Pattern Formations and Oscillatory Phenomena

Pattern Formations and Oscillatory Phenomena

Edited by

Shuichi Kinoshita

AMSTERDAM • BOSTON • HEIDELBERG • LONDON • NEW YORK • OXFORD
PARIS • SAN DIEGO • SAN FRANCISCO • SYDNEY • TOKYO

Elsevier
225, Wyman Street, Waltham, MA 02451, USA
The Boulevard, Langford Lane, Kidlington, Oxford OX5 1GB, UK
Radarweg 29, PO Box 211, 1000 AE Amsterdam, The Netherlands

© 2013 Elsevier Inc. All rights reserved

No part of this publication may be reproduced, stored in a retrieval system or transmitted in any form or by any means electronic, mechanical, photocopying, recording or otherwise without the prior written permission of the publisher. Permissions may be sought directly from Elsevier's Science & Technology Rights Department in Oxford, UK: phone (+44) (0) 1865 843830; fax (+44) (0) 1865 853333; email: permissions@elsevier.com. Alternatively you can submit your request online by visiting the Elsevier web site at http://elsevier.com/locate/permissions, and selecting Obtaining permission to use Elsevier material

Notice
No responsibility is assumed by the publisher for any injury and/or damage to persons or property as a matter of products liability, negligence or otherwise, or from any use or operation of any methods, products, instructions or ideas contained in the material herein. Because of rapid advances in the medical sciences, in particular, independent verification of diagnoses and drug dosages should be made

British Library Cataloguing in Publication Data
A catalogue record for this book is available from the British Library

Library of Congress Cataloging-in-Publication Data
Application submitted

ISBN: 978-0-12-397014-5

For information on all Elsevier publications
visit our web site at store.elsevier.com

Printed and bound in the U.S.A.
13 14 15 16 17   10 9 8 7 6 5 4 3 2 1

Contents

Preface   ix

1 Introduction to Nonequilibrium Phenomena   1
Shuichi Kinoshita
  1.1 Overview   1
  1.2 Oscillatory Phenomena   2
    1.2.1 Oscillation Models   2
    1.2.2 Belousov–Zhabotinsky Reaction   12
    1.2.3 Relaxation Oscillation   17
  1.3 Order-Formation Process   20
    1.3.1 Colloidal Crystals   21
  1.4 Pattern Formation   28
    1.4.1 Instability in Crystal Growth   29
    1.4.2 Turing Pattern   35
    1.4.3 Color-Producing Nanostructures   41
  1.5 Order and Fluctuation   51
    1.5.1 Phase-Transition Analogy   51

2 Belousov–Zhabotinsky Reaction   61
Jun Miyazaki
  2.1 Introduction   61
  2.2 Oscillation, Wave Propagation, and Pattern Formation   64
    2.2.1 Composition   64
    2.2.2 Oscillation in a Reactor   65
    2.2.3 Wave Propagation and Pattern Formation   66
  2.3 Reaction Mechanism and Numerical Simulation   68
    2.3.1 FKN Mechanism   68
    2.3.2 Oregonator Model and Limit Cycle Oscillation   69
    2.3.3 Wave Propagation, Target Pattern, and Spiral Waves   72
  2.4 Synchronization   75
    2.4.1 Synchronization in Chemical Oscillation   75
    2.4.2 Analytical Method: Phase Model   75
    2.4.3 Method for Determining Coupling Function   76
    2.4.4 Application to the Coupled BZ Oscillators   77

3	**Dynamics of Droplets**	**85**
	Hiroyuki Kitahata, Natsuhiko Yoshinaga, Ken H. Nagai, Yutaka Sumino	
	3.1 Introduction	85
	3.1.1 Active Matter	85
	3.1.2 Surface Tension	86
	3.2 Surface Tension–Driven Spontaneous Motion	88
	3.2.1 Droplet Gliding on a Glass Surface	88
	3.2.2 Droplet Drifting on an Aqueous Surface	92
	3.2.3 Suspended Droplet Swimming in an Aqueous Phase	95
	3.3 Hydrodynamics for Spontaneous Motion	97
	3.3.1 Basic Knowledge	97
	3.3.2 Stokes Flow	99
	3.3.3 Surface Tension in the Frame of Hydrodynamics	101
	3.3.4 Spherical Droplet Moving Under a Concentration Gradient	102
	3.4 Motion Coupled with Pattern Formation	106
	3.4.1 Motion of a BZ Droplet	106
	3.4.2 Numerical Results	107
	3.5 Concluding Remarks	111
	Appendix	116
4	**Density Oscillators**	**119**
	Takeshi Kano	
	4.1 Introduction to Density Oscillators	119
	4.1.1 Self-Oscillatory Phenomena	119
	4.1.2 Relaxation Oscillations	121
	4.1.3 Density Oscillators	123
	4.2 Phenomenological Description	127
	4.2.1 Experimental Procedure and General Oscillation Trend	127
	4.2.2 Hydrodynamic Analysis of Each Upflow and Downflow Branch	131
	4.2.3 Phenomenological Model	137
	4.3 Fundamental Mechanism of Oscillation	140
	4.3.1 Hydrodynamic Analysis of Flow Reversal	141
	4.3.2 Viscosity-Dependent Flow Reversal	143
	4.3.3 Model Including Flow-Reversal Process	149
	4.4 Concluding Remarks	158
	Appendix	162
5	**Colloidal Crystals**	**165**
	Junpei Yamanaka, Tohru Okuzono, Akiko Toyotama	
	5.1 Introduction	165
	5.1.1 Order Induced by Entropy	165
	5.1.2 Structures of Colloidal Dispersions	166
	5.1.3 Interactions between Colloidal Particles	170

	5.2	Samples and Methodology	175
		5.2.1 Colloidal Samples	176
		5.2.2 Characterization of Crystal Structures	177
	5.3	Crystallization of Charged Colloids	181
		5.3.1 Charge-Induced Crystallization	181
		5.3.2 Unidirectional Crystallization	183
		5.3.3 Gel Immobilization	189
		5.3.4 Exclusion of Impurity Particles	189
	5.4	Current Topics	193
		5.4.1 Opal-Type Crystals	193
		5.4.2 Complex Structures	194
6	**Structural Color in Nature: Basic Observations and Analysis**		**199**
	Shinya Yoshioka		
	6.1	Introduction	199
		6.1.1 Multiscale Systems and Structural Color in Nature	199
	6.2	Basic Observations	201
		6.2.1 How to Illuminate	201
		6.2.2 *Morpho* Butterfly	202
		6.2.3 Jewel Beetle	204
		6.2.4 Liquid Immersion Experiment	206
	6.3	Optical Characterization	207
		6.3.1 Angle-Dependent Reflection	207
		6.3.2 θ–2θ Scan Measurement	212
		6.3.3 Integrated Optical Properties	215
		6.3.4 Refractive Index Value	217
	6.4	Analysis I	218
		6.4.1 Fresnel's Equations	218
		6.4.2 Single Thin-Layer Interference	221
		6.4.3 Multilayer Interference	227
		6.4.4 Analysis of Jewel Beetle's Iridescence	234
	6.5	Analysis II	237
		6.5.1 Fraunhofer Diffraction	237
		6.5.2 Diffraction Grating	240
		6.5.3 Analysis of *Morpho* Butterfly's Structural Color	242
		6.5.4 Role of Pigment	246
	6.6	Concluding Remarks	248

References to Each Chapter   **253**
Index   **257**

Preface

Pattern formations and oscillatory phenomena are typical manifestations of nonequilibrium order formations, the research field of which has been rapidly growing with the recent development of nanotechnology. In addition, they are closely related to various biological activities and thus have attracted great attention from a biological viewpoint. In this book we select several interesting phenomena, which are very familiar even to ordinary persons, such as the Belousov–Zhabotinsky reaction, spontaneous motion of a droplet, density oscillator, colloidal crystal formation, and natural photonic structures.

We aim for this book to be an easy guide to this newly growing field. Although a large number of books are now available, they tend to be inclined solely to a theoretical or mathematical part of this field so that it is rather difficult for beginners to tackle it by themselves. Otherwise, only the experimental methods are provided without any detailed scientific explanation. To understand the nonequilibrium order formations as a whole and to be a good guide for this field, one should intuitively feel and actually touch the phenomena, and then understand how they are analyzed from a scientific viewpoint. Furthermore, the up-to-date status of each research area should be also referred to properly.

We achieve this tough work by several young and energetic researchers in physics. Actually, we will first show their attracting points through easy observations or simple experiments, and then explain how to analyze them from a physical viewpoint. Since the observations and experiments are given as a form of column, readers can appreciate them independently of the text, which surely makes them familiar to various interesting phenomena. We hope this book will be a good first book for the beginners and also an excellent guide to students and teachers who wish to study this fascinating research field. Furthermore, we believe it will be valuable for the experts to grasp the recent status of the field.

1 Introduction to Nonequilibrium Phenomena

Shuichi Kinoshita

Chapter Contents
1.1 Overview 1
1.2 Oscillatory Phenomena 2
    1.2.1 Oscillation Models 2
    1.2.2 Belousov–Zhabotinsky Reaction 12
    1.2.3 Relaxation Oscillation 17
1.3 Order-Formation Process 20
    1.3.1 Colloidal Crystals 21
1.4 Pattern Formation 28
    1.4.1 Instability in Crystal Growth 29
    1.4.2 Turing Pattern 35
    1.4.3 Color-Producing Nanostructures 41
1.5 Order and Fluctuation 51
    1.5.1 Phase-Transition Analogy 51

1.1 Overview

The term *nonequilibrium* is the antonym of *equilibrium*. Thus, one may think nonequilibrium indicates peculiar states of materials and phenomena in nature. If so, then let us consider what equilibrium means. Generally, equilibrium is defined as a condition in which transfer of heat and/or particles between two systems does not occur. In such a case, we refer to the two systems as being in thermodynamic equilibrium or in thermal equilibrium. For two systems in contact, if one system is sufficiently large and can be considered as a thermal bath of an absolute temperature T, the energy distribution of the constituents of the other system is known to obey the Maxwell–Boltzmann law for the temperature T. Although the concept of an equilibrium state is familiar in the field of science and technology, it is an ideal state that does not exist in the real world and is approximately attainable only in laboratory conditions. Thus, the equilibrium state is the very state that we should call peculiar.

It seems there is no room to consider the presence of the thermal equilibrium state in our daily life: the sun is pouring down brightly, a breath of fresh air is blowing everywhere, the rivers flow smoothly, and it rains for many a day. Earth receives electromagnetic energy from the sun, which is converted into various types of energy and eventually radiated to outer space. Therefore, Earth as a whole is a nonequilibrium system because it constantly experiences energy transfer. In an equilibrium system, owing to the law of increasing entropy, the transition of the system from an ordered state to a disordered state is a certainty. In contrast, in a nonequilibrium system, various types of order can be achieved by applying the flow of energy and materials, which is directly connected with order formation in the global environment, including the phenomena of life.

Thus, the fundamental issue of the necessity and manner of establishing order in a nonequilibrium state arises. This is a curious issue because it is closely related to another fundamental question: What is life? However, we have been focusing on the tractable problems of the equilibrium system, and thus our methodology is inclined systematically toward the equilibrium system. This is extremely inadequate for dealing with the nonequilibrium system. Through this book, we aim to acquaint readers with the nonequilibrium state by introducing various interesting phenomena with the help of simple experiments. In this chapter, we mention various nonequilibrium phenomena and their corresponding characteristics, such as oscillatory phenomena, order-formation processes, and pattern formation, which will be described in later chapters. We further introduce the typical mathematical methods of dealing with these phenomena. The theory of the phase-transition analogy (Haken, 1975, 1978) has also been introduced, which might provide a clue to the fundamental questions mentioned here.

1.2 Oscillatory Phenomena

1.2.1 Oscillation Models

Oscillatory phenomena are familiar in daily life. It is not too much to say that we are surrounded by them. One typical and important oscillatory phenomenon is the circadian rhythm, which is directly associated with Earth's rotation. We also experience repeated seasonal variations owing to Earth's revolution around the sun. The stability of the human body in a constantly changing environment is maintained by the continuous oscillations of heartbeat and breathing. A clock pendulum is an example of mechanical oscillation; the vibrations of strings and membranes, as well as air vibration in tubes, which are used to compose music, are other examples, while speakers and microphones are applications of electromechanical oscillation. In electrical circuits, various types of oscillators have been used to generate radio waves, which are ubiquitous in the present age of information technology. The laser is an oscillator in the optical region, and its use in DVDs and Blu-ray discs has become common. A significantly broader range of oscillatory phenomena can be found in economics, politics, history, climate, astronomy, ecology, biology, etc.

Although numerous types of oscillatory phenomena can be identified, some of them are realized in an isolated system, while some are realized only in a nonequilibrium system. Thus, oscillatory phenomena are varied in that they are associated with various complicated mechanisms. In this book, we have mentioned a few examples among them, placing special emphasis on nonequilibrium oscillatory phenomena (Kuramoto, 2005). Further, we have demonstrated a simple method of scientifically dealing with oscillations.

Harmonic Oscillator

First, consider the case of a simple harmonic oscillator, as shown in Figure 1.1(a). The equation of motion of a weight attached to a spring is generally expressed as

$$m\ddot{x} + kx = 0, \qquad (1\text{-}1)$$

where x is the displacement of the weight from its equilibrium position, and $\ddot{x}$ is its second time derivative. In this case, m and k are the mass of the weight and the spring constant, respectively, and any friction term, which might appear during actual oscillation, is neglected. The solution of this equation is

$$\begin{aligned} x &= A\sin\omega t + B\cos\omega t \\ &= C\sin(\omega t + \phi), \end{aligned} \qquad (1\text{-}2)$$

where A and B are arbitrary real constants that are required to be determined from the initial conditions, with $\omega = \sqrt{k/m}$, $C = \sqrt{A^2 + B^2}$, and $\phi = \tan^{-1}(B/A)$. The calculation of x against t is shown in Figure 1.1(b), which shows that the solution of this equation actually yields an infinite oscillation with an angular frequency of ω.

If we express the velocity $\dot{x}$ using a new variable y, Eq. (1-1) of motion can be transformed into a set of linear differential equations expressed as

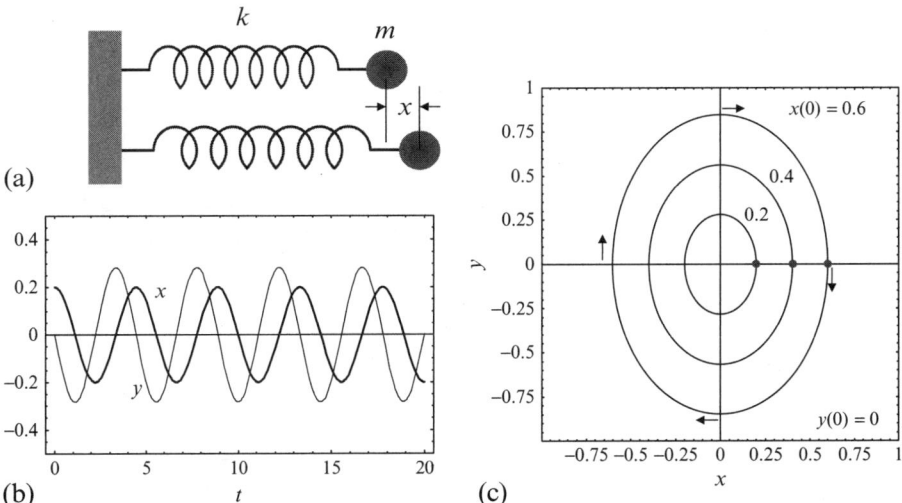

Figure 1.1 (a) Mechanical harmonic oscillator, (b) time variations of x and y ($=\dot{x}$) expressed by Eqs. (1-3) and (1-4) with $k/m=2$, and (c) orbits in phase space for various initial conditions.

$$\dot{x} = y, \tag{1-3}$$

$$\dot{y} = -\left(\frac{k}{m}\right)x. \tag{1-4}$$

Further, dividing each side of the latter equation by that of the former, we obtain

$$\frac{dy}{dx} = -\frac{(k/m)x}{y},$$

which leads to

$$\left(\frac{k}{m}\right)x dx + y dy = 0. \tag{1-5}$$

Integrating both sides with respect to t from 0 to t and multiplying them by m, we obtain

$$\frac{1}{2}kx^2(t) + \frac{1}{2}my^2(t) = U, \tag{1-6}$$

where U is a constant expressed as $U \equiv (1/2)kx^2(0)+(1/2)my^2(0)$. This relation ensures that the motion of the harmonic oscillator is conservative with the constant of motion U, which is generally called mechanical energy. Thus, when the motion of the harmonic oscillator is illustrated in phase space spanned by x and y, concentric elliptical orbits of sizes determined solely by the initial conditions are obtained, as shown in Figure 1.1(c).

Lotka–Volterra Model

A similar situation is realized in a completely different case known as the Lotka–Volterra model (Lotka, 1925; Volterra, 1926). A set of equations for two variables x and y, respectively referring to prey and predator in an ecosystem, are described as

$$\dot{x} = x(a - by), \tag{1-7}$$

$$\dot{y} = -y(c - dx). \tag{1-8}$$

Here, a, b, c, and d are assumed to be positive constants.

We consider the well-known example of a system consisting of rabbits and foxes, where the former and latter correspond to the prey and predator, respectively, as shown in Figure 1.2(a). The rabbits and foxes are assumed to be uniformly distributed in an area, and their numbers are represented by the variables x and y, respectively. The number of rabbits is considered to increase in a self-reproducing way, possibly

Introduction to Nonequilibrium Phenomena

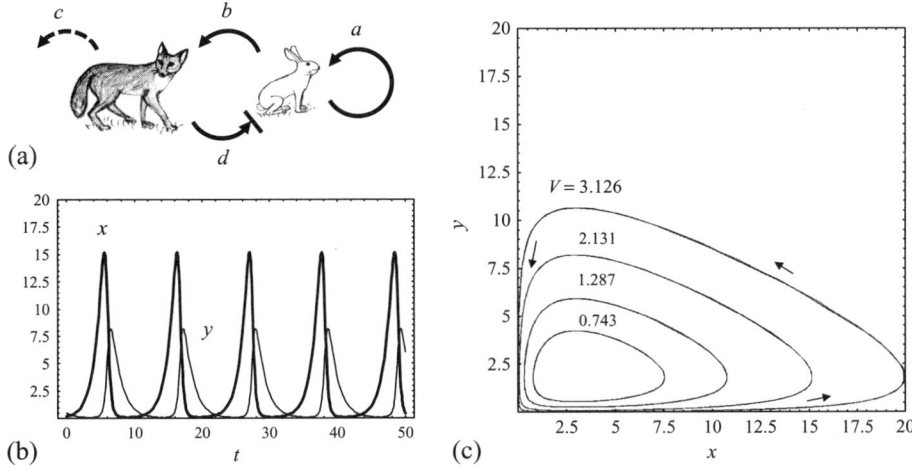

Figure 1.2 (a) Lotka–Volterra model applied to prey–predator (rabbit–fox) dynamics in an ecosystem; (b) time variations in x and y expressed by Eqs. (1-7) and (1-8) for $a=0.9$, $b=0.5$, $c=0.75$, and $d=0.25$ with the initial conditions $x(0)=0.2$ and $y(0)=0.5$; and (c) orbits in phase space for various values of V.

because they eat abundant grass, the growth rate of which is characterized by a parameter a. On the other hand, the number of foxes cannot grow without rabbits, and as a result, the number of foxes will decrease in a pattern characterized by a parameter c, if the number of rabbits decreases. When foxes and rabbits are in proximity, the latter is the food for the former; thus, the number of foxes increases at a rate expressed as a product of their numbers and proportionality constant d, while that of rabbits decreases with a proportionality constant b.

Thus, when the prey–predator interaction is absent, the number of rabbits increases exponentially, while that of foxes decreases exponentially. Therefore, it is not necessary to consider the presence of oscillation. However, in the presence of prey–predator interaction, the concept of oscillation is required to be considered, irrespective of the extent of the interaction. First, we consider a case in which the number of foxes is small and that of rabbits is moderate. In this case, the number of rabbits increases exponentially in a self-reproducing pattern, which causes a gradual increase in the number of foxes. This results in a rapid decrease in the number of rabbits, which eventually causes an exponential decrease in the number of foxes. These processes are illustrated by the simulation shown in Figure 1.2(b); the simple exponential increase and decrease in the number of rabbits and foxes appear in the absence of predator–prey interaction, while the number abruptly varies in the presence of this interaction. Thus, a system without oscillatory characteristics begins to oscillate after the interaction is introduced.

The Lotka–Volterra system is a typical case of an activator–inhibitor system in which the activator and inhibitor correspond to rabbits and foxes, respectively. The following are the conditions corresponding to the roles of activators and inhibitors:

$$f_x > 0, \, f_y < 0, \, g_x > 0, \text{ and } g_y < 0, \tag{1-9}$$

where we assume that the related dynamics are generally expressed by the following two equations of motion:

$$\dot{x} = f(x, y), \tag{1-10}$$

$$\dot{y} = g(x, y), \tag{1-11}$$

with $f_x = (\partial f/\partial x)_{x=x_0, y=y_0}$, and similar expressions can be obtained for g. Here, x_0 and y_0 are stationary points (or fixed points) obtained by setting $\dot{x} = 0$ and $\dot{y} = 0$. In the Lotka–Volterra model, these are given by

$$x_0 = 0, \quad y_0 = 0, \tag{1-12}$$

or

$$x_0 = c/d, \quad y_0 = a/b, \tag{1-13}$$

where Eq. (1-12) corresponds to a trivial case, while Eq. (1-13) gives the condition for nonzero results.

We divide each side of Eq. (1-8) by that of Eq. (1-7) and obtain

$$x(a - by)dy + y(c - dx)dx = 0, \tag{1-14}$$

which yields

$$a\frac{dy}{y} - bdy + c\frac{dx}{x} - ddx = 0 \tag{1-15}$$

after we divide both sides of Eq. (1-14) by xy. Integrating both sides of Eq. (1-15) with respect to t from 0 to t, we obtain

$$a \ln y(t) - by(t) + c \ln x(t) - dx(t) = V, \tag{1-16}$$

where $V \equiv a \ln y(0) - by(0) + c \ln x(0) - dx(0)$. Thus, the dynamics of the Lotka–Volterra model is conservative with respect to V, which becomes a constant of motion in this case. This result is similar to the case of the harmonic oscillator model.

This situation can be easily understood in terms of the motion in phase space, as shown in Figure 1.2(c). This figure shows the solutions of the Lotka–Volterra equations for $a=0.9, b=0.5, c=0.75,$ and $d=0.25$ with changing initial values of $x(0)$, and hence with changing V. The orbits exhibit deformed closed circuits with stationary points of $x_0=c/d=3, y_0=a/b=1.8$. Thus, if an external instantaneous fluctuation or perturbation is applied, the orbit is irreversibly transferred to another orbit. The Lotka–Volterra model has been widely employed to analyze the dynamics of competition and is easily generalized to the case of multiple species and by considering the saturation effect for each species.

Van der Pol Oscillator

The van der Pol oscillator, which was discovered by van der Pol (1926), is a nonconservative stable oscillator, and was termed as a relaxation oscillator on its discovery. This oscillator has been frequently employed for the investigation of the properties of nonlinear oscillators and various oscillatory phenomena in physical and biological fields, such as analyses of electrical circuits and models of the heartbeat, as well as the action potential in neurons.

The basic equation for the van der Pol oscillator is simple and is described as

$$\ddot{x} - \mu(1 - x^2)\dot{x} + x = 0 \tag{1-17}$$

and is of the form of a harmonic oscillator with an angular frequency of unity; the oscillator contains an additional amplitude-dependent amplification/friction term: if we consider μ as a positive constant, the second term on the left side provides amplification for small amplitudes, while it describes friction at large amplitudes. Thus, the oscillation is expected to grow at first and then become saturated owing to the friction.

As is similar to the previous cases, Eq. (1-17) can be reduced to a set of equations by setting $y = \dot{x}$ as follows:

$$\dot{x} = y, \tag{1-18}$$

$$\dot{y} = \mu(1 - x^2)y - x. \tag{1-19}$$

Alternatively, to emphasize the significance of μ, the preceding equation is divided in a different way by setting $y = \dot{x} + F(x)$, where $F(x) = \int_0^x f(\xi)\mathrm{d}\xi$ with $f(\xi) = -\mu(1-\xi^2)$. Then, using the relation $\dot{y} = \ddot{x} + f(x)\dot{x} = -x$, we obtain a set of equations expressed as $\dot{x} = y + \mu(x - x^3/3)$ and $\dot{y} = -x$. Further, replacing y with $-\mu y$, we finally obtain the well-known form

$$\dot{x} = \mu(x - x^3/3 - y), \tag{1-20}$$

$$\dot{y} = (1/\mu)x. \tag{1-21}$$

In Figure 1.3, we show the results of simulations, using Eqs. (1-20) and (1-21) for $\mu = 1$ under various initial conditions. Initially, the orbits in phase space vary with the difference in the initial values; they later merge into a single circuit and apparently rotate along the circuit. Thus, the final circuit does not depend on the initial conditions. This type of orbit is called a limit cycle.

A limit cycle generally originates from a condition that is referred to as Liénard's theorem. This theorem is applicable to the Liénard equation $\ddot{x} + f(x)\dot{x} + g(x) = 0$, and the conditions for generating the limit cycle are as follows: (1) derivatives of $f(x)$ and $g(x)$ are continuous, (2) $g(x)$ is an odd function of x, while $f(x)$ is an even function, (3) for $x > 0$, $g(x) > 0$, and (4) a positive value of α satisfies the conditions $F(x) < 0$ for $0 < x < \alpha$, $F(\alpha) = 0$, and $F(x)$ is positive and monotonically increases with x

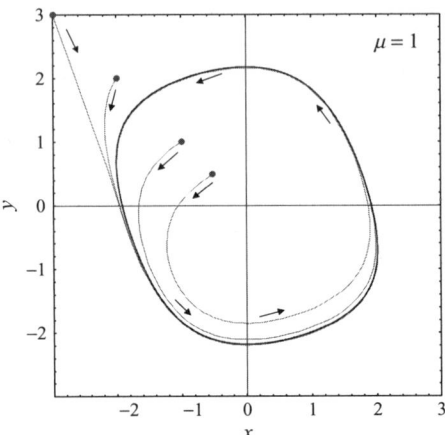

Figure 1.3 Limit-cycle orbits in phase space for the van der Pol oscillator expressed by Eqs. (1-20) and (1-21) with various initial conditions. μ is constant at $\mu = 1$.

for $\alpha < x$, where $F(x) = \int_0^x f(\xi)d\xi$. Comparing these conditions with Eq. (1-17), because $F(x) = -\mu(x - x^3/3)$ and $g(x) = x$, it is easily demonstrated that the van der Pol equation satisfies Liénard's theorem for $\mu > 0$ but not for $\mu < 0$.

In Figure 1.4, we show the orbits calculated for four different values of μ ranging from -1 to 5, considering the time course of the oscillations. The limit cycles are realized for $\mu > 0$, while only damped oscillation is observed for $\mu = -1$. This result is in good agreement with the previous consideration. However, it is found that the orbits change from circular to a parallelogram-like pattern. For small μ, the shape of the orbit and its time course are similar to those of a harmonic oscillator. This is reasonable because the original form of the van der Pol oscillator (Eq. 1-17) reduces to the equation of a harmonic oscillator when $\mu \to 0$.

On the other hand, the time variation in x for large μ is significantly different from harmonic oscillation and is characterized by abrupt increases and decreases, which is similar to the case of a square wave. In Figure 1.5(a), we show the limit-cycle orbit in phase space for $\mu = 10$ with the curve for $y = x - x^3/3$, which is obtained by setting $\dot{x} = 0$ in Eq. (1-20) and is called a nullcline. The orbit completely follows the nullcline on the left and right sides, but it moves transversely from right to left or left to right to form a circuit in the center.

The reason for this peculiar behavior at large μ can be comprehended by considering that a strong attractive force is produced on the left and right sides of the nullcline. For example, if we focus on the right side of the nullcline, the outer area is characterized by an area represented by $dx/dt < 0$, but in the inner area, $dx/dt > 0$ is dominant. For large values of μ, a small deviation from the nullcline causes a strong attraction toward the right side of the nullcline.

This behavior can be clarified by the calculation of the vector field defined by $(\dot{x}, \dot{y})/\sqrt{\dot{x}^2 + \dot{y}^2}$; the direction of this field indicates the movement of a locus at each

Introduction to Nonequilibrium Phenomena 9

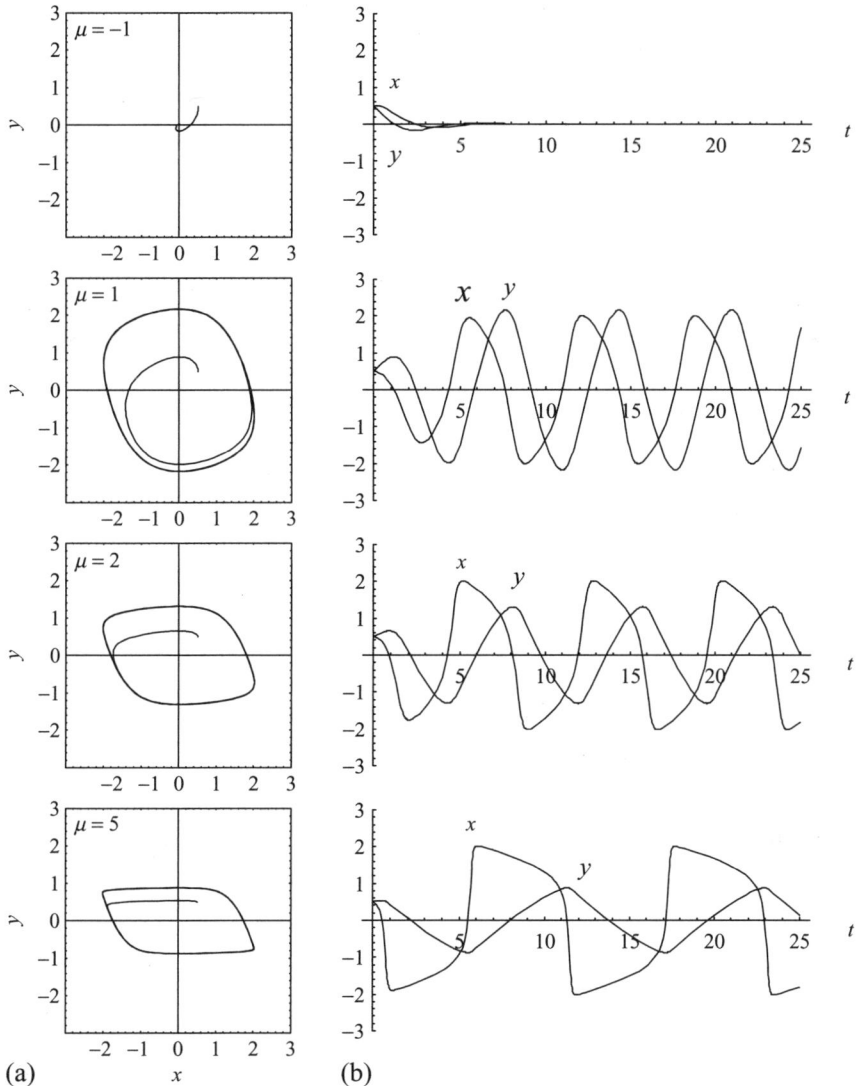

Figure 1.4 (a) Orbits in phase space and (b) time variations in x and y in the van der Pol oscillator for various values of μ.

point in phase space. We show the calculation result in Figure 1.5(b) for $\mu=3$; as shown in the figure, the right and left sides of the nullcline are attractive, while its center is repulsive. Further, the vectors are generally directed along the x axis and point left and right in the upper and lower areas, respectively, while changes in their directions are observable only near the nullcline, which clearly explains the limit-cycle motion in this system.

Thus, because of the condition $dy/dt>0$ for $x>0$, the locus moves upward along the nullcline; when it reaches a maximum, the conditions $dx/dt<0$ and $dy/dt>0$ cause

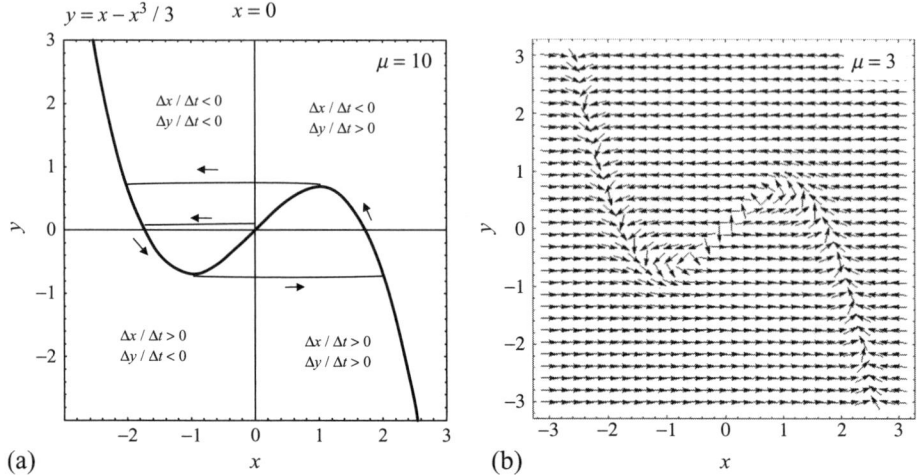

Figure 1.5 (a) Nullclines obtained from $\dot{x} = 0$ and $\dot{y} = 0$, and limit-cycle orbits for $\mu = 10$. (b) Vector fields expressed by $(\dot{x}, \dot{y})/\sqrt{\dot{x}^2 + \dot{y}^2}$ for $\mu = 3$, which explains the origin of the limit-cycle orbit along the nullcline of $\dot{x} = 0$.

the locus to move leftward away from the nullcline. This movement is rapid because of the large absolute value of dx/dt owing to large μ. A similar process continues and eventually closes the circuit, which explains the parallelogram-like shape of the limit cycle and the square-wave-like time response.

The stationary point for the van der Pol equation is obtained as $(x_0, y_0) = (0, 0)$, which is calculated by $\dot{x} = 0$ and $\dot{y} = 0$. The fact that the van der Pol equation shows only oscillatory behavior for $\mu > 0$ instead of converging into the stationary point (x_0, y_0) requires further explanation using the following linear stability analysis.

Consider a general case in which the dynamics of a system are expressed by the following set of equations:

$$\dot{x} = f(x, y), \tag{1-22}$$

$$\dot{y} = g(x, y). \tag{1-23}$$

Assume that a small deviation from the stationary point is imposed on the system from outside and is expressed as $x = x_0 + x_1(t)$ and $y = y_0 + y_1(t)$, where x_0 and y_0 satisfy $f(x_0, y_0) = g(x_0, y_0) = 0$. Here, $x_1(t)$ and $y_1(t)$ indicate small time-dependent deviations from the stationary point. Inserting them into Eqs. (1-22) and (1-23), we obtain

$$\dot{x}_1 = f_x(x_0, y_0)x_1 + f_y(x_0, y_0)y_1, \tag{1-24}$$

$$\dot{y}_1 = g_x(x_0, y_0)x_1 + g_y(x_0, y_0)y_1, \tag{1-25}$$

where $f_x(x_0, y_0) = (\partial f / \partial x)_{x=x_0, y=y_0}$, and relations for the other parameters are similar. Here, we consider only linear terms with respect to x_1 and y_1 because we

Introduction to Nonequilibrium Phenomena

have assumed that they are sufficiently small. These equations are rewritten in a matrix form as

$$\frac{d}{dt}\begin{pmatrix} x_1 \\ y_1 \end{pmatrix} = \begin{pmatrix} f_x(x_0, y_0) & f_y(x_0, y_0) \\ g_x(x_0, y_0) & g_y(x_0, y_0) \end{pmatrix}\begin{pmatrix} x_1 \\ y_1 \end{pmatrix}, \quad (1\text{-}26)$$

which can be easily solved if the eigenvalues for the matrix on the right side are given. The eigenvalues are obtained by solving the following equation with respect to λ:

$$\begin{vmatrix} f_x(x_0, y_0) - \lambda & f_y(x_0, y_0) \\ g_x(x_0, y_0) & g_y(x_0, y_0) - \lambda \end{vmatrix} = 0, \quad (1\text{-}27)$$

which yields the characteristic equation

$$\lambda^2 - (f_x(x_0, y_0) + g_y(x_0, y_0))\lambda + (f_x(x_0, y_0)g_y(x_0, y_0) - f_y(x_0, y_0)g_x(x_0, y_0)) = 0. \quad (1\text{-}28)$$

We denote the solutions of this equation as λ_1 and λ_2. Then, the general solutions for x_1 and y_1 are given as

$$\begin{pmatrix} x_1 \\ y_1 \end{pmatrix} = \begin{pmatrix} c_{11} \\ c_{21} \end{pmatrix}e^{\lambda_1 t} + \begin{pmatrix} c_{12} \\ c_{22} \end{pmatrix}e^{\lambda_2 t}, \quad (1\text{-}29)$$

where c_{ij} is an arbitrary constant and should be determined from the initial deviation. If both $\mathrm{Re}\{\lambda_1\}$ and $\mathrm{Re}\{\lambda_2\}$ are negative, a small initial deviation converges to the stationary point, and the system becomes stable. In contrast, if at least one of them is positive, a small deviation from the stationary point causes the orbit to grow divergently, and the system eventually becomes unstable.

We apply this analysis to the stationary point $(x_0, y_0) = (0, 0)$ of the van der Pol equation. The linearized equations in this case are given as

$$\frac{d}{dt}\begin{pmatrix} x_1 \\ y_1 \end{pmatrix} = \begin{pmatrix} \mu & -\mu \\ 1/\mu & 0 \end{pmatrix}\begin{pmatrix} x_1 \\ y_1 \end{pmatrix}, \quad (1\text{-}30)$$

which leads to the characteristic equation $\lambda^2 - \mu\lambda + 1 = 0$. On solving this equation, we obtain

$$\lambda = \frac{\mu \pm \sqrt{\mu^2 - 4}}{2} \equiv \lambda_1, \lambda_2. \quad (1\text{-}31)$$

Because we have assumed that μ is positive, $\mathrm{Re}\{\lambda_{1,2}\}$ is positive; therefore, the stationary point $(x_0, y_0) = (0, 0)$ is absolutely unstable, which partially explains the appearance of limit-cycle oscillation irrespective of the initial conditions. On the other hand, if μ is considered to be negative, $\mathrm{Re}\{\lambda_{1,2}\}$ is always negative, so the orbit converges to a stationary point. Thus, the characteristics of the orbit in phase space are significantly affected by the sign of μ. This type of change with respect to λ, which appears at $\mu = 0$ as a border, is called Hopf bifurcation.

1.2.2 Belousov–Zhabotinsky Reaction

Most studies on oscillatory behavior in nonequilibrium systems were developed after the discovery of a peculiar chemical reaction now known as the Belousov–Zhabotinsky (BZ) reaction (Kuramoto, 1984; Miike et al., 1997) (see details in Chapter 2). This reaction was discovered by Belousov in 1951, who found that when citric acid and bromate were reacted in the presence of Ce ions, the color of the solution oscillated between yellow and colorless (Belousov, 1959). Zhabotinsky (1964) rediscovered this reaction and found that the reaction occurred even when malonic acid ($CH_2(COOH)_2$) was used instead of citric acid, and Fe or Mn ions instead of Ce ions.

Various experimental improvements have been performed in the BZ reaction since its rediscovery. Among them, the invention of the continuous-flow stirred tank reactor, which enabled the BZ reaction to be performed under stationary conditions, improved the reproducibility of the phenomenon, and thus increased the accuracy of the experiment. Thus, the BZ reaction has become a typical nonequilibrium oscillatory phenomenon with appropriately controlled experimental conditions, and it has been studied intensively from the experimental and theoretical viewpoints, particularly in various phenomena concerning coupled oscillation, pattern formation, light-controlled reactions, and small-volume experiments.

Brusselator Model

The bromination of malonic acid using metallic compound ions and bromine ions as catalysts is an essential step in the BZ reaction using malonic acid as a reactant, which is expressed by the following reaction formula (Field et al., 1972):

$$5CH_2(COOH)_2 + 3BrO_3^- + 3H^+ \rightarrow 3CHBr(COOH)_2 + 2HCOOH + 4CO_2 + 5H_2O.$$

However, the actual reaction processes are complex, and more than 10 elementary processes are known to be involved in this reaction.

Thus, the BZ reaction consists of a series of extremely complicated reaction pathways, which gave a good opportunity to test the theoretical models of oscillatory phenomena. The Brusselator model, one of the models explaining the origin of the chemical oscillations, was proposed by Prigogine's group (Prigogine and Lefever, 1968; Nicolis and Prigogine, 1977).

The Brusselator model is based on a series of the following hypothetical reactions:

$$A \rightarrow X,$$
$$2X + Y \rightarrow 3X,$$
$$B + X \rightarrow Y + D,$$
$$X \rightarrow E,$$

Introduction to Nonequilibrium Phenomena

where A, B, D, E, X, and Y indicate the chemical species involved in the reactions. The second reaction indicates that the product X is also a reactant, which means that Y is converted into X in the presence of X as a catalyst, and this process is called autocatalysis. Assuming that the quantities of A and B are sufficient for maintaining constant concentrations during the reactions and the elimination of products D and E is sufficiently rapid, we can construct rate equations for X and Y as

$$\dot{x} = a + x^2 y - bx - x, \tag{1-32}$$

$$\dot{y} = bx - x^2 y, \tag{1-33}$$

where a, b, x, and y indicate the concentrations of the chemical species. Furthermore, we have set all the kinetic constants to unity, which is possible after appropriate transformations of variables.

The stationary points corresponding to Eqs. (1-32) and (1-33) can be obtained by setting $\dot{x} = 0$ and $\dot{y} = 0$, leading to $a + x^2 y - bx - x = 0$ and $x(b - xy) = 0$, respectively. These relations yield the following stationary points: $(x_0, y_0) = (a, b/a)$. The simulation of Eqs. (1-32) and (1-33) for various parameter values and initial conditions suggests that a locus actually moves around $(a, b/a)$. However, its behavior significantly depends on the parameter values employed. This behavior is indicated in Figure 1.6, which shows a variety of orbits in phase space with varying b, while a is constant at $a=1$. The orbit apparently converges to a stationary point below $b=2$, while it shows a limit cycle above.

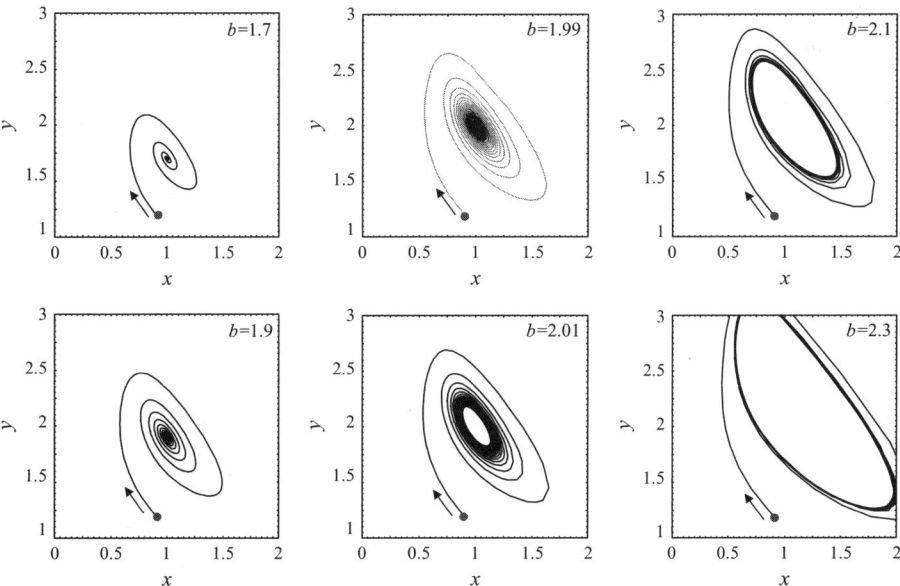

Figure 1.6 Orbits in phase space for a Brusselator model with various values of b while a is kept constant at $a=1$.

This peculiar behavior can be clarified by the investigation of linear stability around the stationary point (x_0, y_0). By using Eq. (1-26), the linearized equations are obtained as follows:

$$\frac{d}{dt}\begin{pmatrix} x_1 \\ y_1 \end{pmatrix} = \begin{pmatrix} 2x_0y_0 - b - 1 & x_0^2 \\ b - 2x_0y_0 & -x_0^2 \end{pmatrix}\begin{pmatrix} x_1 \\ y_1 \end{pmatrix} = \begin{pmatrix} b - 1 & a^2 \\ -b & -a^2 \end{pmatrix}\begin{pmatrix} x_1 \\ y_1 \end{pmatrix}, \quad (1\text{-}34)$$

where we put $x = x_0 + x_1$ and $y = y_0 + y_1$ with x_1 and y_1 expressing small deviations from the stationary point. Equation (1-34) yields the characteristic equation

$$\lambda^2 + (a^2 - b + 1)\lambda + a^2 = 0. \quad (1\text{-}35)$$

Solving this equation with respect to λ, we obtain

$$\lambda = \frac{-(a^2 - b + 1) \pm \sqrt{(a^2 - b + 1)^2 - 4a^2}}{2}. \quad (1\text{-}36)$$

Thus, the sign of λ significantly depends on the values of a and b. When $b < a^2 + 1$, Re$\{\lambda\}$ is always negative, and a stable orbit converging to the stationary point is obtained. On the other hand, when $b > a^2 + 1$, a positive solution is obtained, which leads to destabilization of the stationary point. Thus, Hopf bifurcation occurs at $b = a^2 + 1$ even in chemical oscillation, which is similar to the case of the van der Pol oscillator.

Oregonator Model

Field et al. (1972) devised a more feasible model that comprised a reaction system consisting of Ce^{4+}, BrO_3^-, $CH_2(COOH)_2$, and H_2SO_4, which is often referred to as the FKN (Field, Körös, and Noyes) mechanism. The complicated reaction pathways were further condensed into the following set of simple reactions (Field and Noyes, 1974):

$$A + Y \rightarrow X + P,$$
$$X + Y \rightarrow 2P,$$
$$A + X \rightarrow 2X + 2Z,$$
$$2X \rightarrow A + P,$$
$$Z + B \rightarrow hY,$$

where A, B, P, X, Y, and Z represent BrO_3^-, $CH_2(COOH)_2 + BrCH(COOH)_2$, HOBr, $HBrO_2$, Br^-, and Ce^{4+}, respectively, and h is a stoichiometric parameter.

Using this reaction scheme, a series of differential equations was constructed, which were further made dimensionless as

Introduction to Nonequilibrium Phenomena

$$\varepsilon \dot{x} = qy - xy + x(1-x), \qquad (1\text{-}37)$$

$$\varepsilon' \dot{y} = -qy - xy + fz, \qquad (1\text{-}38)$$

$$\dot{z} = x - z, \qquad (1\text{-}39)$$

where ε and ε' are small dimensionless constants, while q and f are dimensionless parameters characterizing the reaction sequences. This series of equations is called the three-variable Oregonator. In an actual case, $\varepsilon' \dot{y}$ is significantly smaller than the other terms; therefore, the preceding equations are occasionally condensed further into two variables by setting $\varepsilon' \dot{y} = 0$. The resulting two-variable Oregonator is expressed as

$$\varepsilon \dot{x} = x(1-x) - fz \frac{x-q}{x+q}, \qquad (1\text{-}40)$$

$$\dot{z} = x - z. \qquad (1\text{-}41)$$

Hereafter, we analyze the two-variable Oregonator to investigate the origin of the chemical oscillation. We first show the result of simulations using Eqs. (1-40) and (1-41) in Figure 1.7(a), for which we varied f from 1 to 3 and maintained a constant q at $q = 2 \times 10^{-4}$ with $\varepsilon = 10^{-2}$. The orbit in phase space shows a clear limit cycle at $f = 1$ and 2, while at $f = 3$, the orbit is incomplete, which indicates that the system converges to a stationary point after traveling for a certain period of time.

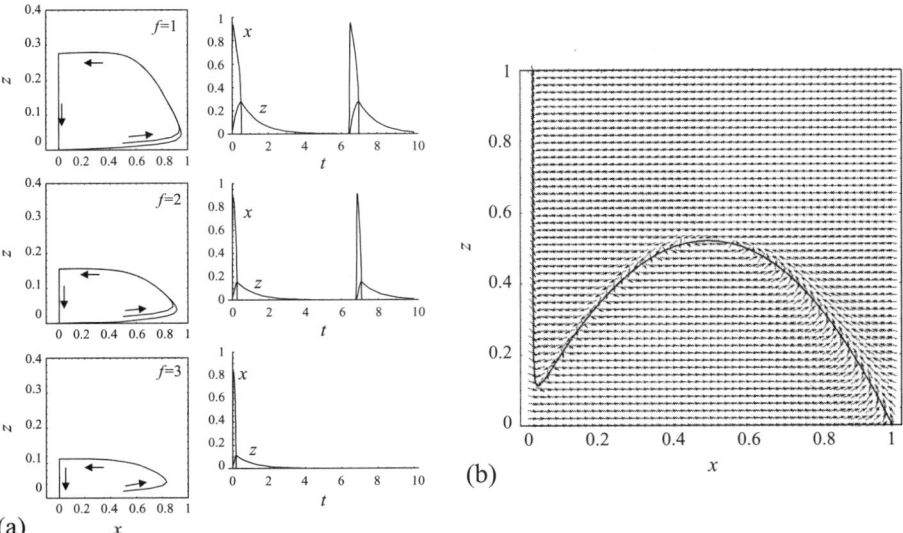

Figure 1.7 (a) Orbits in phase space and time variations of x and z in two-variable Oregonator model for various values of f with $q = 2 \times 10^{-4}$ and $\varepsilon = 10^{-2}$. (b) Vector fields (arrows) expressed by $(\dot{x}, \dot{z})/\sqrt{\dot{x}^2 + \dot{z}^2}$ and a nullcline (solid curve) corresponding to $\dot{x} = 0$ with $q = 0.01$ and $f = 0.5$.

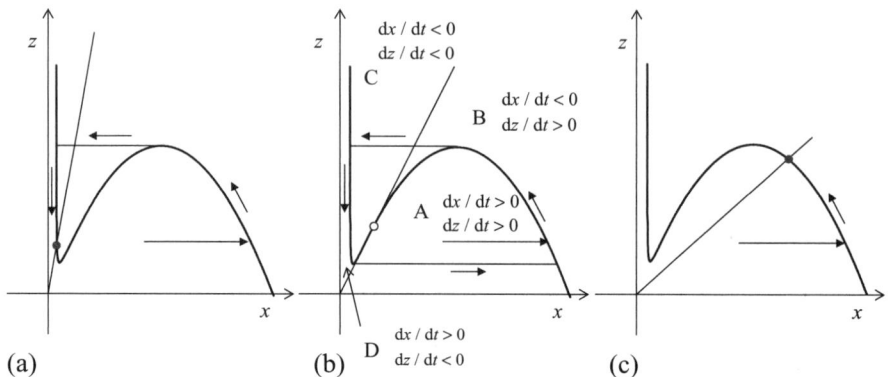

Figure 1.8 (a) Three cases for obtaining a stationary point from a point of intersection of two nullclines corresponding to $\dot{x} = 0$ and $\dot{z} = 0$. (a) and (c) yield stable stationary points, while (b) yields an unstable one, which leads only to a limit cycle. A–D indicate areas divided by the nullclines.

Because ε is usually extremely small, the use of nullcline is effective in this case. The nullclines in the present case are obtained by setting $\dot{x} = 0$ and $\dot{z} = 0$ in Eqs. (1-40) and (1-41), providing $z = x(1-x)/f \cdot \{(x+q)/(x-q)\}$ and $z = x$, respectively. The nullclines obtained are shown in Figure 1.8. The former nullcline is similar to a cubic curve having one minimum and one maximum for $x > q$. The latter nullcline is a straight line passing through the origin of the coordinate system. Thus, a stationary point can be obtained from the crossing point of these two curves, and the following three cases can be considered, as shown in Figure 1.8: the stationary points are located (a) on the left side of the nullcline, (b) at the center, and (c) on the right side.

We can estimate the stability of each stationary point from the signs of $\dot{x}$ and $\dot{z}$. For this purpose, we divide the area for $x > q$ into four regions, as shown in Figure 1.8(b). A and B are characterized by $\dot{z} > 0$, while C and D are areas of $\dot{z} < 0$. On the other hand, A and D are characterized by $\dot{x} > 0$ and B and C by $\dot{x} < 0$. Thus, if we consider the example in Figure 1.8(b) in which the stationary point is located at the center of the nullcline, area A is characterized by $\dot{x} > 0$, while C is characterized by $\dot{x} < 0$. Thus, the center of the nullcline is unstable in any case. Whereas, for left and right sides of the nullclines we obtain a stable stationary point.

In Figure 1.8(b), because the stationary point is located in the unstable region, a locus leaves the stationary point and draws a stable limit-cycle orbit, which is indicated by the vector field shown in Figure 1.7(b) for $q = 0.01$ and $f = 0.5$, where at each position in phase space the direction of a vector with its components expressed by $(\dot{x}, \dot{z})/\sqrt{\dot{x}^2 + \dot{z}^2}$ is plotted. The right and left sides of the nullcline are clearly attractive, while the central regions are repulsive, thus causing a locus in the limit cycle to move as follows. First, it follows the right part of the nullcline and rises to its maximum; then, it immediately moves in the transverse direction, follows the left part of the nullcline, descends to the minimum, and eventually moves to the right part, forming a circuit. This behavior indicates the shape of the orbits in Figure 1.7(a) at $f = 1$ and 2.

Introduction to Nonequilibrium Phenomena

On the other hand, when the stationary points are located on the right or left side of the nullcline (Figure 1.8a and c), the locus converges to these stable points as it travels. If we start near the right side of the nullcline, only a single pulse will be observed in the case of Figure 1.8(a), which is occasionally referred to as excitability, while in the case of Figure 1.8(c), only damping is observed. Thus, the choice of the proper parameter values, which are mainly determined by the rate constants, can help in the realization of various situations, such as excitability, limit cycle, and damping in the reaction system.

1.2.3 Relaxation Oscillation

Relaxation oscillation is an entirely different oscillation mechanism frequently observed in daily life. Relaxation oscillation described here is an oscillation consisting of repeated relaxation processes under the application of a stationary external force. When the external force is applied, the system shifts to a new equilibrium state through the relaxation process. During this stage, the sudden variation in the external force causes the system to shift to another equilibrium state through the relaxation process. These processes are repeated many times. This phenomenon is commonly found in mechanical and electrical systems.

Several examples of relaxation oscillation are found in daily life:

- "Drinking bird," a toy using the thermodynamic function of liquid.
- Shishi-odoshi, a water-filled bamboo tube that clacks against a stone when emptied.
- A neon-lamp oscillator, which continuously charges a condenser, causing an intermittent discharge in a neon lamp.
- A saltwater oscillator, which is the up-and-down motion of salt water in a cup with a small hole at the bottom held in a water-filled vessel.

All these produce regular oscillations characterized by two different relaxation processes exhibiting sudden and periodic variations owing to a type of switching mechanism.

Neon-Lamp Oscillator

First, we present the simple example of a neon-lamp oscillator. This phenomenon was first reported by Pearson and Anson and is now known as the Pearson–Anson effect (Pearson and Anson, 1921a, 1921b). In this study, the electrical circuit shown in Figure 1.9(a) was used, and it was found that continuous charging of a condenser by a direct-current power source caused intermittent current in a neon lamp. They observed that at the beginning, when the voltage v was increased to a value lower than a (see Figure 1.9b), a sudden discharge occurred in the neon lamp, generating a large transient current, but stable oscillations began after the neon lamp was warmed. During the stable oscillation, the voltage–current curve exhibited a peculiar sequential cycle of A→P→Q→B→A. The condenser was charged during (I) B→A, while it was discharged through the neon lamp, which occurred during (II) P→Q. In contrast, processes A→P and Q→B were instantaneous, which implies the presence of a switching mechanism.

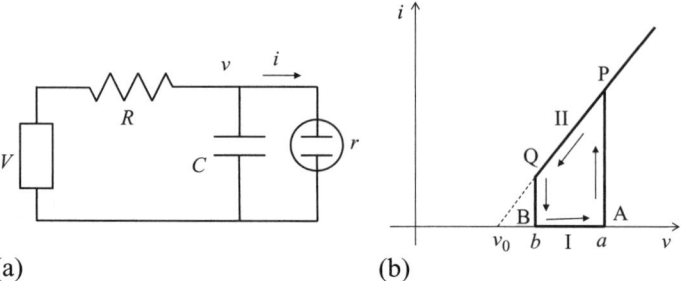

Figure 1.9 (a) Electrical circuit for a neon-lamp oscillator. V: voltage of DC power source, R: resistance, C: capacitance, v: voltage applied to neon lamp, and r: resistance of the neon lamp during discharge. (b) i–v curve for the neon lamp. During period B→A a condenser is charged, while during P→Q the condenser is discharged through the neon lamp.

Thus, the dynamics of the neon-lamp oscillator are expressed using an exponential function:

(I) $v = (V-b)(1-e^{-t/\tau}) + b$,
(II) $v = q + (a-q)e^{-t/\tau'}$,

where t is the time measured from each switching time. Further, we set $\tau = RC$, $\tau' = Rr'C/(R+r')$, and $q = (Vr' + v_0 R)/(R+r')$ with $r' = r(v-v_0)/v$. R and C are the resistance and capacitance, respectively, and r' expresses the gradient of a straight line passing through P and Q in Figure 1.9(b), while r is the apparent resistance of the neon lamp. Because the operating mechanisms for processes (I) and (II) are essentially different, the neon-lamp oscillator can be operated by combining these two processes through switching mechanisms, which are presumably associated with the initiation and termination processes of electric discharge within the neon lamp.

Saltwater Oscillator

Another interesting example is the saltwater oscillator in which oscillations are initiated by holding a cup containing salt water in a water vessel with a small hole or a narrow pipe attached to the bottom of the cup (see the details in Chapter 4). Suppose we fill the cup with salt water while holding a finger over the hole and move it into a vessel filled with fresh water. Then what happens when we release the finger? One will find that the level of the salt water decreases vigorously at first and slowly thereafter. However, for a certain period of time, the saltwater level suddenly increases. This oscillatory movement continues until the salt water is diluted.

A schematic illustration of this experiment is shown in Figure 1.10; the dark area in the figure indicates the region occupied by salt water. This oscillatory phenomenon is peculiar, because during the downflow (I) the level of the salt water continues to decrease until the hydrodynamic balance of $\rho_s x_l = \rho_w h$ is satisfied, where ρ_s and ρ_w are the densities of salt water and fresh water, respectively. However, the switching from downflow to upflow occurs before this balance is attained. In the same way, during

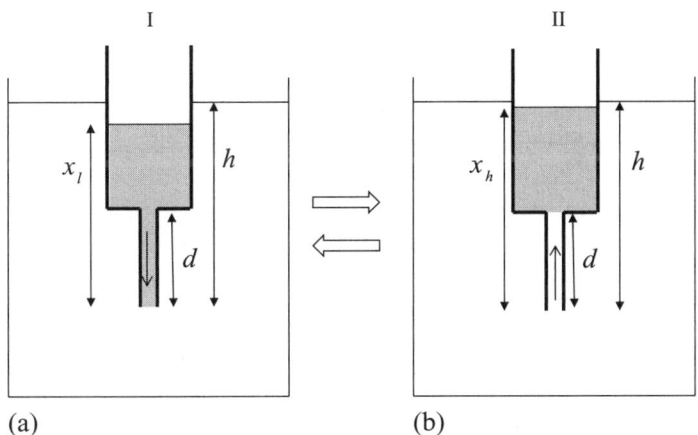

Figure 1.10 (a) Downflow and (b) upflow in saltwater oscillator. Dark region indicates salt water.

the upflow (II), the level of the salt water increases initially. However, before $\rho_s(x_h - d) = \rho_w(h - d)$ is satisfied, where d is the length of the pipe, flow reversal occurs; therefore, the saltwater level continues to oscillate for a long time.

This phenomenon is similar to that of the neon-lamp oscillator, because apart from the switching periods, the phenomenon is appropriately described by a set of simple hydrodynamic equations that express two independent dynamical processes. This experiment was first conducted by Martin (1970), who constructed a model based on the Navier–Stokes equation, assuming that Poiseuille's law was operative inside the pipe. The equations obtained for the downflow and upflow were further made dimensionless and were given as

$$\text{(I)} \ \ddot{y} + \frac{2\sqrt{\sigma_i}}{\beta}\left[\frac{v_i}{v}\dot{y} - \frac{\sqrt{\sigma_i}\beta}{4}\dot{y}^2\right] + \sigma_i y = 0,$$

$$\text{(II)} \ \ddot{y} + \frac{2\sqrt{\sigma_i}}{\beta}\left[\frac{v_i}{v}\dot{y} + \frac{\sqrt{\sigma_i}\beta}{4}\dot{y}^2\right] + \sigma_i(y - 1) = 0,$$

where y is a dimensionless quantity corresponding to the level of salt water, and the other quantities are dimensionless parameters corresponding to the ith oscillation. The second term in parentheses expresses the effect of head loss, which is associated with the abrupt change in the cross-section of the flow at the end of the pipe.

For $\beta \ll 1$, the viscous damping regime, the first term on the left side and the head-loss term can be neglected because $\sigma_i < 1$ is generally satisfied. Therefore, the dynamics are expressed solely by an exponential function such that

$$y_I = (y_1 - y_{I\infty})e^{-t/\tau} + y_{I\infty},$$
$$y_{II} = (y_{II\infty} - y_2)(1 - e^{-t/\tau}) + y_2,$$

where t is again a time measured from each of the switching times, and $y_{I\infty}$ and $y_{II\infty}$ are the equilibrium levels of the salt water for the downflow and upflow with $\tau = 2v_i/(v\sqrt{\sigma_i}\beta)$. Here, y_1 and y_2 are the saltwater levels at which switching from upflow to downflow and from downflow to upflow occurs. This result is similar to the case of the neon-lamp oscillator.

The saltwater oscillator has become widely known, and its mechanism has been analyzed experimentally and theoretically (Yoshikawa et al., 1990; Steinbock et al., 1998). However, its switching mechanism has not been intensively studied. Our group clarified its mechanism experimentally and theoretically (Kano and Kinoshita, 2007, 2008, 2009) (see the details in Chapter 4). These studies have revealed that a small intrusion into the pipe plays an important role in the flow reversal. This peculiar effect occurs when the flow becomes significantly slow, and the diameter of the flow exactly outside the end of the pipe is drastically decreased. It is also concluded that the sum of the viscous drag force, hydrostatic force, and a force associated with the detachment of the flow from the pipe may play a crucial role for the flow reversal. The process of relaxation oscillation is simple in either case, but its switching mechanism is complicated in all cases. Thus, the mechanism of relaxation oscillation cannot be understood until its switching mechanism is clarified.

1.3 Order-Formation Process

Order formation is another type of nonequilibrium phenomena, although order is generally considered in the framework of equilibrium physics. Crystal growth is a typical example; studies on order formation in crystal growth have notably progressed, particularly in the field of epitaxial growth, owing to the development of modern technology. Although significant developments in nanoscale technology have been achieved in recent decades, in situ observation of the formation of atomic and/or molecular crystals remains a difficult task.

We show an interesting phenomenon observed in so-called colloidal crystals, which are macroscopic crystals consisting of colloidal particles. Although the characteristic size of this system is typically 10^3 times greater than that of atomic crystals, the phenomena exhibited by the crystal structure, such as Bragg reflection, domain structure, and crystal growth, are similar to atomic crystals (see the details in Chapter 5). In addition, the crystal structure and its orientation can be observed, using visible light instead of X-rays; thus, the particle fluctuations associated with the formation process are directly observable under a microscope, and various optical methodologies are also applicable to the examination. Therefore, this phenomenon is extremely suitable for investigating the order-formation process in real time. In the following section, we first briefly describe the general properties of colloidal crystals and then present a novel two-dimensional crystal growth process observed in our laboratory.

1.3.1 Colloidal Crystals

General Properties of Colloidal Crystals

When particles of sizes ranging from nanometers to microns are dispersed in fluid, they are called colloids or colloidal particles. Because an ensemble of colloidal particles acquires various forms in accordance with interparticle interactions, they have attracted significant attention, which has set colloid chemistry up as one of the traditional research fields in chemistry. Colloidal particles of the same size are at times arranged simultaneously in solution to form a macroscopic-scale crystal, which is commonly called a colloidal crystal. In 1935, the tobacco mosaic virus was found to form a crystalline shape in liquid, which was the first observation of colloidal crystals (Stanley, 1935). In 1947, colloidal spheres of uniform sizes became commercially available, which facilitated research on colloidal crystals.

Because the lattice constant of colloidal crystals is of the order of the wavelength of light, light illumination on a colloidal crystal causes Bragg reflection in the visible region, which produces iridescent color, as shown in Figure 1.11(a). When a colloidal particle is in solution, the functional groups existing on its surface generally dissociate to form surface charges. For example, for polystyrene spheres, sulfuric groups present at the surface almost completely dissociate into sulfuric ions with negative charges, as illustrated schematically in Figure 1.12(a). For silica spheres, silanol groups on the surface are partly dissociated to form weakly negative surface charges. Even in a solution that is sufficiently deionized, counterions resulting from residual ions are inevitably present; therefore, some of these ions are firmly adsorbed on the surface

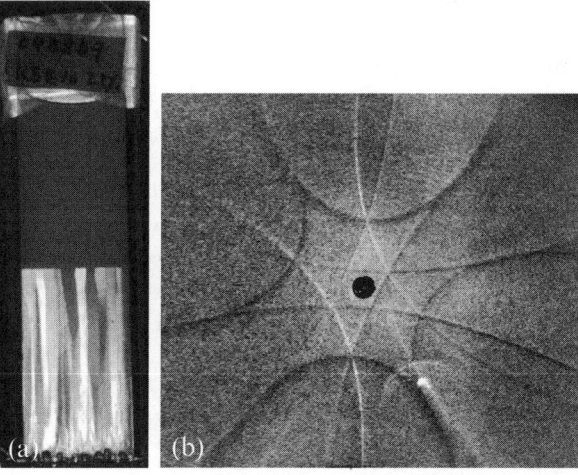

Figure 1.11 (a) Colloidal crystals and (b) Kossel lines. Columnar crystals seen in (a) are single crystals grown from the bottom of the cell where ion exchange resins are placed; the Bragg reflections from these differently directed crystals produce a colorful appearance. (Courtesy of Dr. H. Yamada.) (For color version of this figure, the reader is referred to the online version of this chapter.)

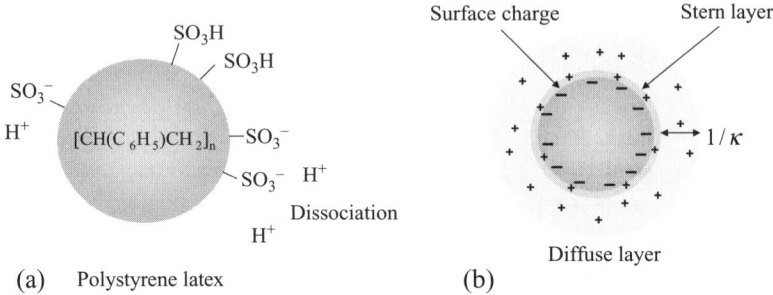

Figure 1.12 (a) Dissociation of functional groups on the surface of a polystyrene sphere and (b) Debye screening due to Stern and diffuse layers, where $1/\kappa$ is the Debye screening length. (Courtesy of Ms. R. Nakazawa.) (For color version of this figure, the reader is referred to the online version of this chapter.)

of the sphere to form the Stern layer, while the other ions surround the sphere to screen the surface charges and form a diffuse layer (Figure 1.12b). The surface charges and charges in the surrounding layers possess opposite signs and constitute the electrical double layer. The screening due to the counterions is called Debye screening, and its effective distance is called Debye screening length.

When the concentration of residual ions increases, Debye screening length generally decreases because only a thin layer is sufficient to screen the surface charges. Thus, particles can approach each other to reach an interparticle distance at which van der Waals' attractive force is operational. Therefore, the particles tend to aggregate with each other (see Figure 1.13a). When the ion concentration decreases, Debye's screening length increases. Even when two such screened particles approach each other, a repulsive force is naturally generated, preventing the overlapping of their diffuse layers owing to osmotic pressure. In this sense, a colloidal particle is considered as a particle covered with a diffuse layer of counterions. Thus, the colloidal particles exhibit independent movement, which is referred to as the liquid state (Figure 1.13b).

When the concentration of residual ions decreases much more, the apparent volume of a particle significantly increases. When the volume exceeds a certain limit, the particles are in constant contact, and their free motions are strongly restricted

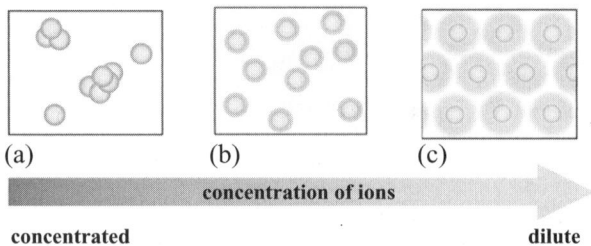

Figure 1.13 Various states of colloidal particles when the ion concentration is varied: (a) aggregation, (b) liquid, and (c) crystal. (Courtesy of Ms. R. Nakazawa.) (For color version of this figure, the reader is referred to the online version of this chapter.)

(see Figure 1.13c). Because the particles are originally subject to random thermal forces, under these conditions and in the presence of slight external forces, such as gravity and a diffusion force due to the concentration gradient, they are arranged into a close-packed structure known as a colloidal crystal. Thus, the physical quantities that determine the conditions of crystal formation are the volume fraction of particles, concentration of residual ions, and magnitude of the surface charges, the latter two of which are related to Debye screening (Yamanaka et al., 1998).

Colloidal crystals are experimentally prepared as follows:

1. Polystyrene or silica spheres of a uniform size are prepared and dispersed in a solution. Thereafter, this solution is dialyzed in ultrapure water to remove residual ions for at least several days.
2. The particles are dispersed in ultrapure water after an ion exchange resin is added and are then stored in a refrigerator.
3. A colloidal crystal is formed after a stored sample is poured into a solution containing a given concentration of ions. To grow large single crystals, ion exchange resins are placed at the bottom of the vessel to produce a gradient in the ion concentration.

The colloidal crystal thus grown generally acquires the form of a face-centered cubic (fcc) structure and shows iridescent color under illumination, as shown in Figure 1.11 (a). In this figure, all columnar crystals are single crystals grown from the bottom of the vessel, where ion exchange resins are placed. The single crystals appear in different colors owing to varying orientations of the crystals. In ordinary atomic crystals, the structure and orientation of a crystal are investigated through Bragg diffraction, using X-rays as a light source. However, in colloidal crystals the scale is required to be increased 10^3 times and visible light is employed to obtain this information.

The difference between atomic and colloidal crystals lies in the forces that maintain their crystal structures. In atomic crystals, the neighboring atoms are bound by a strong binding force, which firmly maintains the crystal structure, while in colloidal crystals, particles float in a solution, and weak electrostatic and chemical forces act between the particles. Thus, only a slight shake of the solution completely destroys a crystal. In addition, within a colloidal crystal, the position of each particle fluctuates owing to Brownian motion, and a clear halo based on this fluctuation is observed around the directional Bragg diffraction under illumination. The halo covers a wide angular range; thus, when its direction coincides with the Bragg condition of the crystal, Bragg diffraction occurs again and deflects part of the halo to a different direction. This phenomenon is clearly observed when the reflected or transmitted light is projected onto a screen. As shown in Figure 1.11(b), the Bragg condition appears as several bright or dark lines on this screen, depending on the observation direction. These lines, called Kossel lines, are used to analyze the crystal structure and orientation.

Although the fundamental question of the simultaneous formation of colloidal crystals has been extensively discussed, it cannot be easily answered, because it has not been confirmed whether only a repulsive force operates between two like-charged particles in a solution. The force operating between two such particles was theoretically analyzed by Derjaguin and Landau (1941) and Verwey and Overbeek

(1948). Their theories are now collectively called the DLVO theory, which has been used as a standard theory to analyze the interaction between colloidal particles. According to this theory, only a repulsive force should operate between two like-charged particles. However, under the presence of the other components, it is still not clear what happens essentially. In fact, an attractive force was reportedly present between two like-charged particles when they were near a glass plate (Larsen and Grier, 1997).

Two-dimensional Colloidal Array Formation

A system of colloidal crystals is particularly suitable for a direct observation of the order-formation process, because it can be observed in real time under a microscope, and various optical techniques are applicable to the examination of this system. We show one example observed in our laboratory (Nakazawa and Kinoshita, 2006), which presumably demonstrates the presence of an attractive force between particles.

Polystyrene latex beads containing sulfate ester were used in the present experiment; their diameter and density were 2.06 ± 0.024 μm and 1.5 g/cm^3, respectively. The particles were suspended in water at a volume fraction of 0.05–0.15 vol%. The suspension was first dialyzed in ultrapure water and then deionized by an ion exchange resin. A given amount of sodium chloride (NaCl) was then added before the suspension was transferred into a sample cell. Each colloidal particle carried a negative charge owing to dissociation of sulfate groups, and the effective charge in the solution was estimated to be 10^5 electron equivalent (-2×10^{-7} C/cm^2) from a measurement of electrophoretic mobility, assuming a constant charge.

The sample cell was made of glass slides; a curved acrylic plate was placed at the bottom of the cell to fix the ion-exchange resins (see Figure 1.14a). After the suspension was transferred to the sample cell, the top of the cell was covered with a cling film to prevent drying and for easy handling of the suspension during the observation. The observation was performed with an inverted microscope equipped with a digital camera.

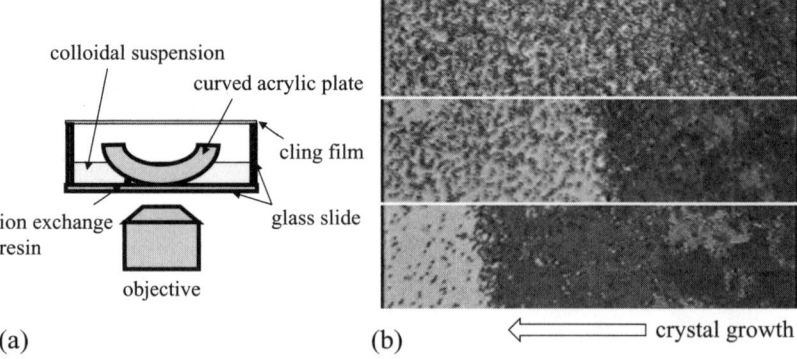

Figure 1.14 (a) Experimental setup and (b) two-dimensional colloidal crystal growth observed using this instrument. (Courtesy of Ms. R. Nakazawa.)

Using this system, we can observe the formation process of the colloidal crystal under a microscope in real time. We show an example in Figure 1.14(b) in which the crystal initially grew at 3.5 μm/min from right to left and then the growth process gradually slowed as the anterior portion of the crystal growth became distinctly visible. The crystal under consideration is not three-dimensional (3D) but is two-dimensional (2D) (we use the word "array" instead of "2D crystal" hereafter). Furthermore, the arrays are generated only under moderate ion concentrations ranging from 10^{-5} to 10^{-6} M, which cannot be classified into any of the categories shown in Figure 1.12.

The array began to grow near the ion exchange resins, which were added to reduce the number of existing ions, and extended within a 2D plane parallel to a glass slide placed at the bottom of the cell. The growth of an array generally begins as a triple layer with a mixture of hexagonal closest-packed (hcp) and fcc lattice structures, as shown in Figure 1.15(a). When the growth rate is reduced sufficiently by a decrease in the particle-number density, a monolayer with a closely packed triangular lattice structure becomes dominant (m in Figures 1-15a and b). This array has an unusually small lattice constant. Furthermore, the array has well-defined interfaces, which are entirely different from those of typical 3D crystals with a sufficiently large lattice constant. The colloidal array thus generated is well-defined and stable. It maintains its form for more than a few days, even though the surrounding ion concentration changes considerably.

Because of Brownian motion, particles fluctuate within a colloidal array under strict motional restriction within a 2D plane; however, the array never dissociates. This is a strong indication of particle–particle attraction within the array. Furthermore, 2D arrays apparently have a characteristic size, and after the stationary state is attained, they neither grow nor bind beyond this size (see Figure 1.15b). Thus, a whole

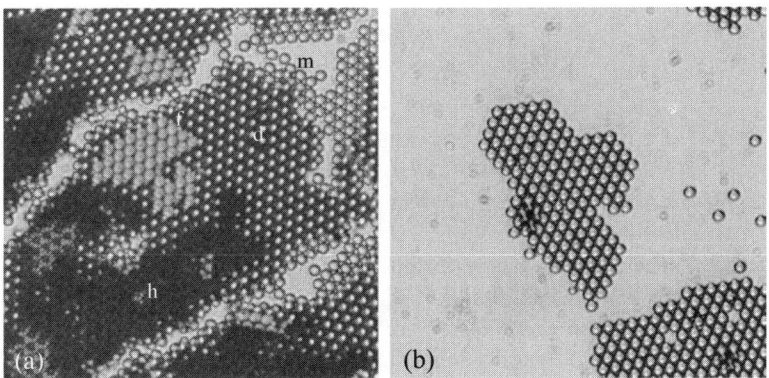

Figure 1.15 Two-dimensional colloidal array formed at an initial NaCl concentration of 10^{-5} M and volume fraction of colloidal particles of 0.05 vol%. (a) Region where mixtures of (m) monolayers, (d) double layers, and (h and f) triple layers are observed, where the triple layer consists of hcp (h) and fcc (f) structures. (b) Well-defined monolayer array observed at the leading edge of the growth. (Courtesy of Ms. R. Nakazawa.)

of the arrays resemble an ensemble of patches. The array is not in direct contact with the glass plate but floats within a distance of 0.5 μm above the plate, because the glass plate carries a negative surface charge that was estimated to be -2×10^{-6} C/cm^2. Therefore, the 2D array actually moves along the plate when the sample cell is slightly inclined.

We investigated the conditions for the formation of the 2D array and found that the initial ion concentration plays an important role under appropriate particle concentrations. However, under our experimental conditions, it was difficult to control the local ion concentration because the ion concentrations decreased continuously owing to the function of the ion exchange resins. This problem was solved as follows: First, we prepared a suspension containing a given concentration of NaCl between 10^{-5} and 10^{-6} M. Thereafter, we transferred the suspension into a cell containing only anion exchange resins. In addition to the extra Na$^+$ and Cl$^-$ ions, the suspension contained excess H$^+$ ions produced by the dissociation of sulfate groups on the surfaces of the particles. After the addition of the anion exchange resins, the H$^+$ ion concentration decreased owing to the following reaction: resin-OH + H$^+$ + Na$^+$ + Cl$^-$ → resin-Cl + Na$^+$ + H$_2$O. Because the Na$^+$ ion concentration remained constant during the reaction, the cation concentration asymptotically approached a constant value determined by the Na$^+$ ion concentration. Thus, the anion exchange resin plays an important role in the formation of the colloidal array: it produces a spatial gradient of the ion concentration, thus assuring crystallization, while maintaining a sufficient ion concentration to form the array.

Next, we investigated whether the attractive force that brings colloidal particles into a 2D array originates in the van der Waals interaction appearing at high ion concentrations. For this purpose, we added a concentrated NaCl solution to the colloidal suspension containing the 2D array to increase the surrounding local ion concentration. Figure 1.16 shows an example of this behavior. In this figure, a 10^{-3} M NaCl solution was added to a suspension prepared 20 h before the addition of extra ions. The NaCl solution was added at a distance of 1 cm from the array with special care not to cause a flow in the suspension. When the extra solution was uniformly distributed, the initial Na$^+$ ion concentration of 5×10^{-6} M eventually increased to 2.5×10^{-5} M. As shown in Figure 1.16, the array split into two small pieces after 34 min and then into separate particles within 45 min.

If the attractive force within the array originates from the van der Waals attraction, the interparticle binding force would increase, and condensation may eventually occur. However, the result in Figure 1.16 is inconsistent with this hypothesis. This experiment strongly suggests that the array exhibits electrostatic attraction instead of van der Waals attraction. However, this type of attraction cannot be explained in terms of the conventional DLVO theory, which is applicable only to a particle pair. We performed additional experiments by adding NaCl at various concentrations and found that the colloidal array collapses only when the concentration of the NaCl solution is greater than 10^{-5} M. Thus, an ion concentration of $\sim 10^{-5}$ M is the upper limit for maintaining the array.

In contrast, the colloidal array is stable against a decrease in the surrounding ion concentration. To investigate this quantitatively, we performed the following

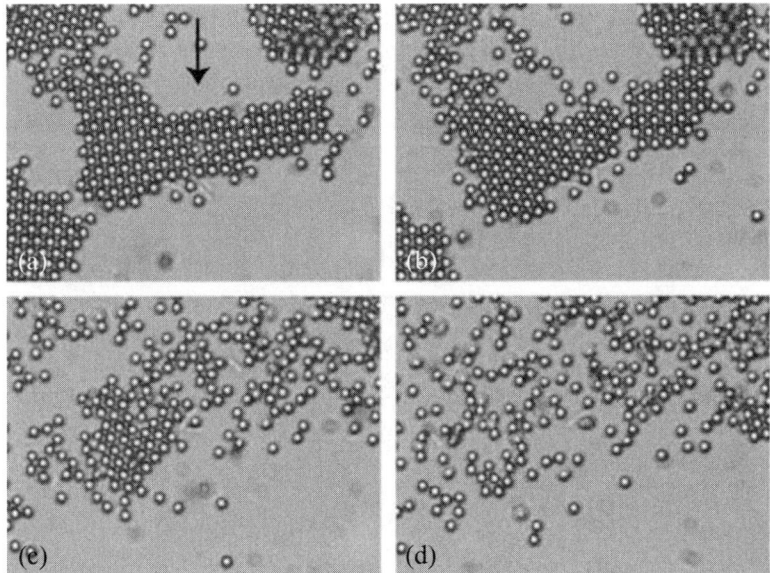

Figure 1.16 Collapse of a monolayer colloidal array caused by the addition of NaCl solution to a sample solution. NaCl solution (24 μl of 10^{-3} M) was added to a 1200 μl sample solution 1 cm separated from the monolayer colloidal array: (a) 30 min., (b) 34 min., (c) 44 min., and (d) 49 min. after addition of NaCl. The arrow indicates direction of diffusion flow of Na^+ ions. (Courtesy of Ms. R. Nakazawa.)

experiment. First, a colloidal suspension containing 10^{-5} M NaCl with anion exchange resins is prepared, and a 2D colloidal array is grown in the sample cell. After one day, a mixture of anion and cation exchange resins is added to the suspension, and the mixture is arranged in a line 5 mm from the array, as shown in Figure 1.17(a). Thus, the ion concentration around the resins begins to decrease, and this decrease extends to the entire cell through ionic diffusion.

During this experiment, the particle–particle distance estimated for a monolayer array hardly changes or decreases only slightly. Therefore, the colloidal array is not thermodynamically equilibrated. Nevertheless, this array is stable and maintains its form for more than 100 h. Thereafter, the colloidal array is divided into small fractures (Figures 1.17b–d), while the particle–particle distance remains nearly constant; it subsequently collapses into separate particles. On the other hand, in a double- or triple-layer region, the array does not collapse, and both particle–particle distance and size remain constant.

The extremely high stability of the array against the decrease in the surrounding ion concentration suggests that the entire system of the 2D array—which consists of the colloidal particles, ionic distribution, electrostatic field distribution, and colloidal surface reaction—is thermodynamically metastable. The remarkable stability of this array is apparently closely related to its dimensionality, which presumably includes the effect of the electrostatic and hydrodynamic interactions with the underlying glass plate. However, to clarify this, more sophisticated experiments and calculations are necessary.

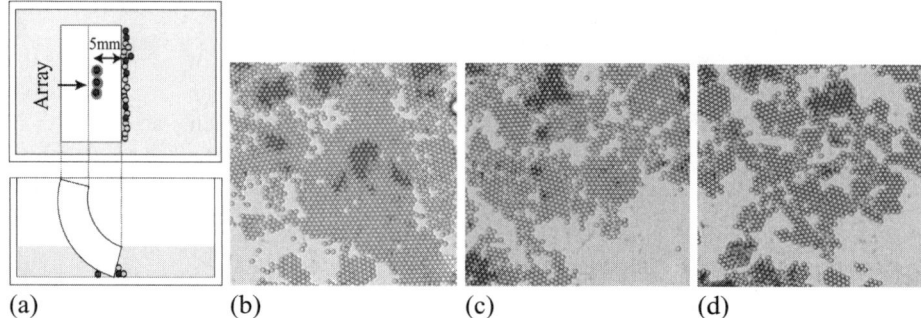

Figure 1.17 (a) Experimental setup for observing 2D colloidal array as the ion concentration decreases. A mixture of anion and cation exchange resins is placed 5 mm separated from the 2D array. The colloidal array is shown during the removal of Na^+ ions at (b) 0 h, (c) 24 h, and (d) 76 h after the addition of the ion exchange resins. (Courtesy of Ms. R. Nakazawa.)

The preceding experiment clarifies that the colloidal crystal is suitable for this type of experiment because the order-formation process is observable in real time under a microscope. In addition, direct measurement of the fluctuations of each particle under a microscope is easy. The 2D array is particularly promising for investigating particle–particle interactions in the presence of other components, such as a glass plate. The presence of such components during self-organization is inevitable; therefore, this system is particularly promising for its future development as an excellent tool for clarifying the fundamental mechanism of the self-organization process in a crystal.

1.4 Pattern Formation

Living and nonliving systems in nature exhibit a wide range of patterns, and therefore the formation processes of these patterns are expected to differ significantly. Here, we will describe only a few cases that are, at least partially, mathematically analyzable and/or cases in which the functions of the resulting patterns can be explained from a viewpoint of physics.

The first case involves pattern formation during crystal growth. The patterns formed during crystal growth are familiar in daily life. We observe a dendritic pattern of ice on a glass window in cold weather. However, questions of how and why such a pattern is formed simultaneously need scientific considerations. The second case involves biological patterns that presumably occur as a kind of marking; such patterns are widely found in the animal and plant kingdoms. This type of pattern has been described as a manifestation of the Turing pattern. Therefore, we briefly explain the essential part of the Turing instability and show some examples observed in chemical and biological systems. Finally, we focus on nanostructures observed in biological systems, which presumably represent a type of nanoscale biological pattern and exhibit coloration mechanisms on their interaction with light (see the details in Chapter 6).

1.4.1 Instability in Crystal Growth

We first show an essential part of the Mullins–Sekerka instability, which explains the initial stage of pattern formation during crystal growth (Mullins and Sekerka, 1963, 1964). The growth form of isotropic crystals is one of the comprehensible examples of pattern formation; it is mathematically analyzable and has been quantitatively evaluated in experiments.

Instability in crystal growth can be observed using an experimental configuration called directional solidification (Langer, 1980) in which a sample is sandwiched between two thin glass plates, applying different temperatures on both sides, as shown in Figure 1.18(a), so that the melting point approximately coincides with the temperature at the center of the glass plates. By solely moving the glass plates under a constant temperature gradient, the time course of one-dimensional (1D) crystal growth can be observed under a microscope. In Figure 1.18(b), we show an example of such experiments conducted using succinonitrile as a sample, which is known to display almost isotropic crystal growth. This figure shows the time course of the experiment at three different growth speeds. At a slow growth speed (C), the liquid–crystal interface is subject to slightly irregular sinusoidal modulation, which acquires a columnar shape with time. On the other hand, during rapid growth (A), a fine projection-like modulation pattern is initially generated and appropriate projections are selected from the pattern, which promotes the growth of only specific columnar projections at nearly constant intervals. Further, the sides of each projection are modulated, which eventually produces a dendritic pattern.

Although such an experiment clearly reveals the presence of various types of instabilities, which depend on the growth speed and lapsed time, we focus on the simplest case of the initial modulation observed in the preceding experiment. Consider an isotropic crystal growing uniformly in the x direction within a 2D plane spanned by the x

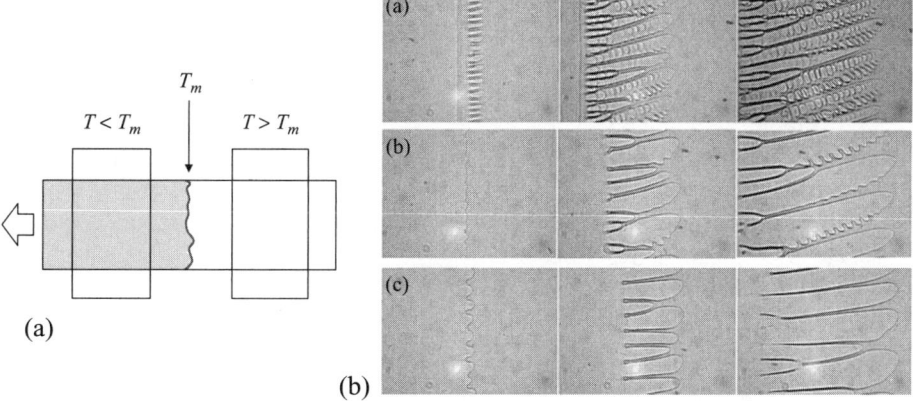

Figure 1.18 (a) Schematic illustration of directional solidification experiment and (b) actual experiments on succinonitrile. Speed with which cell is pulled is (A) fast, (B) medium, and (C) slow. (Courtesy of Mr. Y. Sato.)

and z coordinate axes. The governing equations in this case are expressed by the following three relations (Langer, 1980; Caroli et al., 1991; Takaki, 1992). The first relation is the heat conduction equations

$$D\nabla^2 T^{(l)} = \frac{\partial T^{(l)}}{\partial t}, \tag{1-42}$$

$$D'\nabla^2 T^{(c)} = \frac{\partial T^{(c)}}{\partial t}, \tag{1-42'}$$

where $T^{(l)}$ and $T^{(c)}$ denote the temperatures in the liquid and crystal phases, respectively, which are functions of space and time, and D and D' are the corresponding diffusion constants.

The second relation is known as the Gibbs–Thomson equation, which states that the melting point of a crystal at the interface depends on its curvature:

$$T_{int} = T_m(1 - \gamma\kappa/L), \tag{1-43}$$

where T_{int} and T_m are the melting points at the interface and a flat surface, respectively, and γ, κ, and L are the surface tension coefficient, curvature of the interface, and latent heat, respectively.

The third relation is associated with energy conservation during crystal growth and is expressed by

$$Lv_n = \left(D'c'_p \nabla T_c - Dc_p \nabla T_l\right) \cdot \mathbf{n}, \tag{1-44}$$

where c_p and c'_p are the values of specific heat at constant pressure for the liquid and crystal, while ∇T_l and ∇T_c are the temperature gradients within the liquid and crystal phases, respectively, at their interface. Here, $\mathbf{n}$ is a unit vector normal to the interface, and v_n is the growth speed in this direction. The left side of this equation represents the latent heat production during growth, while the first and second terms on the right side indicate the heat flow measured at the interface in the crystal and liquid phases, respectively.

The previous three equations can be made nondimensional by setting $u \equiv (T - T_m)/(L/c_p)$:

$$D\nabla^2 u^{(l)} = \frac{\partial u^{(l)}}{\partial t} \text{ and } D'\nabla^2 u^{(c)} = \frac{\partial u^{(c)}}{\partial t}, \tag{1-45}$$

$$u_{int} = -d_0\kappa, \tag{1-46}$$

$$v_n = D\{\beta\nabla u_c - \nabla u_l\} \cdot \mathbf{n}, \tag{1-47}$$

where $d_0 = \gamma T_m c_p / L^2$, and $\beta = D'c'_p/(Dc_p)$. ∇u_c and ∇u_l correspond to ∇T_c and ∇T_l, respectively.

We consider crystal growth in the x direction with a constant speed v and assume that a stationary pattern is formed during this process. We observe the growth surface of the crystal in a moving coordinate system at the same speed v. Then, introducing the new variables x' and t' as $x = x' + vt'$ and $t = t'$, and setting $u(x, z, t) = \bar{u}(x', z, t')$, we obtain $\partial \bar{u}/\partial x' = \partial u/\partial x$ and $\partial \bar{u}/\partial t' = v\, \partial u/\partial x + \partial u/\partial t$ for which we have used the following relation:

$$\left(\frac{\partial \bar{u}}{\partial x'} \;\; \frac{\partial \bar{u}}{\partial t'} \right) = \left(\frac{\partial u}{\partial x} \;\; \frac{\partial u}{\partial t} \right) \begin{pmatrix} \frac{\partial x}{\partial x'} & \frac{\partial x}{\partial t'} \\ \frac{\partial t}{\partial x'} & \frac{\partial t}{\partial t'} \end{pmatrix} = \left(\frac{\partial u}{\partial x} \;\; \frac{\partial u}{\partial t} \right) \begin{pmatrix} 1 & v \\ 0 & 1 \end{pmatrix}.$$

Thus, transforming $\partial u/\partial t \to \partial \bar{u}/\partial t' - v\, \partial \bar{u}/\partial x'$ and $\partial u/\partial x \to \partial \bar{u}/\partial x'$ and further setting $\partial \bar{u}/\partial t' = 0$, because stationary growth is assumed, the diffusion equations associated with the moving coordinate system are expressed as

$$\nabla^2 u^{(l)} + (2/l)\partial u^{(l)}/\partial x = 0, \tag{1-48}$$

$$\nabla^2 u^{(c)} + (2/l')\partial u^{(c)}/\partial x = 0, \tag{1-49}$$

where we put $\bar{u} \to u$ and $x' \to x$ with $l = 2D/v$ and $l' = 2D'/v$ for simplicity.

First, we consider a case in which the growth surface is uniform in space and time; a stationary temperature gradient is formed in both liquid and crystalline phases. Setting $u(x, z) \to u(x)$, the solutions of Eqs. (1-48) and (1-49) can be easily obtained by modifying either of the equations into $\partial/\partial x\, (\partial u/\partial x + 2u/l) = 0$, where l represents l or l' for the concerned case. Thus, a first-order linear differential equation $\partial u/\partial x + 2u/l = \text{const}$ is obtained, which is solved to obtain a general solution for $u = A - B\exp[-(2/l)x]$, where A and B are arbitrary constants.

We consider an ideal case of $u(-\infty) = 0$ and $u(\infty) = \text{const}$ for directional solidification, which means that the temperature in the crystalline phase is uniformly maintained at T_m, while that in the liquid phase, which is distant, is set at a low temperature to guarantee stationary crystal growth. Under this condition, $A = B = 0$ in the crystalline phase, while $A - B = 0$ in the liquid phase, which is obtained from the boundary condition at $x = 0$. Therefore, the $u(x)$ values in the crystalline and liquid phases are obtained as $u(x) = 0$ and $u(x) = A(1 - \exp[-2x/l])$, respectively. Further, by inserting these solutions into Eq. (1-47), the relation $v = -DA(2/l)$ is obtained at $x = 0$; consequently, $A = -1$. Thus, the final result is

$$u(x) = \begin{cases} e^{-2x/l} - 1, & (x > 0 \text{ in liquid phase}) \\ 0, & (x < 0 \text{ in crystal phase}) \end{cases} \tag{1-50}\tag{1-51}$$

The temperature gradient obtained is shown as a solid line in Figure 1.19(b), which indicates the presence of a supercooled state in the liquid phase.

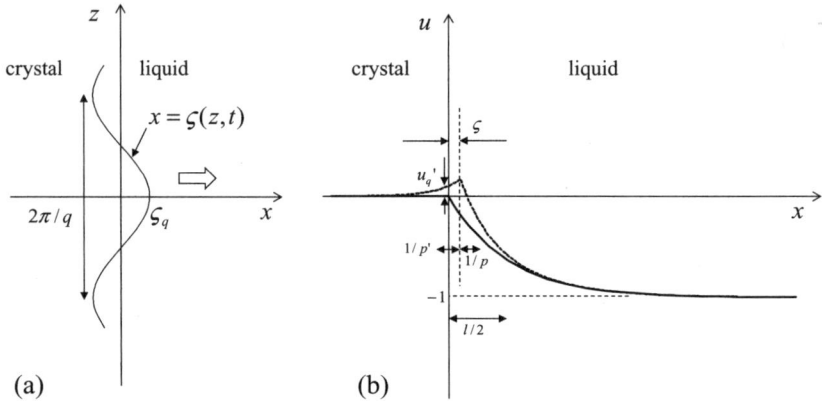

Figure 1.19 (a) Schematic view of modulation of liquid–crystal interface, and (b) its influence on the temperature distribution. Solid and dashed curves are the initial and resulting temperature distributions, respectively.

Next, we consider a case in which the growth surface of the crystal is spatially and temporally modulated during growth. This modulation deforms the interfacial shape, which is denoted by $\zeta(z,t)$; as a result, the temperature distribution is varied in both the liquid and crystalline phases. This situation is considered to be simply introduced into $u^{(l)}$ and $u^{(c)}$ as follows:

$$\begin{cases} \zeta = \zeta_q \exp[iqz + \omega_q t], & (1\text{-}52) \\ u^{(l)} = e^{-2x/l} - 1 + u_q \exp[iqz - px + \omega_q t], & x > \zeta, \quad (1\text{-}53) \\ u^{(c)} = u'_q \exp[iqz + p'x + \omega_q t], & x < \zeta, \quad (1\text{-}54) \end{cases}$$

where we consider a periodic deformation along the z axis, which is characterized by a wavenumber q, as shown in Figure 1.19(a). Here, ζ_q, u_q, and u'_q are constants that take extremely small values, while p, p', and ω_q are parameters that are required to be determined to satisfy the basic equations. The effect of the deformation on the temperature distribution is schematically illustrated as a dashed line in Figure 1.19(b). To avoid divergence, it is implicitly assumed that p, $p' > 0$, while ω_q can take both positive and negative values. When $\omega_q < 0$, the deformation decreases with time, and the stationary and uniform growth surface reappears. In contrast, when $\omega_q > 0$, the deformation grows exponentially with time, resulting in instability with respect to the interfacial shape.

First, inserting Eqs. (1-53) and (1-54) into Eqs. (1-50) and (1-51), respectively, we obtain the following relations:

$$-(2p/l) + p^2 - q^2 = 0, \quad (1\text{-}55)$$

$$(2p'/l') + p'^2 - q^2 = 0. \quad (1\text{-}56)$$

Next, we consider the effect on the Gibbs–Thomson equation Eq. (1-46). For this purpose, we should express the curvature in terms of ζ. The curvature κ is generally defined as

$$\kappa \equiv \frac{d^2x}{dz^2} \bigg/ \left(1 + \left(\frac{dx}{dz}\right)^2\right)^{3/2},$$

where we consider $x=x(z)$ as a function that indicates the shape of the interface. Inserting $x = \zeta = \zeta_q \exp[iqz + \omega_q t]$, we obtain

$$\kappa = \frac{-q^2\zeta}{(1 - q^2\zeta^2)^{3/2}} \approx -q^2\zeta.$$

Further, because u_{int} is expressed as $u_{int} = (u^{(l)})_{x=\zeta} = (u^{(c)})_{x=\zeta}$, the corresponding relation becomes

$$u_{int} = e^{-2\zeta/l} - 1 + u_q \exp[iqz - p\zeta + \omega_q t] = u_q' \exp[iqz + p'\zeta + \omega_q t],$$

which results in

$$-d_0 q^2 \zeta_q \approx -2\zeta_q/l + u_q \approx u_q', \tag{1-57}$$

assuming small values of ζ, u_q, and u_q'.

Finally, we consider Eq. (1-47) approximated as

$$\frac{\partial \zeta}{\partial t} + v \approx D\left\{\beta\left(\frac{\partial u^{(c)}}{\partial x}\right)_{x=\zeta} - \left(\frac{\partial u^{(l)}}{\partial x}\right)_{x=\zeta}\right\},$$

assuming that the deformation is so small that **n** nearly lies along the x axis. This relation leads to

$$\omega_q \zeta_q = D\left\{\beta p' u_q' - (2/l)^2 \zeta_q + u_q p\right\}, \tag{1-58}$$

where we again assume small values of ζ, u_q, and u_q'.

Because Eqs. (1-57) and (1-58) are simultaneous homogeneous equations with respect to ζ_q, u_q, and u_q', nonzero solutions can be obtained only when the determinant of the coefficient matrix is zero. Thus, we finally obtain

$$\omega_q = v(p - 2/l) - D(p + \beta p')d_0 q^2. \tag{1-59}$$

Further, if we assume that l is sufficiently larger than the wavelength of spatial modulation, then $l \gg 1/q$, and because $p > 0$ and $p' > 0$, the solutions of Eqs. (1-55) and (1-56) with respect to p and p' provide $p \approx p' \approx q$, considering $q \geq 0$. Thus, replacing p or p' with q and neglecting the term containing $1/l$, which is extremely smaller than q, we arrive at a simple expression corresponding to Eq. (1-59):

$$\omega_q = vq\left\{1 - \frac{1}{2}l(1+\beta)d_0q^2\right\}. \tag{1-60}$$

Figure 1.20 shows the functional form of Eq. (1-60) for various values of l, indicating that $\omega_q > 0$ is realized within a particular region of q from 0 to $q_c (= ((1+\beta)d_0 l/2)^{-1/2})$, while ω_q is maximum at $q_m = q_c/\sqrt{3}$. This type of instability is known as the Mullins–Sekerka instability. By decreasing l further and thus increasing the growth speed v, the wavenumber, which provides a peak in the curve, increases, indicating the occurrence of fine spatial modulation at the beginning of the instability, which qualitatively explains the experiment shown in Figure 1.18(b).

Chou and Cummins (1988) and Qian et al. (1989) evaluated the Mullins–Sekerka instability quantitatively, using succinonitrile as the sample. Their directional solidification experiments showed that the growth process on the surface began with a sinusoidal fluctuation, but a coarsening phenomenon shortly occurred, which eventually produced a dendritic pattern. A spatial Fourier transform was applied to the interfacial curve obtained from the time-course experiment and the average growth rates for various wavenumbers were deduced by fitting the curve with Eq. (1-52). A curve similar to that predicted by the Mullins–Sekerka theory was obtained, but the q_m thus obtained was considerably smaller than predicted. They attributed the difference to the presence of impurities and differences in the experimental conditions, such as the temperature gradient and the finite thickness of the sample. Furthermore, the deviation from the linear analysis and the presence of residual anisotropy have not been discussed thus far. Therefore, further study is clearly needed for the experimental verification in a true sense.

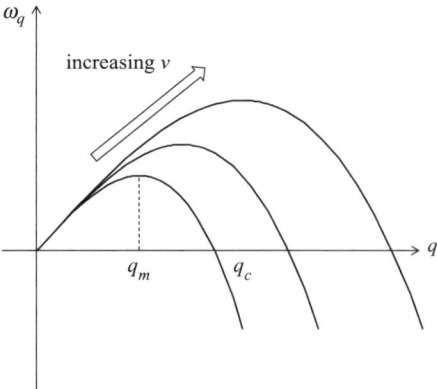

Figure 1.20 Mullins–Sekerka instability occurs within a region of $\omega_q > 0$, which corresponds to a range of q from 0 to q_c. The region is extended by increasing v, which increases q_m, where q_m is the wavenumber at which ω_q is maximum.

1.4.2 Turing Pattern

Turing Instability

Patterns generated in reaction–diffusion systems have received considerable attention in the field of biology. The diffusion process is generally considered to level out any structure and make the system homogeneous. However, by coupling it with the reaction process, it is possible to destabilize the homogeneous system and generate a temporally stationary and spatially inhomogeneous pattern. This idea was first proposed by Turing (1952), and the pattern created is now known as a Turing pattern. The Turing pattern is thought to be widely distributed in nature and to be the origin of various patterns observed in living things. Various theoretical analyses and computer simulations have demonstrated its significance; however, this was not experimentally verified until 1990, when the stationary pattern was first observed experimentally in the reaction–diffusion process (Castets et al., 1990).

Before describing the details of this experiment, we will consider the instability observed in the reaction–diffusion system (Ohta, 2000). Consider a simple 1D model system consisting of an activator and an inhibitor; concentrations of the two components are denoted by u and v, respectively, which are generally expressed by the following reaction–diffusion equations:

$$\frac{\partial u}{\partial t} = D_u \frac{\partial^2 u}{\partial x^2} + f(u,v), \tag{1-61}$$

$$\frac{\partial v}{\partial t} = D_v \frac{\partial^2 v}{\partial x^2} + g(u,v), \tag{1-62}$$

where D_u and D_v are the diffusion coefficients for u and v, respectively, and $f(u,v)$ and $g(u,v)$ are their reaction terms. We investigate the stability of the solutions of these equations using a linearized stability analysis in which their concentrations are spatially modulated with the wavenumber q, considering $u = \bar{u} + a\exp(\lambda t + iqx)$ and $v = \bar{v} + b\exp(\lambda t + iqx)$, where $\bar{u}$ and $\bar{v}$ are stationary homogeneous solutions that satisfy $f(\bar{u},\bar{v}) = g(\bar{u},\bar{v}) = 0$, and a and b are sufficiently small. For example, if we insert $u = \bar{u} + a\exp(\lambda t + iqx)$ into Eq. (1-61) and differentiate it with respect to the spatial and temporal coordinates, we obtain the following expression:

$$a\lambda e^{\lambda t+iqx} = -D_u a q^2 e^{\lambda t+iqx} + f\left(\bar{u} + ae^{\lambda t+iqx}, \bar{v} + be^{\lambda t+iqx}\right)$$
$$\approx -D_u a q^2 e^{\lambda t+iqx} + f_u a e^{\lambda t+iqx} + f_v b e^{\lambda t+iqx},$$

where $f_u \equiv (\partial f/\partial u)_{u=\bar{u},v=\bar{v}}$, and $f_v \equiv (\partial f/\partial v)_{u=\bar{u},v=\bar{v}}$. Because we assumed that the spatial modulation and period of temporal evolution are sufficiently small, we can consider only the first-order partial derivatives of $f(u,v)$. Thus, the preceding relation is reduced to

$$a\lambda = -D_u a q^2 + f_u a + f_v b. \tag{1-63}$$

A similar procedure yields

$$b\lambda = -D_v b q^2 + g_u a + g_v b, \tag{1-64}$$

where $g_u \equiv (\partial g/\partial u)_{u=\bar{u},v=\bar{v}}$, and $g_v \equiv (\partial g/\partial v)_{u=\bar{u},v=\bar{v}}$. Equations (1-63) and (1-64) are simultaneous homogeneous equations in two unknowns, a and b, which have non-zero solutions only when the following relation holds:

$$\begin{vmatrix} \lambda + D_u q^2 - f_u & -f_v \\ -g_u & \lambda + D_v q^2 - g_v \end{vmatrix} = 0.$$

Then, we obtain the following characteristic equation to evaluate the stability of the system:

$$\lambda^2 + A(q)\lambda + B(q) = 0, \tag{1-65}$$

where

$$A(q) = q^2(D_u + D_v) - (f_u + g_v), \tag{1-66}$$

and

$$B(q) = D_u D_v q^4 - q^2(D_u g_v + D_v f_u) + f_u g_v - f_v g_u. \tag{1-67}$$

Generally, in the activator–inhibitor model, the relations $f_u > 0$, $f_v < 0$, $g_u > 0$, and $g_v < 0$ are assumed; the first relation guarantees an autocatalytic reaction such that an increase in u causes a further increase in u, while the second relation ensures the inhibition of this autocatalytic reaction by the presence of the inhibitor v. On the other hand, the third and fourth relations represent the generation of the inhibitor by an increase in the activator concentration and damping of the inhibitor. Solving Eq. (1-68) with respect to λ, we obtain

$$\lambda = \frac{-A(q) \pm \sqrt{A(q)^2 - 4B(q)}}{2}.$$

Linearized stability is obtained when the condition $\text{Re}\{\lambda\} < 0$ is satisfied. For $q=0$, this condition is appropriately satisfied when $A(0) > 0$ and $B(0) > 0$, which are equivalent to $f_u + g_v < 0$ and $f_u g_v - f_v g_u > 0$ from Eqs. (1-66) and (1-67), and correspond to the condition that $\bar{u}$ and $\bar{v}$ are linearly stable solutions.

For $q \neq 0$, if $B(q)$ is negative in a particular range of q, λ acquires a positive value, which results in the instability of the stationary homogeneous solutions. To estimate the value of q, we consider q_c, which gives the minimum value of $B(q)$. This is obtained by differentiating Eq. (1-70) with respect to q^2 and setting it to zero, which is given as

$$q_c^2 = \frac{D_u g_v + D_v f_u}{2 D_u D_v}, \tag{1-68}$$

with

$$B(q_c) = -\frac{(D_u g_v + D_v f_u)^2}{4 D_u D_v} + f_u g_v - f_v g_u. \tag{1-69}$$

The large negative value of the first term on the right side of Eq. (1-69) results in $B(q_c) < 0$ because the sum of the second and third terms gives a positive value, as previously described.

When we focus on the diffusion constants for the activator and inhibitor, the preceding situation is attained when $D_u \ll D_v$ or $D_u \gg D_v$, because in each case, the first term on the right side of Eq. (1-69) reduces to $-D_v f_u^2/(4 D_u)$ or $-D_u g_v^2/(4 D_v)$, respectively. Further, because $f_u > 0$ and $g_v < 0$, the latter condition is inappropriate owing to the relation in Eq. (1-68). Thus, D_u takes a much smaller value than D_v to generate the spatial instability in a finite wavenumber range, which is called the Turing instability. When $-D_u g_v \ll D_v f_u$, Eq. (1-68) reduces to

$$q_c \approx (f_u/2D_u)^{\frac{1}{2}}, \tag{1-70}$$

which gives the relationship between the pattern size and reaction rate/diffusion coefficient.

Turing Patterns in Chemical Systems

The presence of this type of instability was predicted in 1952, but the generation of the stationary pattern was only recently observed experimentally in a laboratory. Castets et al. (1990) employed a chlorite–iodide–malonic acid (CIMA) system to observe the stationary spatial pattern. A flat piece of polyacrylamide gel 3×20 mm^2 in size and 1 mm thick was used as a reactor, which was placed in contact with two chemical reservoirs containing malonic acid (MA)/sulfuric acid on one side and ClO_2^-/I^- on the other. To observe the reaction process, a nondiffusive soluble starchlike color indicator was loaded into the gel, which changed from yellow to blue with the change in the concentration ratio $[I_3^-]/[I_2]$. A periodic pattern with a pitch of 0.2 mm within the gel was found; this pattern was sustained for more than 20 h. They also found that the pattern was sensitive to temperature and depended on the range of the concentrations of the chemicals.

Another elaborate experiment was performed, using the same CIMA system proposed by Ouyang and Swinney (1991), who employed a 2.0 mm–thick gel plate sandwiched between 0.4 mm–thick porous glass disks. Stationary 2D hexagonal pattern and stripes, which were dependent on the concentrations of the chemicals, were found to generate. The temperature dependence of the pattern formation was investigated, and it was found that the pattern formed only below 18°C, and its amplitude depended on the temperature. The pitch of the pattern varied from 0.14 to 0.33 mm, depending on the parameters employed. Despite the significance of these pioneering studies, such patterns are not ubiquitous and are limited to special cases, such as a single phase of the CIMA system and a mixed phase of the BZ system (Vanag and Epstein, 2001).

This limitation is attributed mainly to the difficulty of selecting the parameters. For the Turing instability described earlier, it is suggested that the diffusion constant of the activator should be much smaller than that of the inhibitor. Because both activator and inhibitor generally comprise small molecules, it is difficult to induce a difference between these values. In the preceding CIMA system, this condition was presumably fulfilled through a special mechanism. In this system, three major reactions are believed to take place such that

$$MA + I_2 \rightarrow IMA + I^- + H^+,$$
$$ClO_2 + I^- \rightarrow ClO_2^- + \frac{1}{2}I_2,$$
$$ClO_2^- + 4I^- + H^+ \rightarrow Cl^- + 2I_2 + 2H_2O.$$

Because the concentrations of $[ClO_2]$, $[I_2]$, and $[MA]$ are slowly varying functions of time, Lengyel and Epstein (1991) assumed these parameters to be independent of time and derived a two-variable model equivalent to Eqs. (1-61) and (1-62), where $[I^-]$ and $[ClO_2^-]$ acted as the activator and inhibitor, respectively. Using the appropriate parameter values, the reactions were successfully simulated, and it was concluded that the Turing pattern was obtainable when $D_v/D_u > 10$. Although the actual reason for this large difference in the diffusion constants was unknown, I^- ions were assumed to be specifically trapped with gel or starch for a period of time, which delayed the diffusion process of I^- ions and apparently reduced the diffusion constant. Although various experimental designs have been proposed to generate chemical Turing patterns, the experimental development of these patterns is even now in progress (Horváth et al., 2009).

Turing Patterns in Biological Systems

Turing's concept of instability in a homogeneous system originally aimed to explain the mechanism of various spatial patterns generated in a biological system (Turing, 1952). Turing assumed a set of chemical substances called morphogens, which acted as form producers to create such patterns by reacting chemically with each other and diffusing across the cell membrane. For mathematical convenience, he considerably simplified the problem by assuming a system consisting of an isolated ring of cells in contact with its neighbors or of a continuous ring of tissues, and considered only the fluctuations in morphogen concentration within a region that did not distantly deviate from the original homogeneous condition. He considered that the system was initially in a stable homogeneous condition, which was then disturbed by Brownian motion or the irregularity of the form.

His conclusions are summarized as describing the following six types of patterns: (1) Instability occurs in one direction and uniformly changes the entire system. He cited the example of dappled color patterns in this case. (2) The pattern is uniform, as in (1), but is oscillatory. (3) The spatial scale of the drift from a homogeneous equilibrium condition is of the order of a cell, so the drifts in adjacent cells carry opposite signs. (4) A stationary wave pattern is produced on the ring with no time variation, which corresponds to the so-called Turing pattern. Five to ten tentacles found in the hydra and woodruff leaf whorls were used as samples for this case. All patterns just described involve two morphogens, but when the number of morphogens is increased further, other patterns are generated: (5) a pair of waves traveling in opposite directions and (6) an oscillation with neighboring cells in opposite phases.

Turing's pioneering study considerably promoted the theoretical studies, and various biological patterns (e.g., see Figure 1.21) were simulated by assuming the appropriate reaction–diffusion equations (see Koch and Meinhardt, 1994). However, these studies mainly compared the simulated patterns with actual biological ones to determine the similarities between the patterns; thus, rational explanations for the validity of the Turing pattern could not be provided.

Recently, Kondo and Asai (1995) investigated the time evolution of the stripe pattern in the skin of the marine angelfish *Pomacanthus*. Unlike mammalian skin patterns, during body growth, the stripes of this species maintained spaces between the lines, which was achieved by continuous rearrangement of the patterns. The results

Figure 1.21 Animals exhibiting spotted and striped patterns. (For color version of this figure, the reader is referred to the online version of this chapter.)

of a laser ablation experiment performed on the stripe patterns of the zebra fish suggested that a clear stripe pattern was regenerated when all the pigment cells were ablated, while a bell-shaped deformation of the nearest black stripe was observed when only the melanophores were selectively ablated (Yamaguchi et al., 2007). The black stripes comprised melanophores, while the yellow stripes between the black ones consisted of xanthophores; therefore, these two types of pigment cells interact to a certain extent during the stripe formation. In this study, the time evolutions of the patterns were simulated using the reaction–diffusion equations; on the basis of the similarity between the actual and simulated patterns, they insisted that the Turing instability is a viable mechanism for producing these patterns. Similar experiments and discussions on various biological systems, such as zebra fish stripes (Asai et al., 1999; Maderspacher and Nüsslein-Volhard, 2003; Nakamasu et al., 2009), marine angelfish stripes (Painter et al., 1999), murine hair follicles (Sick et al., 2006), feather buds (Jiang et al., 1999; Baker et al., 2009), mouse limbs (Miura et al., 2006), and the mammalian palate (Economou et al., 2012), have frequently appeared.

It is really fascinating that the pattern formation can be explained on the basis of the Turing instability without the knowledge of its details. However, the direct application of this theory to such complicated systems requires more careful investigations. Although a comparison of the actual and simulated patterns suggests many similarities between the patterns, including those in their time evolution, such a comparison is clearly not quantitative. Furthermore, because a rational explanation is not available for the form of the reaction–diffusion equation, it is generally difficult to discuss the physical validity of the instability and physical significance of the parameter values.

As an example, we discuss the simple equations employed by Kondo and Asai (1995):

$$\frac{\partial u}{\partial t} = D_u \frac{\partial^2 u}{\partial x^2} + a_u u + b_u v + c_u, \qquad (1\text{-}71)$$

$$\frac{\partial v}{\partial t} = D_v \frac{\partial^2 v}{\partial x^2} + a_v u + b_v v + c_v, \qquad (1\text{-}72)$$

where u and v are the concentrations of the activator and inhibitor, respectively. The first term on the right side of each equation corresponds to the diffusion term, while the remaining terms should be assigned to the reaction term. In addition to this simplification, an artificial cutoff was introduced for each reaction term to avoid divergence. Although rationalization of these calculation procedures has not been explained, the patterns simulated using these equations were considerably similar to the actual patterns.

The origin of Eqs. (1-71) and (1-72) can be easily deduced from the general reaction-diffusion equations of Eqs. (1-61) and (1-62); the reaction terms $f(u, v)$ and $g(u, v)$, which are generally nonlinear with respect to u and v, are expanded up to the first order of u and v to linearize the equations, and the omitted nonlinearity is supplemented in the form of the cutoff. Thus, they would be applicable only when the system does not significantly deviate from homogenous equilibrium.

Introduction to Nonequilibrium Phenomena

At present, it is quite suspicious whether the Turing instability actually dominates over the biological pattern formation or not. In fact, biologists themselves have recently claimed the necessity of considering a basic process of "local self-enhancement" with "long-range inhibition" (Koch and Meinhardt, 1994; Yamaguchi et al., 2007), or "short-range positive feedback" with "long-range negative feedback" (Kondo and Miura, 2010), which is not always limited to the Turing instability. In addition, they are discussing the necessity of reevaluating alternative mechanisms, such as the clock-and-wavefront model and cell-contact-mediated lateral inhibition (Economou et al., 2012). Thus, before judging the validity of the Turing instability in biological pattern formation, further intensive investigations using well-controlled experimental systems are required.

1.4.3 Color-Producing Nanostructures

When the characteristic size of a structural pattern is of the order of the wavelength of light, special functions emerge through the interaction of the pattern with light. The coloration produced by such a mechanism, known as structural color, is widely distributed in the biological world and plays an important role in mating, camouflage, deceptive behavior, and generating warning and threat signals (Ghiradella, 1991; Srinivasarao, 1999; Vukusic and Sambles, 2003; Kinoshita and Yoshioka, 2005a, 2005b; Kinoshita, 2008; Kinoshita et al., 2008; details in Chapter 6).

The peculiar structures that strongly interact with light are usually categorized into the following four types, as shown in Figure 1.22 (Kinoshita et al., 2008):

1. Thin films and multiple layers causing thin-layer interference.
2. Diffraction gratings.
3. Photonic crystals.
4. Random particles causing light scattering.

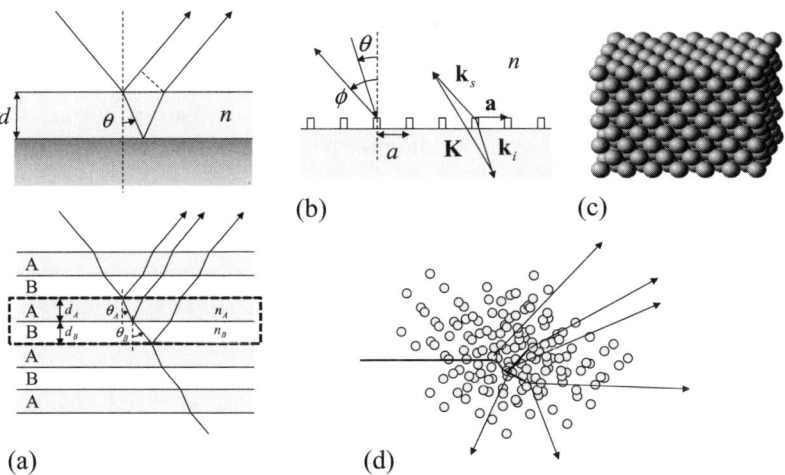

Figure 1.22 Typical mechanisms that produce structural colors: (a) thin-film and multilayer interference, (b) diffraction grating, (c) photonic crystal, and (d) light scattering.

Because interference and scattering of light are major coloration mechanisms, the characteristic sizes of these structures are generally of the order of the wavelength of visible light and thus are on a submicron scale or nanoscale.

Although the optical processes described above are typical origins of biological structural colors, they are generally modified in various ways; they are combined with other optical processes to produce a special optical effect, or their basic structures are deformed and coupled with macroscopic structures. Furthermore, visual effects are introduced by using the viewer's physiological and cognitive processes. Thus, the actual structures and their functions are generally complicated.

Thin-Layer Interference

We briefly explain the principles of the preceding optical processes, placing special emphasis on their structures. First, we choose a thin film, which is one of the simplest structures producing specific coloration. Its principle is explained as follows. Light incident on a thin film is reflected and transmitted at its surface and is reflected again at the other side of the film (see the top illustration in Figure 1.22a). The two reflected light rays with different phases interfere with each other. The interference condition for enhanced reflectivity is expressed as $2nd\cos\theta = (m - 1/2)\lambda$ when the film is attached to the low-index medium (the soap-bubble case), while it is given by $2nd\cos\theta = m\lambda$ when the film is attached to the high-index medium (antireflection-coating case), where n, d, λ, θ, and m are the refractive index and thickness of the film, the wavelength of light that satisfies the interference condition, the refraction angle of the film, and a natural integer, respectively. The two expressions here differ because the reflected light undergoes a phase change of π when it is incident from a lower- to a higher-index medium, while the phase remains constant in the opposite case.

Typical examples of thin-film interference are soap bubbles, oil films on water, and a single-layer antireflective (AR) coating on glasses; these materials generally exhibit low reflectivity, and thus the effect appears faint to the viewers. However, when the refractive index of the film increases or the substrate to which the film is attached is replaced by a metal, the reflectivity considerably increases, and the coloration effect is prominent to the viewers. This characteristic is used to fabricate specific metallic pigments called interference pearl pigments and optically variable pigments®.

Coloration based on thin-film interference is seldom observed in nature, probably for the reasons described above. However, the coloration in the neck feathers of rock dove (Figure 1.23a) gives a special effect on the eyes because the film is extraordinarily thick (500–700 nm, Figure 1.23b), which is sufficient to produce two reflection peaks in the visible region (Yoshioka et al., 2007). Consequently, the feather exhibits a nonspectral color, the color associated with visual sensation, and changes between green and purple as the viewing angle is changed. Such dichroism is commonly experienced in daily life, for example, color change in the abalone.

Stacking of many layers with their structural units repeated causes multilayer interference. A typical example is shown in the bottom illustration in Figure 1.22(a), which

Introduction to Nonequilibrium Phenomena 43

Figure 1.23 (a) Rock dove and (c) jewel beetle, and (b, d) the microstructures producing their respective structural colors. Scale bars: (b) 3 μm and (d) 0.4 μm. (Courtesy of Professor T. Hariyama and reproduced from Yoshioka et al. [2007] with permission.) (For color version of this figure, the reader is referred to the online version of this chapter.)

consists of two repeating layers labeled A and B. Let us focus on the pair of A and B layers enclosed by a dashed line and assume that the refractive indices of these layers satisfy the condition $n_A > n_B$. Then, light incident on this pair is transmitted both from the B layer to the A layer. Thus, the AR coating–type condition can be applied, which is expressed as $2(n_A d_A \cos\theta_A + n_B d_B \cos\theta_B) = m\lambda$, where $n_{A,B}$, $d_{A,B}$, and $\theta_{A,B}$ denote the refractive index, thickness, and refraction angle with respect to A or B, respectively.

On the other hand, if we solely consider layer A, light incident from B to A is transmitted from A to B. Thus, the soap bubble–type condition can be applied, which gives $2n_A d_A \cos\theta_A = (m' - 1/2)\lambda$, where m' is a natural integer satisfying $m' \leq m$. When both conditions are satisfied at a single wavelength λ, the multilayer interference produces maximum reflectivity, which is usually called ideal multilayer interference. On the other hand, when the refractive index or thickness deviates from the ideal condition, the light reflected from each layer interferes destructively, and the reflectivity decreases with decreasing bandwidth, which is the case of a nonideal multilayer interference condition.

Multilayer interference generally offers high reflectivity with a broad bandwidth; therefore, it is often used in dielectric mirrors in laser technology. However, even in ideal multilayer interference, the maximum reflectivity and the bandwidth strongly depend on the refractive index contrast between the two layers. When the difference between their refractive indices is sufficiently large, the reflection spectrum is characterized by reflection bands with high reflectivity and broad bandwidth, while when the difference is sufficiently small, the spectrum shows narrow peaks with low reflectivity. Ideal multilayer interference is rarely found in nature, and most cases of multilayer interference are nonideal, possibly because the perception of color is distinct when the reflection bandwidth is narrow.

The structures causing multilayer interference are widely distributed in the natural world in both living things, such as insects, fish, reptiles, and amphibians, and nonliving things, such as minerals. We discuss the elytra of beetles as a typical example. Figure 1.23(d) shows the cross-section of the elytron of a jewel beetle (Hariyama et al., 2005), which shows green metallic reflection (see Figure 1.23c) and was occasionally used to decorate traditional craftwork in the olden days. Electron microscopy reveals that the surface layer of the elytron, called the epicuticle, contains ~ 20 black-and-white layers, which correspond to layers with and without small granules of melanin pigments, respectively. Thus, typical multilayer interference occurs in this system.

However, even in this apparently simplest case, various structural modifications appearing in the form of surface modification and spacing between the layers can be observed. The elytral surface is doubly modulated at sizes of ~ 50 μm and ~ 10 μm (Kinoshita, 2008). The former appears as depressions, while the latter is a polygonal structure covering the elytral surface; both contribute to the mirror-ball-like diffuse reflection.

On the other hand, two or three layers near the elytral surface are thicker than the inner layers (Yoshioka et al., 2012). A detailed investigation clarified that this feature contributes to the phase matching that produces constructive interference between the surface reflection and the inner multilayer reflection because the refractive index contrast of the inner multilayer is rather small (Yoshioka and Kinoshita, 2011). Therefore, a large contrast between air and the elytral surface, which causes strong phase-inverted reflection, adversely affects multilayer reflection. This effect is prevented in the jewel beetle because the spacing between the layers is appropriately adjusted. In addition, the absorption due to the melanin-containing layers enhances the color contrast by absorbing the unwanted background reflection. Thus, we can find various multifunctional structures even in the elytron of the beetle.

Diffraction Grating

If microstructures are arranged regularly in a line or on a plane, which is generally called diffraction grating, a significant optical effect emerges (Figure 1.22b). If $N \times N'$ microstructures consisting of small particles designated as j are regularly arranged on a plane to form a 2D diffraction grating, the amplitude of the diffracted light in the far field is expressed as

$$\tilde{u} \propto \sum_{m=0}^{N-1} \sum_{n=0}^{N'-1} \sum_{j} e^{-i\mathbf{K} \cdot (\mathbf{R}_{mn} + \mathbf{r}_j)}, \tag{1-73}$$

where $\mathbf{R}_{mn} = m\mathbf{a} + n\mathbf{b}$ with unit lattice vectors $\mathbf{a}$ and $\mathbf{b}$. $\mathbf{K}$ is the scattering vector and is defined as $\mathbf{K} = \mathbf{k}_s - \mathbf{k}_i$, where $\mathbf{k}_i$ and $\mathbf{k}_s$ are the wave vectors of the incident and scattered light waves, respectively. Further, $\mathbf{r}_j$ is a position vector for particle j in a unit cell. The resulting diffraction intensity is then given as

$$|\tilde{u}|^2 \propto \frac{\sin^2(N\mathbf{K} \cdot \mathbf{a}/2)}{\sin^2(\mathbf{K} \cdot \mathbf{a}/2)} \cdot \frac{\sin^2(N'\mathbf{K} \cdot \mathbf{b}/2)}{\sin^2(\mathbf{K} \cdot \mathbf{b}/2)} \left| \sum_j e^{-i\mathbf{K} \cdot \mathbf{r}_j} \right|^2. \quad (1\text{-}74)$$

Thus, diffraction spots appear when both $\mathbf{K} \cdot \mathbf{a}/2 = m\pi$ and $\mathbf{K} \cdot \mathbf{b}/2 = m'\pi$ are satisfied for a scattering vector $\mathbf{K}$, while the last term expresses the scattering due to a single microstructure.

Consider the simplest case of one dimension, which is obtained by setting $N' = 1$. Equation (1-74) is reduced to

$$|\tilde{u}|^2 \propto \frac{\sin^2(N\mathbf{K} \cdot \mathbf{a}/2)}{\sin^2(\mathbf{K} \cdot \mathbf{a}/2)} \left| \sum_j e^{-i\mathbf{K} \cdot \mathbf{r}_j} \right|^2. \quad (1\text{-}75)$$

Diffraction spots are obtained from the relation $\mathbf{K} \cdot \mathbf{a}/2 = m\pi$, which is reduced to

$$na(\sin\theta + \sin\phi) = m\lambda, \quad (1\text{-}76)$$

where we define θ and ϕ as shown in Figure 1.22(b) and replace $m \rightarrow -m$ for convenience. Here, n is the refractive index of the medium.

The coloration produced by a diffraction grating is peculiar, because under a fixed incident angle, the wavelength that satisfies the interference condition varies with the viewing angle ϕ. Thus, the diffraction grating appears rainbow-colored, which is often observed in CDs or DVDs. In nature, diffraction grating seldom appears as the coloration mechanism of living things because it does not show the individual colors acting as important biological signals.

Instead, it has a special function when a 2D diffraction-grating-like structure is produced on the cornea of the insect eye. Consider a case in which the spacing between the microstructures considerably decreases. Because the maximum value of the left side of the interference condition becomes $2na$, the condition of Eq. (1-76) cannot be satisfied even for $m = 1$ for a wavelength $\lambda > 2na$. In this case, only the condition $m = 0$ is satisfied when $\theta = -\phi$, where only specular reflection occurs for all the wavelengths. This special case is called zeroth-order grating.

The moth-eye structure is a famous structure found in the corneal surfaces of moths, butterflies, and flies (Bernhard and Miller, 1962), as well as in the wing surfaces of cicadas and moths with transparent wings (Yoshida et al., 1997). Each microstructure in this case is known to be bell-shaped. Thus, if we slice the two-dimensionally aligned bell-shaped projections along the surface into thin layers, the averaged refractive indices of the layers become a smooth function of depth, and change from that of air to that of the inner material. Thus, it is possible to prevent most surface reflection, which presumably contributes to the defense of these insects against predators. The moth-eye structure has recently been considered as an excellent AR device, particularly because of its broad angular and wavelength expandability, and it is now expected to be a strong candidate for reducing surface reflection in solar cell panels.

Photonic Crystals

Microstructures arranged regularly with a period of the order of the wavelength of light are generally called photonic crystals (Figure 1.22c). In this sense, the multilayer and 1D diffraction grating should be classified as 1D photonic crystals, while 2D diffraction grating can be described as a 2D photonic crystal. Sanders (1964) first discovered photonic crystals in precious opal. Using an electron microscope, he found that the opal comprised amorphous silica spheres of diameter $\sim$200 nm arranged in the fcc configuration. Later, the inner structure of the sphere was further investigated (Darragh et al., 1966) and a patent was sought to fabricate synthetic opal.

The first 2D photonic crystals in the biological world were observed in a peacock feather (Durrer, 1962), in which cylindrical melanin granules of diameter $\sim$170 nm and length $\sim$700 nm were arranged in a square lattice (see Figure 1.24b). Later, a more complete 3D photonic crystal was found in the scale of a small green butterfly (Morris,

Figure 1.24 (a) Indian peafowl and (c) common kingfisher, and (b, d) the microstructures producing their respective structural colors (scale bar: 1 μm). (Reproduced from Yoshioka and Kinoshita [2002] and Kinoshita and Yoshioka [2005a].) (For color version of this figure, the reader is referred to the online version of this chapter.)

1975). Similar structures have been discovered in various butterfly and beetle scales since then. Thus, the case of photonic crystals is one of the basic coloration mechanisms in the natural world. These photonic structures are mostly divided into small domains of size ~ 10 µm and are known to contain considerable irregularity. It has also been suggested that the photonic crystals in butterfly scales acquire the form of a gyroid, an infinitely connected minimal surface, which is known as a mechanism for microphase separation in block copolymers (Michielsen and Stavenga, 2008). Thus, studies on their structures and detailed optical properties are in progress.

Light Scattering

Light scattering (Figure 1.22d) is the simplest coloration mechanism and exploits the fact that the scattering cross-section depends on the wavelength of light. Rayleigh scattering, which scatters light in inverse proportion to the fourth power of the wavelength, occurs for small particle sizes. In contrast, when the size of the particle becomes of the order of the wavelength, Mie scattering, which scatters light in the visible region depending on the size and refractive index of the sphere, occurs. The feathers of birds and the bodies of dragonflies occasionally turn blue without any noticeable color variation with the varying viewing angle; this color is usually referred to as Tyndall blue. This is because under an electron microscope, only a random network structure or random ensemble of particles is observed (e.g., see Figure 1.24d).

Using the spatial Fourier transform of the electron micrograms, Prum et al. (1998) clarified that these seemingly random structures actually involve a short-range order, and the interference effect plays an important role in their coloration. They investigated the structures of various living organisms, such as avian barbs, butterfly scales, the bodies of damselflies and dragonflies, and avian and mammalian skins, and found that similar short-range order was ubiquitous. These observations were supported by the fact that these seemingly random structures showed definite peaks in the reflection spectra, while Rayleigh scattering is characterized by a smooth increase in the reflectivity toward shorter wavelengths. Thus, it is now generally believed that eventually no structure will be found in the biological world, of which the origin of the color is purely due to light scattering.

Developmental Studies on Nanostructures

At present, the issue how the previously mentioned submicron or nanostructures are created under natural conditions remains unclear. Actually, detailed microscopic investigations of the developmental stages of these structures are very few in number. In the following section, we discuss a few such exceptional examples.

The first example is the multilayer structure found in tiger beetles. Schultz and Rankin (1985) conducted a developmental study on festive tiger beetle (*Cicindela scutellaris*) and splendid tiger beetle (*C. splendida*). A color change was noticed in the wing before and after ecdysis; the color was initially pale white and became blue and pigmented after ecdysis. Electron microscopic observations during this period revealed that the presumptive epicuticle was already deposited four days before

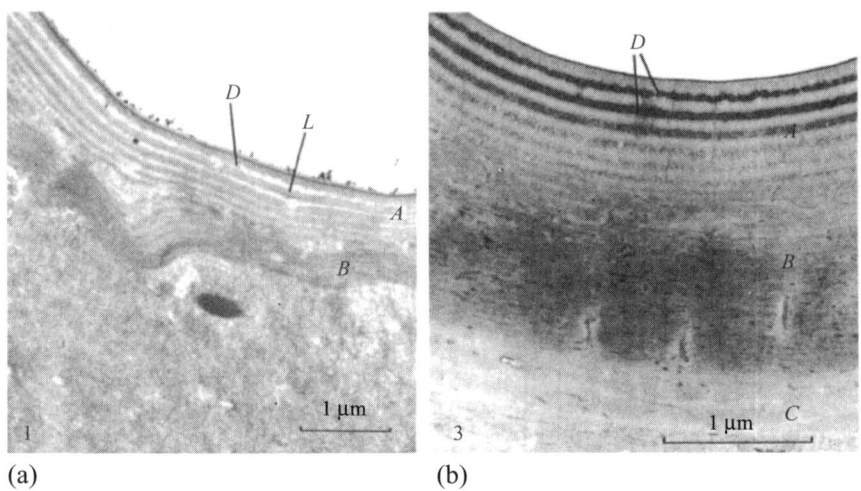

Figure 1.25 Cross-sections of developing elytral cuticle of *Cicindela scutellaris*: (a) 10 days after pupation and (b) 12 hours after ecdysis. A: epicuticle, B: outer exocuticle, C: inner exocuticle, and D and L: presumptive electron-dense and -lucent layers, respectively. (Reproduced from Schultz and Rankin [1985] with permission.)

ecdysis. The number of layers was completely prepared for later growth, but the presumptive electron-dense bands were electron lucent, while the presumptive electron-lucent layers were slightly electron dense, as shown in Figure 1.25(a). Twelve hours after ecdysis, the electron-dense bands became distinct, and their densities were stronger toward the outer side (Figure 1.25b). On the basis of these observations, it was concluded that the change in the electron density was entirely caused by the chemical change and possibly by in situ melanoprotein formation, which accounted for the change in the apparent color of the wing and for the emergence of interference during development.

Butterflies in their pupal stages represent another example. Ghiradella (1989) conducted developmental studies on two Lycaenid species with different microstructures in the lower part of the scale. One was the olive hairstreak (*Mitoura grynea*), which exhibits a crystalline structure, while the other was the spring azure (*Celastrina ladon*), which exhibits a lamellar structure consisting of a fibrous network. Electron microscopic observations were conducted during the pupal periods of these butterflies. Their scales initially exhibited architectures similar to those in ordinary scales, but after the eighth day in *M. grynea* and the seventh day in *C. ladon*, internal specialization began.

In *M. grynea*, the membrane and cuticle together formed a unit (see Figure 1.26). The enclosing membranes were continuous with each other and also with the outer space and essentially had an extracellular character. These units aggregated into lattices, forming small crystallites that grew until the adult morphology was achieved. The lattice formation continued for two days, and the structure became firmly fixed at that position as the cell died back and disappeared. In *C. ladon*, the stack of

Introduction to Nonequilibrium Phenomena

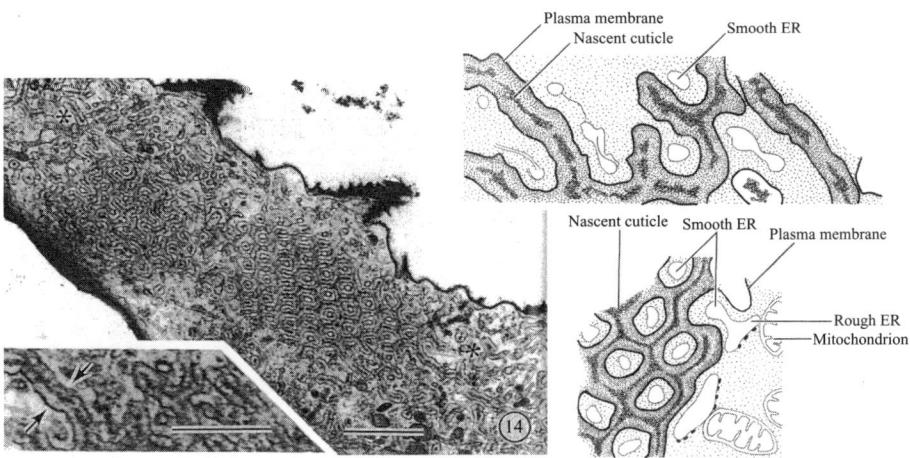

Figure 1.26 Transverse section (left) of developing scale nine days after pupation of *Mitoura grynea*, showing two crystallites in a cell (scale bar: 1 μm). Each crystallite consists of membrane–cuticle units (lower right), which first appear as nascent cuticle enclosing a smooth endoplasmic reticulum as a nucleus (upper right: schematic illustration of inset of photograph on left; scale bar: 0.5 μm). (Reproduced from Ghiradella [1989] with permission).

endoplasmic reticulum began from the lower end of the scale cell. In addition, electronically dense pale vesicles appeared. The fibrils and tubes appeared to be lined up in a stack and were closely associated with the plasma membrane. These processes apparently adopted completely different methods of organization, although these species were phylogenetically close, and intermediate species having lamellar structure with regular holes were observed.

In addition, Ghiradella (1974) extensively studied the development of UV-reflecting scales, using the orange sulfur butterfly (*Colias eurytheme*), which is distributed in North America. The scales of this butterfly are decorated by many lines (called ridges), each of which exhibits a well-developed lamellar structure consisting of alternating air and cuticle layers with thicknesses of 84.8 nm and 51 nm, respectively (see Figure 1.27). Its pupal stage lasts for approximately seven days, and scale formation continues throughout this period. In Ghiradella's study, the ridge formation occurred before the third day, when the bundles of microfilaments were arranged near the upper and lower surfaces of a flattened scale. Then, ridge formation occurred between the upper bundles such that cuticulin deposition proceeded during early ridge formation along the entire surface of the scale, and the height and width of the ridges continued to increase. The first lamellae appeared as buckling of the cuticulin at the top of the ridge, and the others appeared more basally. Then, the cuticle quickly began to fill in the lamellae. The plasma membrane was initially close to the ridge but retreated during the later stage except for the close-contact region, which became a window at a later stage. Finally, the trabeculae (stems) and ellipsoidal beads (pigment granules) were laid down.

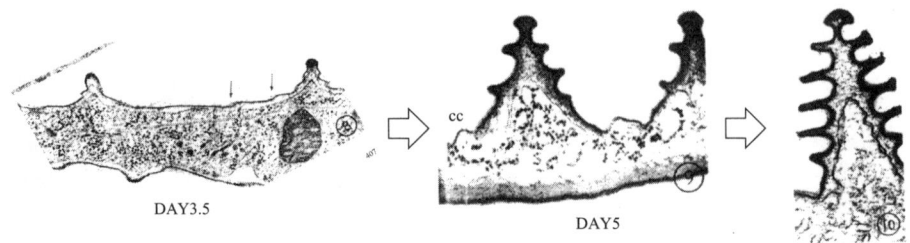

Figure 1.27 Developmental study of UV-reflecting scale of *Colias eurytheme*. Cross-sections of a developing scale at various stages after pupation are shown. (Reproduced from Ghiradella [1974] with permission).

Ghiradella proposed a model of elastic buckling (sinusoidal modulation) for the formation of the ridge-lamella system. In this model, the lamellae are formed from either compression or stretching in the longitudinal direction of the scale owing to the contractile element of the microfilaments. Thus, the elastic stress due to the function of the microfilaments causes buckling on the surface of the nascent ridge. However, once the lamellae begin to fold, an additional mechanism, which is probably an interaction between adjacent parts of a fold, is necessary for the final formation of the ridge-lamella system.

Gemne (1971) investigated the developmental process of the moth-eye structure by detailed observations of a pupae of the tobacco hornworm moth (*Manduca sexta*), using optical and electron microscopes. In this study, it was observed that after pupation, the processes on the corneagenous cell surface became slender and was separated from each other. With a further increase in length, the processes became true microvilli of length 400–500 nm and width 70–100 nm separated by 170 nm. The microvilli were arranged in a hexagonal pattern and were finally separated by 200 nm. At the tip of the microvilli, a patched lamellar element appeared forming a dome. After 5.5–6 days, the dome assumed the shape of a higher cupola with an amplitude of 150 nm, and after 6–8 days, the height of the cupola increased, eventually reaching 200–250 nm. The corneal material was then deposited to fill the evaginations created by evaginating the epicorneal membrane above the tip of the microvilli, which was followed by consolidation. Then, a hexagonal array of well-developed corneal projections was formed.

Thus, the biological mechanisms to create such elaborated structures are scattered, depending on the species and on the microstructures to be made. However, the resulting structures seem to have something in common. Namely, they have the periodic structures of which the periods always lie on the order of ~ 200 nm, which is inevitable to produce the coloration in a visible region. In this respect, it is fascinating to consider that the mechanisms to create such structures are somehow related to the Turing's mechanism as described earlier.

Applying $q_c \approx (f_u/2D_u)^{1/2}$ (see Eq. (1-70)) to the color-producing structures, we can estimate the necessary condition to create such microstructures. Because the reaction rate is unknown, we estimate only the diffusion coefficient with the help of recently reported reaction–diffusion systems. For example, employing the values of $D_u \approx 10^{-9}$

m^2/s and $q_c \approx 10^4$ m^{-1} to a gel reactor (Vigil et al., 1992), we estimated the diffusion constant as $D_u \approx 10^{-15}$ m^2/s for $q_c \approx 10^7$ m^{-1} assuming the same reaction rate. Thus, the diffusion coefficient required for creating the color-producing pattern is estimated to be seven orders of magnitude smaller than that in water. Although the difference in the reaction rates may significantly affect the result, this estimation partially explains the reason for the use of complex chemical structures, such as the cuticle (in insects) and keratin (in birds) for creating their microstructures. Thus, in the near future, it will be extremely important to construct a model system for investigating the formation process in a controlled laboratory environment.

1.5 Order and Fluctuation

1.5.1 Phase-Transition Analogy

We have presented many examples of oscillatory phenomena, order-formation processes, and pattern formations, and clarified their characteristics using various mathematical treatments. In each case, naturally complicated phenomena can be condensed into a few differential equations with few degrees of freedom. By solving these equations, we can appropriately describe the temporal and spatial behaviors. In addition, by varying the parameter values appearing in the equations, we can understand the origins and characteristics of the phenomena within the framework of mathematics; on the basis of information on the stability of the solution, we can assume the conditions for the related order-formation process. Therefore, we might have conveyed to readers that all mechanisms for various phenomena occurring in nonequilibrium systems have been solved, using the methods described so far.

However, these methods only provide the mathematical characteristics of given differential equations and suggested analogies with natural phenomena. They do not give any answer to a fundamental question of why various temporal and spatial orders are spontaneously created in a nonequilibrium system. In this sense, these equations are sometimes called phenomenological equations. In addition, the reason for the small number of degrees of freedom being sufficient for describing a complex system has not been clarified. However, the answers to these questions are extremely difficult because the latter is directly connected with a fundamental problem of physics regarding the relationship of microscopic degrees of freedom with macroscopic quantities—that is, the association of quantum mechanics with statistical mechanics.

Instead of dealing with these issues, we refer to a study by Haken (1975, 1978), who conducted to clarify order formation in nonequilibrium systems. Haken focused on an order parameter that characterized the degree of order and established an analogy between the phase-transition phenomena observed in equilibrium and nonequilibrium systems. He suggested laser oscillation as an example and developed a method to analyze it mathematically. He selected three quantities as the basic variables of laser oscillation: electric field of light, atomic polarization, and population difference between the excited and ground states of the atom. The other quantities, such as those related to external pumping and reflection loss of mirrors, were treated as phenomenological parameters.

The Hamiltonian in this system is expressed as

$$H = \hbar\omega b^\dagger b + \frac{1}{2}\hbar\omega_0 \sum_\mu \sigma_\mu + \hbar \sum_\mu \left(g_\mu^* b \alpha_\mu^\dagger + \text{h.c.}\right), \tag{1-77}$$

where ω and ω_0 are the angular frequency of light and that corresponding to the energy difference between the excited and ground states of an atom, respectively. Further, b and $b^\dagger$ are the annihilation and creation operators of a photon, respectively, where we consider only a single-mode case for light. In addition, $\alpha_\mu^\dagger$ and σ_μ denote a polarization operator and an operator expressing the population difference of an atom μ, which are explicitly expressed as $\alpha_\mu^\dagger = a_{2\mu}^\dagger a_{1\mu}$ and $\sigma_\mu = a_{2\mu}^\dagger a_{2\mu} - a_{1\mu}^\dagger a_{1\mu}$, where the annihilation and creation operators of an electron at level j in the atom μ are denoted by $a_{j\mu}$ and $a_{j\mu}^\dagger$, respectively. Finally, g_μ^* is a coupling constant between the electric field of light and atomic polarization.

The equation of motion for $b^\dagger$ is then obtained, using the Heisenberg equation of motion as

$$\dot{b}^\dagger = \frac{i}{\hbar}[H, b^\dagger] = (i\omega - \kappa)b^\dagger + i\sum_\mu g_\mu^* \alpha_\mu^\dagger + F^\dagger(t). \tag{1-78}$$

Similarly, we obtain the equations of motion for $\alpha_\mu^\dagger$ and σ_μ as follows:

$$\dot{\alpha}_\mu^\dagger = (i\omega_0 - \gamma)\alpha_\mu^\dagger - ig_\mu b^\dagger \sigma_\mu + \Gamma_\mu^\dagger(t), \tag{1-79}$$

$$\dot{\sigma}_\mu = \gamma_\|(d_0 - \sigma_\mu) + 2i(g_\mu b^\dagger \alpha_\mu - \text{h.c.}) + \Gamma_{\sigma,\mu}(t), \tag{1-80}$$

where κ, γ, $\gamma_\|$, and d_0 are phenomenological parameters expressing the cavity loss, phase and population relaxation rates, and amount of pumping, respectively. $F^\dagger(t)$, $\Gamma_\mu^\dagger(t)$, and $\Gamma_{\sigma,\mu}(t)$ indicate the fluctuation force for each quantity. Further, their time dependence is eliminated by putting $b^\dagger = \tilde{b}^\dagger \exp[i\omega t]$, $\alpha_\mu^\dagger = \tilde{\alpha}_\mu^\dagger \exp[i\omega_0 t]$ and $F^\dagger = \tilde{F}^\dagger \exp[i\omega t]$, where the tilde is finally dropped.

Haken employed the concept of adiabatic approximation in which the phase relaxation rate γ is sufficiently high, and thus, the polarization of the atom completely follows the changes in the electric field amplitude and population difference. Thus, the atomic polarization is obtained as

$$\alpha_\mu^\dagger = -ig_\mu \int_{-\infty}^t e^{-\gamma(t-\tau)} b^\dagger \sigma_\mu d\tau + \int_{-\infty}^t e^{-\gamma(t-\tau)} \Gamma_\mu^\dagger d\tau$$

$$\approx -(ig_\mu/\gamma)b^\dagger \sigma_\mu + \hat{\Gamma}_\mu^\dagger(t), \tag{1-81}$$

where we set $\hat{\Gamma}_\mu^\dagger(t) \equiv \int_{-\infty}^t e^{-\gamma(t-\tau)} \Gamma_\mu^\dagger d\tau$. Inserting this relation into Eq. (1-80), and further assuming the adiabatic approximation for σ_μ by setting $\dot{\sigma}_\mu = 0$, we obtain

Introduction to Nonequilibrium Phenomena

$$\sigma_\mu = d_0 / \left[1 + 4\left(g^2/\gamma\gamma_\|\right) b^\dagger b \right] \approx d_0 - 4\left(g^2/\gamma\gamma_\|\right) d_0 b^\dagger b, \tag{1-82}$$

where the electric field of a light wave traveling in the x direction is expressed as

$$E(x,t) = i\left\{ (\hbar\omega/2\varepsilon_0 V)^{1/2} e^{ikx} b - \text{h.c.} \right\}, \tag{1-83}$$

and set $g_\mu^* \equiv g\exp[ikx_\mu]$. Here, k is the wavenumber of the light, while ε_0 and V are the permittivity of vacuum and the quantization volume, respectively. If we insert all these relations into Eq. (1-78), the resulting equation of motion for the photon field becomes

$$\begin{aligned}
\dot{b}^\dagger &= -\kappa b^\dagger + \frac{g^2}{\gamma} b^\dagger \sum_\mu \sigma_\mu + \hat{F}(t) \\
&\approx \left(-\kappa + \frac{g^2}{\gamma} D_0 \right) b^\dagger - \frac{4g^2 \kappa}{\gamma\gamma_\|} b^\dagger b^\dagger b + \hat{F}(t),
\end{aligned} \tag{1-84}$$

where we define the population inversion caused by external pumping as $D_0 = \sum_\mu d_0$ and also set $\sum_\mu d_0 \approx \kappa\gamma/g^2$ in the second term on the right side.

The dynamics of the electric field within a laser cavity are expressed as the sum of the linear and cubic terms; the behavior of this type of electric field can be easily analyzed, using the classical picture by replacing the quantum operator b or $b^\dagger$ with a classical quantity q. Thus, the corresponding classical equation of motion is

$$\dot{q} = -aq - bq^3 + F(t), \tag{1-85}$$

where $F(t)$ represents a random force, and its characteristics are assumed to be expressed by a correlation function of the Markovian type as

$$\langle F(t) F(t') \rangle = C\delta(t-t'). \tag{1-86}$$

Here, b is assumed to be a positive constant, while a takes both positive and negative values. In fact, because $a \cong \kappa - g^2 D_0/\gamma$, a is positive for small D_0, while with increasing D_0, a becomes negative. Thus, $\kappa = g^2 D_0/\gamma$ represents a threshold, which is similar to those of the instabilities described in the previous sections.

However, in the present case, the presence of $-bq^3$ prevents the system from being divergent. This can be easily understood if a potential is defined as $V(q) = (1/2) aq^2 + (1/4) bq^4$, and express Eq. (1-85) using $V(q)$ as

$$\dot{q} = -dV(q)/dq + F(t). \tag{1-87}$$

This equation can be interpreted in terms of the overdamped motion of a particle under the presence of the potential $V(q)$ and the fluctuation force, or in terms of the dynamics

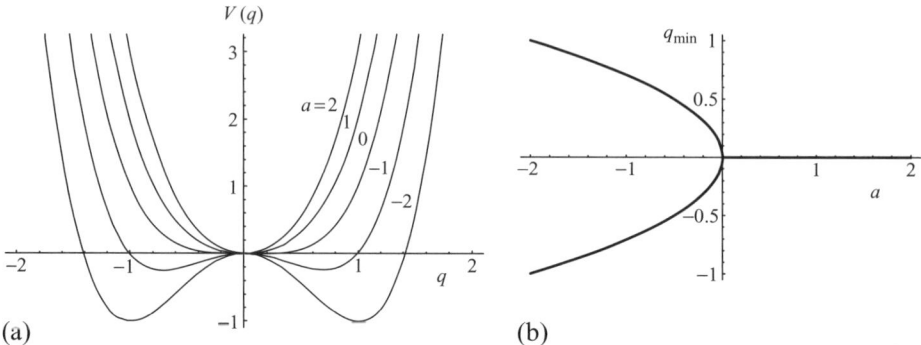

Figure 1.28 (a) Functional shape of $V(q)$ for various values of a, where $b=1$. (b) Variation in $q_{\min}$ with changing a, which gives the stable minimum value of $V(q)$.

based on the second-order phase transition phenomenon with q as an order parameter. In Figure 1.28(a), we show the distribution of the potential $V(q)$ for various values of a. When $a > 0$, the curve closely resembles a parabola with its minimum at $q = 0$, while with decreasing a, the shape is deformed by an extension of the flat area around $q = 0$. Finally, the curve exhibits a double minimum when a is negative. Therefore, the instability at $q = 0$ actually occurs at $a = 0$. This situation can be understood by plotting the value of $q_{\min}$ that yields the minimum $V(q)$, as shown in Figure 1.28(b), which clearly shows the presence of the bifurcation at $a = 0$. This phenomenon is similar to the Hopf bifurcation appearing in the van der Pol oscillator.

Let us apply the preceding behavior to laser oscillation by noting the relation $a \cong \kappa - g^2 D_0/\gamma$. Initially, under the condition of small external pumping, D_0 should be small; thus, $\kappa \gg g^2 D_0/\gamma$, resulting in a parabola-like curve and a minimum at $\langle b^\dagger \rangle = 0$. Therefore, the light field fluctuates around a zero mean value. With a further increase in D_0, and thus $\kappa \gtrsim g^2 D_0/\gamma$, the flat area near the bottom of the potential curve (see the case of $a = 0$ in Figure 1.28a) is considerably extended, causing a large fluctuation in the light field. Then, with a greater increase in D_0, the system enters a double-minimum state after crossing the critical value of $\kappa = g^2 D_0/\gamma$. This means that the laser oscillation begins, and the oscillation is characterized by the presence of a finite electric field value. Although the classical model predicts the presence of electric fields that have opposite directions, $b^\dagger$ is actually a complex quantity. Thus, the potential curve shown in Figure 1.28(a) should be plotted in the complex plane, which yields a curved surface with a donut-shaped minimum.

Although the model described here is a simple one-variable case, it is significant because nonequilibrium order-formation phenomena, such as laser oscillation, can be treated quantitatively by comparing them with the phase-transition phenomena in equilibrium systems. An experimental measurement of the fluctuation of the light intensity near the laser oscillation threshold revealed a considerable increase in the fluctuation amplitude (Arecchi et al., 1967), which was in appropriate agreement with the present prediction. The mathematical models described earlier differ from the present one in that the present model develops the formulations to a point at which it is possible to compare them with the existing model in equilibrium systems, and the fluctuation forces are naturally introduced and can be compared directly with the experimental results.

From Eq. (1-85), we can construct the Fokker–Planck equation, which changes the mechanical motion of a particle into its statistical distribution as follows:

$$\dot{f}(q,t) = -\frac{\partial}{\partial q}\left\{\left(-\frac{\partial V(q)}{\partial q}f(q,t)\right) - \frac{1}{2}C\frac{\partial f(q,t)}{\partial q}\right\}, \tag{1-88}$$

where $f(q, t)$ is a probability function representing the probability that a particle exists at the position $q=q$ and time $t=t$. Further, setting $\dot{f}(q,t) = 0$ and solving the equation with respect to $f(q, t)$, we can obtain the stationary distribution of q, which is comparable to the Boltzmann distribution in a system in thermal equilibrium. The explicit expression is easily obtained as

$$f_{eq}(q) = N\exp[-2V(q)/C], \tag{1-89}$$

where C is given by Eq. (1-86) and corresponds to the magnitude of the fluctuation, and N is a normalization factor. Thus, a result similar to that of the equilibrium system is obtained, even in the nonequilibrium case. Therefore, the spontaneous creation of order in nonequilibrium systems can be clarified by its comparison with existing physical phenomena. Further studies on these lines are strongly required.

Acknowledgments

This study was partly supported by Grants-in-Aid for Scientific Research (nos. 22340121 and 24120004) from the Japanese Ministry of Education, Culture, Sports, Science, and Technology.

References

Arecchi, F.T., Rodari, G.S., Sona, A., 1967. Statistics of the laser radiation at threshold. Phys. Lett. 25 A, 59–60.
Asai, R., Taguchi, E., Kume, Y., Saito, M., Kondo, S., 1999. Zebrafish *Leopard* gene as a component of the putative reaction–diffusion system. Mech. Dev. 89, 87–92.
Baker, R.E., Schnell, S., Maini, P.K., 2009. Waves and patterning in developmental biology: Vertebrate segmentation and feather bud formation as case studies. Int. J. Dev. Biol. 53, 783–794.
Belousov, B.P., 1959. Periodically acting reaction and its mechanism. Collection of Abstracts on Radiation Medicine, Medgiz, Moscow, p.145.
Bernhard, C.G., Miller, W.H., 1962. A corneal nipple pattern in insect compound eyes. Acta Physiol. Scand. 56, 385–386.

Caroli, B., Caroli, C., Roulet, B., 1991. Instabilities of Planar Solidification Fronts. In: Godrèche, C. (Ed.), Solids Far from Equilibrium. Press Syndicate of University of Cambridge, New York, pp. 155–296.

Castets, V., Dulos, E., Boissonade, J., De Kepper, P., 1990. Experimental evidence of a sustained standing Turing-type nonequilibrium chemical pattern. Phys. Rev. Lett. 64, 2953–2956.

Chou, H., Cummins, H.Z., 1988. Evolution of the dendritic instability in solidifying succinonitrile. Phys. Rev. Lett. 61, 173–176.

Darragh, P.J., Gaskin, A.J., Terrell, B.C., Sanders, J.V., 1966. Origin of precious opal. Nature 209, 13–16.

Derjaguin, B.V., Landau, L.D., 1941. Theory of the stability of strongly charged lyophobic sols and of the adhesion of strongly charged particles in solutions of electrolytes. Acta Phys. Chem. URSS 14, 633–662.

Durrer, H., 1962. Schillerfarben beim Pfau (*Pavo cristantus L.*). Verhand Naturf Ges. Basel 73, 204–224.

Economou, A.D., Ohazama, A., Porntaveetus, T., Sharpe, P.T., Kondo, S., Basson, M.A., et al., 2012. Periodic stripe formation by a Turing mechanism operating at growth zones in the mammalian palate. Nat. Genet. 44, 348–352.

Field, R.J., Noyes, R.M., 1974. Oscillations in chemical systems. IV. Limit cycle behavior in a model of a real chemical reaction. J. Chem. Phys. 60, 1877–1884.

Field, R.J., Körös, E., Noyes, R.M., 1972. Oscillations in chemical systems. II. Through analysis of temporal oscillation in the bromate-cerium-malonic acid system. J. Am. Chem. Soc. 94, 8649–8664.

Gemne, G., 1971. Ontogenesis of corneal surface ultrastructure in nocturnal Lepidoptera. Phil. Trans. R. Soc. B 262, 343–363.

Ghiradella, H., 1974. Development of ultraviolet-reflecting butterfly scales: How to make an interference filter. J. Morphol. 142, 395–410.

Ghiradella, H., 1989. Structure and development of iridescent butterfly scales: Lattices and laminae. J. Morphol. 202, 69–88.

Ghiradella, H., 1991. Light and color on the wing: Structural colors in butterflies and moths. Appl. Opt. 30, 3492–3500.

Haken, H., 1975. Cooperative phenomena in systems far from thermal equilibrium and in nonphysical systems. Rev. Mod. Phys. 47, 67–121.

Haken, H., 1978. Synergetics. An Introduction. Nonequilibrium Phase Transitions and Self-Organization in Physics, Chemistry and Biology. Springer-Verlag, Berlin.

Hariyama, T., Hironaka, M., Horiguchi, H., Stavenga, D.G., 2005. The leaf beetle, the jewel beetle, and the damselfy; insects with a multilayered show case. In: Kinoshita, S., Yoshioka, S. (Eds.), Structural Colors in Biological Systems—Principles and Applications. Osaka University Press, Osaka, pp. 153–176.

Horváth, J., Szalai, I., De Kepper, P., 2009. An experimental design method leading to chemical Turing patterns. Science 324, 772–775.

Jiang, T.X., Jung, H.S., Widelitz, R.B., Chuong, C.M., 1999. Self-organization of periodic patterns by dissociated feather mesenchymal cells and the regulation of size, number and spacing of primordial. Develop 126, 4997–5009.

Kano, T., Kinoshita, S., 2007. Viscosity-dependent flow reversal in a density oscillator. Phys. Rev. E 76, 046208.

Kano, T., Kinoshita, S., 2008. Modeling of the flow-reversal process in a density oscillator. J. Korean Phys. Soc. 53, 1273–1279.

Kano, T., Kinoshita, S., 2009. Modeling of a density oscillator. Phys. Rev. E 80, 046217.

Kinoshita, S., 2008. Structural Colors in the Realm of Nature. World Scientific Publishing, Singapore.
Kinoshita, S., Yoshioka, S., 2005a. Structural colors in nature: The role of regularity and irregularity in the structure. ChemPhysChem 6, 1442–1459.
Kinoshita, S., Yoshioka, S. (Eds.), 2005b. Structural Colors in Biological Systems—Principles and Applications. Osaka University Press, Osaka.
Kinoshita, S., Yoshioka, S., Miyazaki, J., 2008. Physics of structural colors. Rep. Prog. Phys. 71, 076401.
Koch, A.J., Meinhardt, H., 1994. Biological pattern formation: From basic mechanisms to complex structures. Rev. Mod. Phys. 66, 1481–1507.
Kondo, S., Asai, R., 1995. A reaction–diffusion wave on the skin of the marine angelfish *Pomacanthus*. Nature 376, 765–768.
Kondo, S., Miura, T., 2010. Reaction–diffusion model as a framework for understanding biological pattern formation. Science 329, 1616–1620.
Kuramoto, Y., 1984. Chemical Oscillations, Waves, and Turbulence. Springer-Verlag, Berlin.
Kuramoto, Y., 2005. The World of Rhythmic Phenomena. University of Tokyo Press, Tokyo (in Japanese).
Langer, J.S., 1980. Instabilities and pattern formation in crystal growth. Rev. Mod. Phys. 52, 1–28.
Larsen, A.E., Grier, D.G., 1997. Like-charge attractions in metastable colloidal crystallites. Nature 385, 230–233.
Lengyel, I., Epstein, I.R., 1991. Modeling of Turing structures in the chlorite–iodide–malonic acid–starch reaction system. Science 251, 650–652.
Lotka, A.J., 1925. Elements of Physical Biology. Williams and Wilkins, Baltimore.
Maderspacher, F., Nüsslein-Volhard, C., 2003. Formation of the adult pigment pattern in zebrafish requires *leopard* and *obelix* dependent cell interactions. Develop 130, 3447–3457.
Martin, S., 1970. A hydrodynamic curiosity: The salt oscillator. Geophys. Fluid Dyn. 1, 143–160.
Michielsen, K., Stavenga, D.G., 2008. Gyroid cuticular structures in butterfly wing scales: Biological photonic crystals. J. R. Soc. Interface 5, 85–94.
Miike, H., Mori, Y., Yamaguchi, T., 1997. Science of Nonequilibrium Systems. III. Dynamics in Reaction–Diffusion System. Kodansha, Tokyo (in Japanese).
Miura, T., Shiota, K., Morriss-Kay, G., Maini, P.K., 2006. Mixed-mode pattern in *Doublefoot* mutant mouse limb—Turing reaction–diffusion model on a growing domain during limb development. J. Theor. Biol. 240, 562–573.
Morris, R.B., 1975. Iridescence from diffraction structures in the wing scales of *Callophrys rubi*, the Green Hairstreak. J. Ent. (A) 49, 149–154.
Mullins, W.W., Sekerka, R.F., 1963. Morphological stability of a particle growing by diffusion or heat flow. J. Appl. Phys. 34, 323–329.
Mullins, W.W., Sekerka, R.F., 1964. Stability of a planar interface during solidification of a dilute binary alloy. J. Appl. Phys. 35, 444–451.
Nakamasu, A., Takahashi, G., Kanbe, A., Kondo, S., 2009. Interactions between zebrafish pigment cells responsible for the generation of Turing patterns. Proc. Natl. Acad. Sci. U. S. A. 106, 8429–8434.
Nakazawa, R., Kinoshita, S., 2006. Self-organization of well-defined two-dimensional colloidal array near a glass plate, unpublished work.
Nicolis, G., Prigogine, I., 1977. Self-Organization in Nonequilibrium Systems: From Dissipative Structures to Order through Fluctuations. Wiley, New York.
Ohta, T., 2000. Physics of Nonequilibrium Systems. Shokabo, Tokyo (in Japanese).

Ouyang, Q., Swinney, H.L., 1991. Transition from a uniform state to hexagonal and striped Turing patterns. Nature 352, 610–612.
Painter, K.J., Maini, P.K., Othmer, H.G., 1999. Stripe formation in juvenile *Pomacanthus* explained by a generalized Turing mechanism with chemotaxis. Proc. Natl. Acad. Sci. U. S. A. 96, 5549–5554.
Pearson, S.O., Anson, H.St.G., 1921a. Demonstration of some electrical properties of neon-filled lamps. Proc. Phys. Soc. Lond. 34, 175–176.
Pearson, S.O., Anson, H.St.G., 1921b. The neon tube as a means of producing intermittent currents. Proc. Phys. Soc. Lond. 34, 204–212.
Prigogine, I., Lefever, R., 1968. Symmetry breaking instabilities in dissipative systems II. J. Chem. Phys. 48, 1695–1700.
Prum, R.O., Torres, R.H., Williamson, S., Dyck, J., 1998. Coherent light scattering by blue feather barbs. Nature 396, 28–29.
Qian, X.W., Chou, H., Muschol, M., Cummins, H.Z., 1989. Role of noise in the initial state of solidification instability. Phys. Rev. B 39, 2529–2531.
Sanders, J.V., 1964. Colour of precious opal. Nature 204, 1151–1153.
Schultz, T.D., Rankin, M.A., 1985. Developmental changes in the interference reflectors and colorations of tiger beetles (*Cicindela*). J. Exp. Biol. 117, 111–117.
Sick, S., Reinker, S., Timmer, J., Schlake, T., 2006. WNT and DKK determine hair follicle spacing through a reaction–diffusion mechanism. Science 314, 1447–1450.
Srinivasarao, M., 1999. Nano-optics in the biological world: Beetles, butterflies, birds, and moths. Chem. Rev. 99, 1935–1961.
Stanley, W.M., 1935. Isolation of a crystalline protein possessing the properties of tobacco-mosaic virus. Science 81, 644–645.
Steinbock, O., Lange, A., Rehberg, I., 1998. Density oscillator: Analysis of flow dynamics and stability. Phys. Rev. Lett. 81, 798–801.
Takaki, R., 1992. Mathematical Sciences of Forms. Asakura, Tokyo (in Japanese).
Turing, A.M., 1952. The chemical basis of morphogenesis. Phil. Trans. Roy. Soc. 237, 37–72.
Vanag, V.K., Epstein, I.R., 2001. Pattern formation in a tunable medium: The Belousov–Zhabotinsky reaction in an aerosol OT microemulsion. Phys. Rev. Lett. 87, 228301.
van der Pol, B., 1926. On relaxation-oscillations. The London, Edinburgh, and Dublin Phil. Mag. J. Sci. 2, 978–992.
Verwey, E.J.W., Overbeek, J.Th.G., 1948. Theory of the Stability of Lyophobic Colloids. Elsevier, New York.
Vigil, R.D., Ouyang, Q., Swinney, H.L., 1992. Turing patterns in a simple gel reactor. Physica A 188, 17–25.
Volterra, V., 1926. Variazioni e fluttuazioni del numero d'individui in specie animali conviventi. Mem. Acad. Lincei Roma 2, 31–113.
Vukusic, P., Sambles, J.R., 2003. Photonic structures in biology. Nature 424, 852–855.
Yamanaka, J., Yoshida, H., Koga, T., Ise, N., Hashimoto, T., 1998. Reentrant solid–liquid transition in ionic colloidal dispersions by varying particle charge density. Phys. Rev. Lett. 80, 5806–5809.
Yamaguchi, M., Yoshimoto, E., Kondo, S., 2007. Pattern regulation in the stripe of zebrafish suggests an underlying dynamic and autonomous mechanism. Proc. Natl. Acad. Sci. U. S. A. 104, 4790–4793.
Yoshida, A., Motoyama, M., Kosaku, A., Miyamoto, K., 1997. Antireflective nanoprotuberance array in the transparent wing of a hawkmoth, *Cephonodes hylas*. Zool. Sci. 14, 737–741.
Yoshikawa, K., Fukunaga, K., Kawakami, H., 1990. A tri-phasic mode is stable when three nonlinear oscillators interact with each other. Chem. Phys. Lett. 174, 203–207.

Yoshioka, S., Kinoshita, S., 2002. Effect of macroscopic structure in iridescent color of the peacock feathers. Forma 17, 169–181.
Yoshioka, S., Kinoshita, S., 2011. Direct determination of the refractive index of natural multilayer systems. Phys. Rev. E 83, 051917.
Yoshioka, S., Nakamura, E., Kinoshita, S., 2007. Origin of two-color iridescence in rock dove's feather. J. Phys. Soc. Jpn. 76, 013801.
Yoshioka, S., Kinoshita, S., Iida, H., Hariyama, T., 2012. Phase-adjusting layers in the multilayer reflector of a jewel beetle. J. Phys. Soc. Jpn. 81, 054801.
Zhabotinsky, A.M., 1964. Periodical process of oxidation of malonic acid solution. Biophysics 9, 306–311.

2 Belousov–Zhabotinsky Reaction

Jun Miyazaki

Chapter Contents
2.1 Introduction 61
2.2 Oscillation, Wave Propagation, and Pattern Formation 64
  2.2.1 Composition 64
  2.2.2 Oscillation in a Reactor 65
  2.2.3 Wave Propagation and Pattern Formation 66
2.3 Reaction Mechanism and Numerical Simulation 68
  2.3.1 FKN Mechanism 68
  2.3.2 Oregonator Model and Limit Cycle Oscillation 69
  2.3.3 Wave Propagation, Target Pattern, and Spiral Waves 72
2.4 Synchronization 75
  2.4.1 Synchronization in Chemical Oscillation 75
  2.4.2 Analytical Method: Phase Model 75
  2.4.3 Method for Determining Coupling Function 76
  2.4.4 Application to the Coupled BZ Oscillators 77

2.1 Introduction

Chemical oscillation, which is a periodic change in the concentrations of reacting reagents, is one of the most typical pattern formation processes maintained under the flux of energy or matter. It occurs far from equilibrium and usually persists for a long time on the way to its equilibrium state. In particular, the Belousov–Zhabotinsky (BZ) reaction is a well-known chemical oscillation. Figure 2.1 shows that a solution undergoing the BZ reaction changes color periodically from red to blue. When the solution is unstirred, one can see blue waves propagating on a red background, forming concentric rings or rotating spirals. The period is typically several minutes, and the reaction occurs for several tens of minutes.

The spatial and temporal patterns of the BZ reaction are often classified as dissipative structures, and have significant similarity with living systems. For example, a rotating spiral pattern is observed in many biological systems (Lechleiterand and Clapham, 1992; Gorelova and Bures, 1983; Pertsov et al., 1993; Davidenko et al., 1991). In contrast to these living systems, an experiment with the BZ reaction can be performed by simply mixing several kinds of reagents such as bromine, organic acid, and metal catalyst. This feature has attracted the interest of researchers because

Figure 2.1 Oscillations and spatiotemporal patterns observed in the BZ reaction. (a) When the BZ solution is continuously stirred, the solution periodically changes color between red and blue. (b) Concentric circles and (c) rotating spirals appear in the thin layer when the solution is unstirred. (For interpretation of the references to color in this figure legend, the reader is referred to the online version of this chapter.)

it is a suitable system to understand the general aspects of pattern formation processes. Indeed, the study of the BZ reaction has contributed considerably to the development of nonequilibrium thermodynamics.

The BZ reaction was discovered serendipitously in 1951 by a Russian chemist, Boris Belousov, when he was investigating the citric cycle (Belousov, 1958, 1985). He employed cerium ions instead of an enzyme as a catalyst for the citric cycle, and found that the color of the solution alternated periodically between yellow

---

Box 2.1 Temporal Pattern

Prepare four water solutions: $[NaBrO_3] = 0.8$ (M), $[H_2SO_4] = 0.8$ (M), $[CH_2(COOH)_2$: Malonic acid$] = 0.8$ (M), and $[Fe(phen)_3^{2+}$ (ferroin)$] = 3.5$ (mM). Ferroin is synthesized by mixing 1,10-phenanthroline monohydrate ($C_{12}H_8N_2 \cdot H_2O$) and iron sulfate ($FeSO_4 \cdot 7H_2O$) with the molar ratio of 3:1. Temporal oscillation can be initiated by mixing the solutions with equal volume in the preceding order. It is convenient to use a magnetic stirrer to maintain spatial uniformity. The experiment should be performed in a well-ventilated space because bromine, which is extremely harmful to humans, is produced through the reaction.

> **Box 2.2 Spatial Pattern**
>
> Prepare five stock solutions: $[NaBrO_3] = 1$ (M), $[H_2SO_4] = 2$ (M), [Malonic acid] $= 2$ (M), $[KBr] = 1$ (M), and [ferroin] $= 25$ (mM). First, mix the solutions of $NaBrO_3$, H_2SO_4, and malonic acid in a petri dish with a volume of 2 ml for each. Then, add 0.5 ml of KBr solution. The color of the solution changes to yellow as Br_2 gas is produced. Wait for a few minutes until the solution becomes transparent. Add 0.2 ml of ferroin solution, and then you can see blue spots appearing on the red background, from which a wave propagates periodically to form a concentric ring. The chemical wave can also be initiated by dipping a silver wire in the solution. A pair of rotating spirals appears if the chemical wave is slightly stirred by a thin rod.

Ce^{4+} and colorless Ce^{3+}. He also observed yellow traveling waves when the solution was unstirred. After this discovery, a graduate student in biophysics, Anatoly Zhabotinsky, confirmed the reaction and investigated it further (Zhabotinsky, 1964a, 1964b; Zaikin and Zhabotinsky, 1970; Zhabotinsky, 1974). He employed ferroin, an iron complex that is red (blue) in the reduced (oxidized) state, as a catalyst instead of cerium to enhance contrast during oscillation, and observed a concentric circle pattern (the target pattern). Since then, this chemical reaction has attracted much attention from scientists and has been named the Belousov–Zhabotinsky reaction.

Although the BZ reaction is exhibited by only a few kinds of reagents, many reaction intermediates appear during the reaction process, and the mechanism is very complicated. In the 1970s, Fields, Körös, and Noyes (FKN) made progress in determining the mechanism of the BZ reaction (Field et al., 1972; Noyes et al., 1972). They proposed a kinetic model for the BZ reaction, now known as the FKN mechanism. Later, they simplified the mechanism and proposed the Oregonator model, which has only three variable concentrations (Field and Noyes, 1974; Gyorgyi and Field, 1992). This model remains essential for studying the BZ reaction and allows us to reproduce the oscillation behavior in numerical simulations. Since then, many numerical investigations have been performed based on this model.

So far, various spatiotemporal patterns have been observed in the BZ reaction. Soon after the discovery of target pattern, Zhabotinsky discovered a rotating spiral pattern in the BZ medium. Winfree (1980) independently discovered this two-dimensional (2D) pattern formed in a thin layer, and then studied the three-dimensional (3D) spiral structure. The spatiotemporal pattern formed in the BZ reaction is often compared with the Turing pattern. In 1952, a British mathematician, Alan Turing, suggested in his paper "The Chemical Basis of Morphogenesis," that a steady periodic structure spontaneously arises in a reaction diffusion system with an activator–inhibitor mode (Turing, 1952). The Turing pattern was suggested to appear when the inhibitor diffuses much faster than the activator. In the BZ reaction, the diffusion coefficients of the reactants that correspond to the activator and inhibitor are of nearly the same magnitude. Therefore, the dynamical patterns observed in the BZ reaction are basically different from the Turing pattern. However, in the 1990s, De Kepper and colleagues experimentally observed the Turing pattern in the chloride–iodide–malonic acid (CIMA) reaction, which is similar to the BZ reaction (Castets et al., 1990). They performed the CIMA reaction in acrylamide gel, in which the inhibitor diffuses faster than the activator, and consequently, a clear steady Turing pattern was observed.

To date, many studies have been performed based on the BZ reaction for applications such as the control of pattern formation by light irradiation (Sekiguchi et al., 1994; Srivastava et al., 1992), image processing (Kuhnert et al., 1989), chemical circuits (Toth and Showalter, 1995; Agladze et al., 1996; Motoike and Yoshikawa, 1999), and chemical motors based on the conversion of chemical energy into mechanical energy under an isothermal condition (Kitahata et al., 2002).

2.2 Oscillation, Wave Propagation, and Pattern Formation

2.2.1 Composition

Several kinds of reagent, such as bromate ions, organic substrate, and sulfuric acid, are employed to initiate the BZ reaction. Malonic acid is frequently used as an organic substrate. Cerium ions (Ce^{3+}), iron–phenanthroline complex (ferroin: $Fe(phen)_3^{2+}$), and ruthenium–bipyridine complex $\left(Ru(bipyridine)_3^{2+}\right)$ are typically used as catalysts. The color of the oscillating solution depends on the catalyst employed. For cerium ions, the solution oscillates between colorless Ce^{3+} and yellow Ce^{4+}. On the other hand, when ferroin is employed, the solution oscillates between red $Fe(phen)_3^{2+}$ and blue $Fe(phen)_3^{2+}$, and when the ruthenium–bipyridine complex is employed, the solution oscillates between orange $Ru(bipyridine)_3^{2+}$ and light green $Ru(bipyridine)_3^{3+}$.

Among these catalysts, ferroin has been widely employed because the light absorption band in the visible region has a high contrast between the reduced state $\left(Fe(phen)_3^{2+}\right)$ and the oxidized state $\left(Fe(phen)_3^{3+}\right)$. Figure 2.2 shows the typical

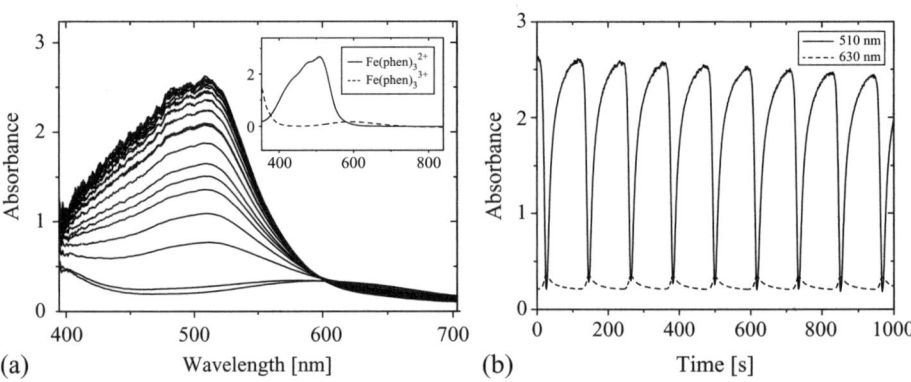

Figure 2.2 (a) Periodic change in the absorption spectrum of the ferroin-catalyzed BZ reaction. The inset shows the absorption spectra of $Fe(phen)_3^{2+}$ and $Fe(phen)_3^{3+}$, respectively. (b) Time traces of absorbance at 510 nm (black line) and 630 nm (gray line).

periodic change in the absorption spectrum of the ferroin-catalyzed BZ reaction. $Fe(phen)_3^{2+}$ and $Fe(phen)_3^{3+}$ show their absorption maxima at 510 nm and 630 nm, respectively; the former shows red, while the latter shows blue. Figure 2.2(b) shows the time courses of absorbance at 550 nm and 630 nm, which show complementary behaviors. The color of the solution changes quickly from red to blue due to the self-catalytic reaction of $HBrO_2$, where $Fe(phen)_3^{2+}$ is oxidized to $Fe(phen)_3^{3+}$ as described in the following section. This rapid color change is often called *firing* in an analogy to neural firing. It subsequently changes from blue to red, as $Fe(phen)_3^{3+}$ is reduced to $Fe(phen)_3^{2+}$. It is known that iron complex undergoes decomposition in the presence of strong acid, and therefore, the contrast of the oscillating BZ solution gradually decreases. However, the decomposition rate is slow compared with the oscillation period, and thus, this usually does not cause a problem. Ferroin is synthesized by simply mixing 1,10-phenanthroline monohydrate and iron sulfate in water in a molar ratio of 3:1.

The ruthernium–bipyridine complex is also widely used for catalysis. The periodic change in its absorption spectrum is shown in Figure 2.3. It changes color between orange and light green during oscillation. One of the most important features of the BZ reaction catalyzed by the ruthernium–bypyridine complex is its photosensitivity (Sekiguchi et al., 1994; Srivastava et al., 1992; Kuhnert et al., 1989). The photoexcited triplet state of $Ru(bipyridine)_3^{2+}$ induces a photochemical reaction and produces Br^-, which serves as an inhibitor in the BZ reaction; consequently, the oscillation is usually suppressed. This allows us to control the spatial as well as temporal patterns by photo irradiation (Kuhnert et al., 1989).

2.2.2 Oscillation in a Reactor

In principle, nonlinear chemical phenomena occur in open systems where matter and energy flow into and out of the system. Oscillation occurs on the way to equilibrium. Therefore, when a closed system such as a batch reactor is considered, the initial reagents are consumed gradually, and the oscillation period and amplitude change.

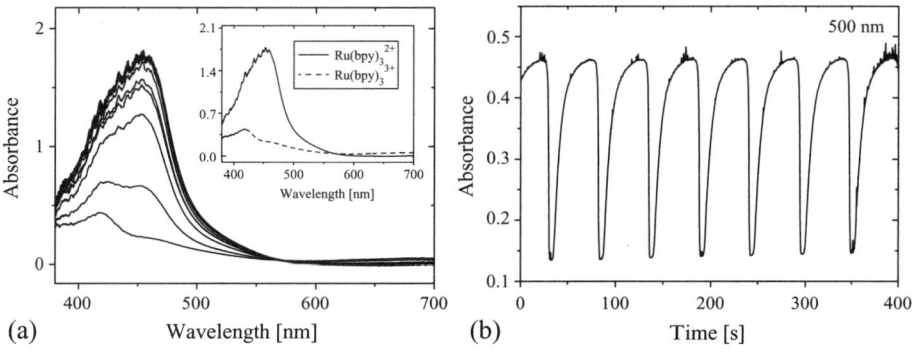

Figure 2.3 (a) Periodic change in the absorption spectrum of the $Ru(bipyridine)_3^{2+}$-catalyzed BZ reaction. The inset shows the absorption spectra of $Ru(bipyridine)_3^{2+}$ and $Ru(bipyridine)_3^{3+}$, respectively. (b) Time traces of absorbance at 500 nm.

To maintain stable oscillation, it is necessary to supply fresh reagents to the reactor. To this end, a continuous-flow stirred tank reactor (CSTR) is often employed, in which fresh reagents are continuously supplied to the reactor by a pump, and the solution undergoing the reaction flows out to maintain a constant volume (Epstein and Pojman, 1998). This type of reactor allows us to maintain the system far from equilibrium for a long time. The principle of the CSTR is similar to that of living organs that take energy from nutrition to maintain vital activity. Figure 2.4(a) shows a schematic illustration of a typical CSTR, where reactants are supplied separately through the two inlets, and reaction takes place by mixing the solution. Reacting solution then overflows from the outlet. Figure 2.4(b) shows that the oscillation period is kept constant for a long time by using the CSTR. Taking advantage of this, the CSTR is often employed for quantitative measurement (Epstein and Pojman, 1998; Miyazaki and Kinoshita, 2006b). However, when a batch reactor is employed, the oscillation period increases with time, and oscillation eventually stops.

The BZ solution is stirred often to maintain spatial uniformity for observation of temporal behavior. An interesting point is that oscillatory behavior strongly depends on the stirring condition (Ali and Menzinger, 1991; Noszticzius et al., 1991; Pojman et al., 1992; Ruoff, 1993; Vanag and Melikhov, 1995). In general, the oscillation period increases with the stirring rate. The explanation is that in a weak mixing condition, oscillation periods depend on position due to concentration fluctuations or defects on the surface of the reactor. In this case, fast oscillating regions serve as *pacemakers*, and the slow regions are entrained to these oscillations so that the oscillation frequency usually increases.

2.2.3 Wave Propagation and Pattern Formation

When the ferroin-catalyzed BZ solution is unstirred and remains in a glass dish, one can observe blue spots of the oxidized state randomly appearing on the red background of the reduced state, from which a wave propagates to form a concentric ring. When the central region oscillates periodically, multiple concentric rings (target pattern) are

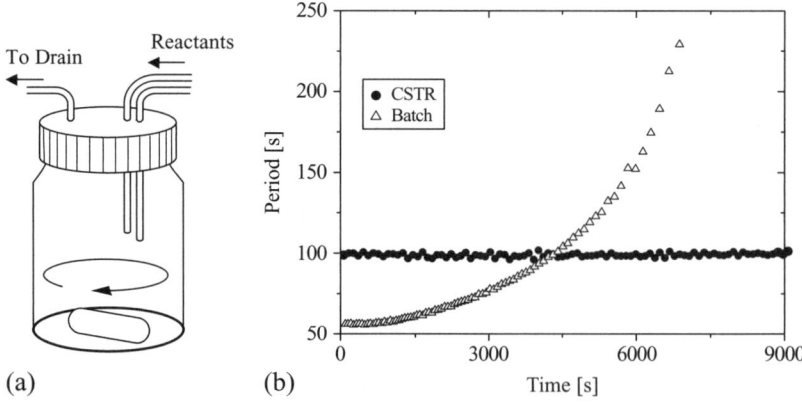

Figure 2.4 (a) Schematic view of a CSTR. Reactants are continuously supplied by a pump and mixed in the reactor. (b) Time courses of the oscillation periods in a CSTR (filled circle) and a batch reactor (open triangles).

formed, as shown in Figure 2.5. The number of the concentric rings corresponds to the number of cycles. It is assumed that there are some defects on the glass surface or dust at the center of the target pattern, where the oscillation frequency becomes faster than in the surrounding areas. The wave is initiated in such a way that the excited region of the solution further excites the neighboring region. Since the period of the propagating wave is determined by the period at the center, the central point is called the pacemaker. The wave can be also initiated by dipping a silver wire in the BZ solution, which reacts with Br to form AgBr, leading to a decrease in the local concentration of the Br^- inhibitor. An interesting feature is that when the two waves collide with each other, they annihilate. Therefore, only a single target pattern that exhibits the fastest oscillation period can finally remain in the glass dish, as shown in Figure 2.5.

Another interesting feature is that when the solution is slightly perturbed after a wave is initiated, rotating spiral waves emerge, as shown in Figure 2.6. Contrary to the target pattern, the spiral wave is maintained without a pacemaker. The trajectory of the tip of the spiral wave exhibits various meandering patterns, such as circle and flower shapes, depending on the experimental condition (Müller et al., 1985). The oscillating frequency of the spiral wave is generally faster than that of the target pattern, and consequently the pitch of the wave is shorter.

The catalysts can be immobilized in a gel such as silica or acrylamide. When the gel sheet with catalysts is immersed in the BZ solution, traveling waves, target patterns, and spiral waves are observed in the gel as well. The gel system enables us to control

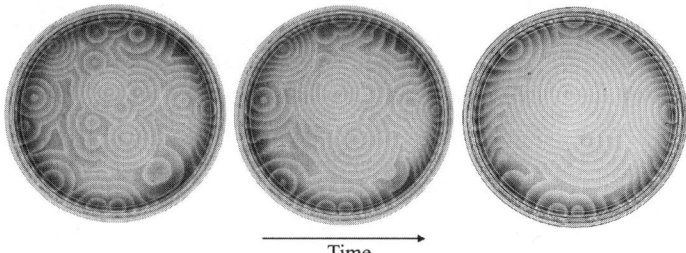

Figure 2.5 Time course of target patterns. A single target pattern that exhibits the fastest oscillation period eventually dominates. (For color version of this figure, the reader is referred to the online version of this chapter.)

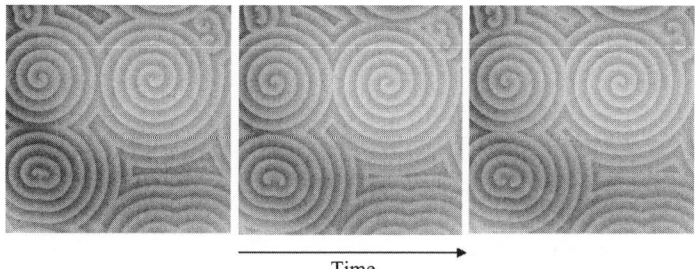

Figure 2.6 Rotating spiral waves. (For color version of this figure, the reader is referred to the online version of this chapter.)

2.3 Reaction Mechanism and Numerical Simulation

2.3.1 FKN Mechanism

Although only a few kinds of reagents are employed to initiate the BZ reaction, the detailed reaction mechanism is complicated because many intermediates appear during the reaction process. The FKN mechanism proposed by Fields, Körös, and Noyes consists of 10 elementary processes, as shown in Table 2.1, where Ce^{2+} is used as a catalyst (Field et al., 1972; Miike et al., 1997).

These reactions are summarized as the bromination of malonic acid in (R1)–(R10), where bromine is supplied from bromated ion through two reaction processes. The two processes proceed alternately; when one of the processes dominates, the other process is suppressed. Each process takes place repeatedly, and therefore, oscillation is observed.

Figure 2.7 shows the reaction scheme of the FKN mechanism. Process 1 (P1) consists of (R1)–(R3) in Table 2.1. First, the initial reagent, BrO_3^-, consumes Br^- and changes into $HBrO_2$ (R1). $HBrO_2$ also consumes Br^- and changes into 2HBrO (R2). HBrO reacts with Br^-, and then Br_2 is produced (R3). Since Br^- is consumed in (R1)–(R3), (P1) is dominant under high concentrations of Br^-.

Process 2 (P2) consists of (R1) and (R4)–(R6), which is dominant at low concentrations of Br^-. $HBrO_2$ produced through (R1) changes into a BrO_2 radical ($\cdot BrO_2$) through (R5) instead of (R2) at a low concentration of Br^-. $HBrO_2$ reacts with BrO_2 through (R5), and two radicals of BrO_2 are produced. The BrO_2 radical is deoxidized by a Ce^{3+} and changes into two $HBrO_2$, through which Ce^{3+} is oxidized into Ce^{4+} (R6). This process can be expressed explicitly as

$$(R5) + 2(R6)\ HBrO_2 + BrO_3^- + 3H^+ + 2Ce^{3+} \rightarrow 2HBrO_2 + H_2O + 2Ce^{4+}.$$

Table 2.1 FKN Mechanism

(R1) $Br^- + BrO_3^- + 2H^+ \leftrightarrow HOBr + HBrO_2$
(R2) $Br^- + HBrO_2 + H^+ \leftrightarrow 2HOBr$
(R3) $Br^- + HOBr + H^+ \leftrightarrow Br_2 + H_2O$
(R4) $2HBrO_2 \leftrightarrow HOBr + BrO_3^- + H^+$
(R5) $HBrO_2 + BrO_3^- + H^+ \leftrightarrow 2\cdot BrO_2 + H_2O$
(R6) $\cdot BrO_2 + Ce^{3+} + H^+ \leftrightarrow HBrO_2 + Ce^{4+}$
(R7) $\cdot BrO_2 + Ce^{4+} + H_2O \leftrightarrow BrO_3^- + Ce^{3+} + 2H^+$
(R8) $Br_2 + CH_2(COOH)_2 \rightarrow BrCH(COOH)_2 + Br^- + H^+$
(R9) $6Ce^{4+} + CH_2(COOH)_2 + 2H_2O \rightarrow 6Ce^{3+} + HCOOH + 2CO_2 + 6H^+$
(R10) $4Ce^{4+} + BrCH(COOH)_2 + 2H_2O \rightarrow 4Ce^{3+} + HCOOH + Br^- + 2CO_2 + 5H^+$

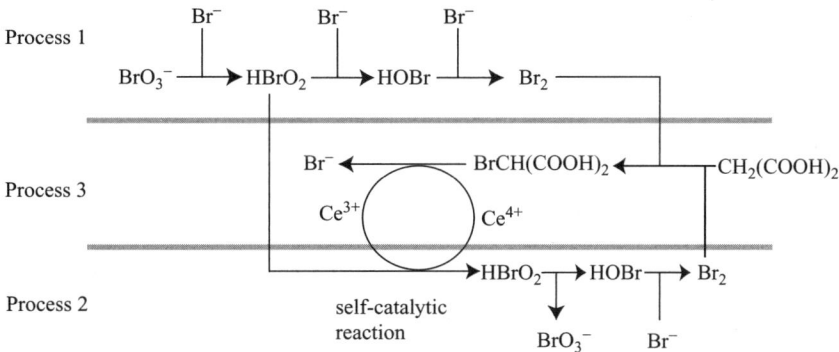

Figure 2.7 Three processes in the FKN mechanism.

Here, two $HBrO_2$ molecules are produced from one $HBrO_2$ through a self-catalytic reaction. This is a positive-feedback reaction, where the reaction occurs more rapidly as the concentration of $HBrO_2$ increases, and the reaction proceeds explosively. The self-catalytic reaction is eventually suppressed as Ce^{3+} is consumed. $HBrO_2$ is subsequently deoxidized through (R4) and (R3), and 10 Br_2 molecules are produced.

The switching mechanism between (P1) and (P2) can be understood as follows. Suppose that the concentration of Br^- ($[Br^-]$) is initially large. In this case, (P1) is dominant because $HBrO_2$ is consumed by Br^-, while the self-catalytic reaction in (P2) is suppressed. $[Br^-]$ decreases gradually through (P1), and eventually (P1) is suppressed. Then (P2) becomes dominant instead of (P1) because Br^- is not involved in the self-catalytic reaction. Therefore, switching from (P1) to (P2) occurs. Ce^{3+} is oxidized to Ce^{4+} through (P2). To induce switching from (P2) to (P1), Ce^{4+} must be deoxidized to Ce^{3+}, and Br^- must be generated through the subset process involving organic reaction (P3). This process consists of (R9) and (R10):

$$(R9) + (R10) \ 10\ Ce^{4+} + CH_2(COOH)_2 + BrCH(COOH)_2 + 4H_2O$$
$$\rightarrow 10Ce^{3+} + 2HCOOH + Br^- + 2CO_2 + 6H^+,$$

where Ce^{4+} is deoxidized to Ce^{3+}, and Br^- is produced; therefore, switching from (P2) to (P1) occurs. The reduction of the catalyst and the production process of Br^- are expressed as (P3) in the FKN mechanism. However, the details of this subset process have not been fully understood yet.

2.3.2 Oregonator Model and Limit Cycle Oscillation

The FKN mechanism is still too complicated for numerical calculation as it includes many elementary steps and chemical species. Field and Noyes (1974) simplified the FKN mechanism and proposed the Oregonator model, which had only three variable concentrations but retained the essential characteristics of the BZ reaction.

Table 2.2 Oregonator Model Reactions

(R1) $Br^- + BrO_3^- + 2H^+ \rightarrow HOBr + HBrO_2$
(R2) $Br^- + HBrO_2 + H^+ \rightarrow 2HOBr$
(R4) $2HBrO_2 \rightarrow HOBr + BrO_3^- + H^+$
(R5) + 2(R6) $HBrO_2 + BrO_3^- + 3H^+ + 2Ce^{3+} \rightarrow 2HBrO_2 + H_2O + 2Ce^{4+}$
(R10) $4Ce^{4+} + BrCH(COOH)_2 + 2H_2O \rightarrow 4Ce^{3+} + HCOOH + Br^- + 2\,CO_2 + 5H^+$
(M1) $A + W \rightarrow U + P$
(M2) $U + W \rightarrow 2P$
(M3) $A + U \rightarrow 2U + 2V$
(M4) $2U \rightarrow A + P$
(M5) $B + V \rightarrow hW$

The Oregonator model consists of five elementary reactions, as shown in Table 2.2; in this model, the inverse reactions in the FKN mechanism are not taken into account, and therefore, only the essence of the FKN mechanism is retained.

The top section of Table 2.2 is summarized in the bottom section by setting $A = [BrO_3^-]$, $B = [BrCH(COOH)_2]$, $P = [HOBr]$, $U = [HBrO_2]$, $V = [Ce^{4+}]$, and $W = [Br^-]$. The value A represents the concentration of the initial reagent, BrO_3^-, which is treated as a constant parameter by assuming that it is consumed slowly. Because Br^- is produced through complicated reaction processes that are not included in (R10), a stoichiometric factor h that reflects the number of Br^- ions generated through the organic oxidation reaction is introduced in (M5). The values U, V, and W are treated as variables. The rate equations are then given by

$$\frac{dU}{dt} = k_1 AW - k_2 UW + k_3 AU - k_4 U^2, \tag{2-1}$$

$$\frac{dV}{dt} = 2k_3 AU - k_5 BV, \tag{2-2}$$

$$\frac{dW}{dt} = -k_1 AW - k_2 UW + k_5 hBV, \tag{2-3}$$

where k_i ($i = 1, 2, \ldots, 5$) are the rate constants for (M1)–(M5). Equations (2-1), (2-2), and (2-3) can be expressed in a nondimensional form as

$$\varepsilon \frac{du}{d\tau} = qw - uw + u - u^2, \tag{2-4}$$

$$\frac{dv}{d\tau} = u - v, \tag{2-5}$$

$$\varepsilon' \frac{dw}{d\tau} = -qw - uw + fv, \tag{2-6}$$

where $u = [2k_4/(k_3A)]U, v = [k_4k_5B/(k_3A)^2]V, w = [k_2/(k_3A)]W, t = k_5Bt, \varepsilon = k_5B/(k_3A)$, $\varepsilon' = 2k_4k_5B/(k_2k_3A), q = 2k_1k_4/(k_2k_3), f = 2h$.

The set of Eqs. (2-4), (2-5), and (2-6) is widely employed for numerical calculations to reproduce the dynamical behavior of the BZ reaction. Since ε is much larger than ε', w can be regarded as a fast variable. This allows us to eliminate w adiabatically by setting $dw/dt = 0$, and the Oregonator model is expressed by the two variables of u and v:

$$\varepsilon \frac{du}{d\tau} = u(1-u) - \frac{fv(u-q)}{u+q}, \tag{2-7}$$

$$\frac{dv}{d\tau} = u - v. \tag{2-8}$$

Equations (2-7) and (2-8) exhibit a limit cycle when appropriate parameters are employed. A limit cycle is a closed trajectory in a phase space. It is different from a harmonic oscillation, the amplitude of which depends on the initial position and velocity.

For any initial point, the trajectory eventually converges to the limit cycle. The amplitude of the limit cycle does not depend on the initial condition. Such a behavior can be seen in a wide variety of nonlinear systems. In terms of thermodynamics, the oscillation is induced through dissipation of energy and is often called a self-sustained oscillator.

To obtain a qualitative explanation of the oscillation using Eqs. (2-7) and (2-8), it is convenient to plot nullclines on the 2D uv surface. The nullclines are curves where $du/dt = 0$ (u-nullcline) and $dv/dt = 0$ (v-nullcline), which are respectively given by

$$v = \frac{u(1-u)(u+q)}{(u-q)f}, \tag{2-9}$$

$$v = u. \tag{2-10}$$

The dynamic of the system is expressed as a trajectory on the surface. Figure 2.8 shows the nullclines and the trajectory of the limit cycle. Arrows show the direction

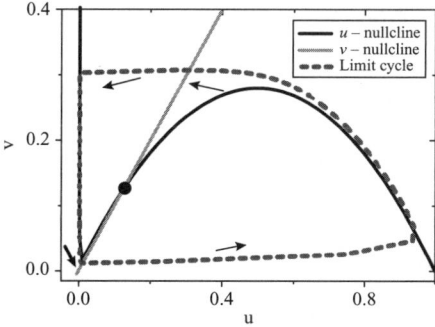

Figure 2.8 The trajectory of the system exhibits limit-cycle oscillation (broken line) on the uv surface. The solid back and gray lines are u- and v-nullclines, respectively.

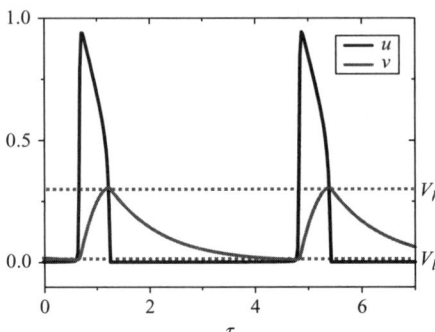

Figure 2.9 Time courses of u and v. v_h and v_l are the upper and lower thresholds of v, respectively.

of the vectors $(du/dt, dv/dt)$ in the regions separated by the nullclines. One can see that the state point is attracted to the limit cycle, irrespective of the initial position, and therefore, the system exhibits a stable oscillation. Since the parameter ε is usually much larger than unity, u evolves faster than v. Therefore, the state point is quickly attracted to the u-nullcline and moves along it for most of the time during one period.

Figure 2.9 shows the time evolution of u and v. When v reaches the lower threshold value v_l, u abruptly increases. This corresponds to the self-catalytic reaction. Subsequently, v begins to increase gradually. This corresponds to the Br^- production process (M5). During that time, u decreases as it subjects to v. After v reaches the upper threshold v_u, u decreases quickly, and then v begins to decrease. In this system, the variable u serves to promote the production of v, while v suppresses the production of u. Therefore, the variables u and v are often called activator and inhibitor, respectively, and the system is called the activator–inhibitor model.

The dynamical system governed by Eqs. (2-7) and (2-8) also exhibits excitability under an appropriate parameter condition. Figure 2.10 shows the nullclines and the trajectory of the system with several values of f. The intersection point of u- and v-nullclines in Figure 2.10(a), to which the state point is attracted, is stable. This corresponds to the reduced state of the BZ solution. In this case, when perturbation is applied, the state point exhibits a large deviation from the stable point. This transient behavior is called excitability, and the traveling wave occurs in such a way that the excited region of the solution further excites (perturbs) the neighboring region, and therefore, the wave propagates. Figure 2.10(b) shows that the system exhibits a limit cycle for $f = 1$, where the intersection point of the nullclines is unstable. For a small value of f, the state point is attracted to the intersection point corresponding to the oxidized BZ solution, as shown in Figure 2.10(c).

2.3.3 Wave Propagation, Target Pattern, and Spiral Waves

When the BZ solution is unstirred, the dynamics of the system should be described as a function of time and spatial coordinates. The substances distributed in space undergo Brownian motion, and the concentration changes due to diffusion as well as chemical

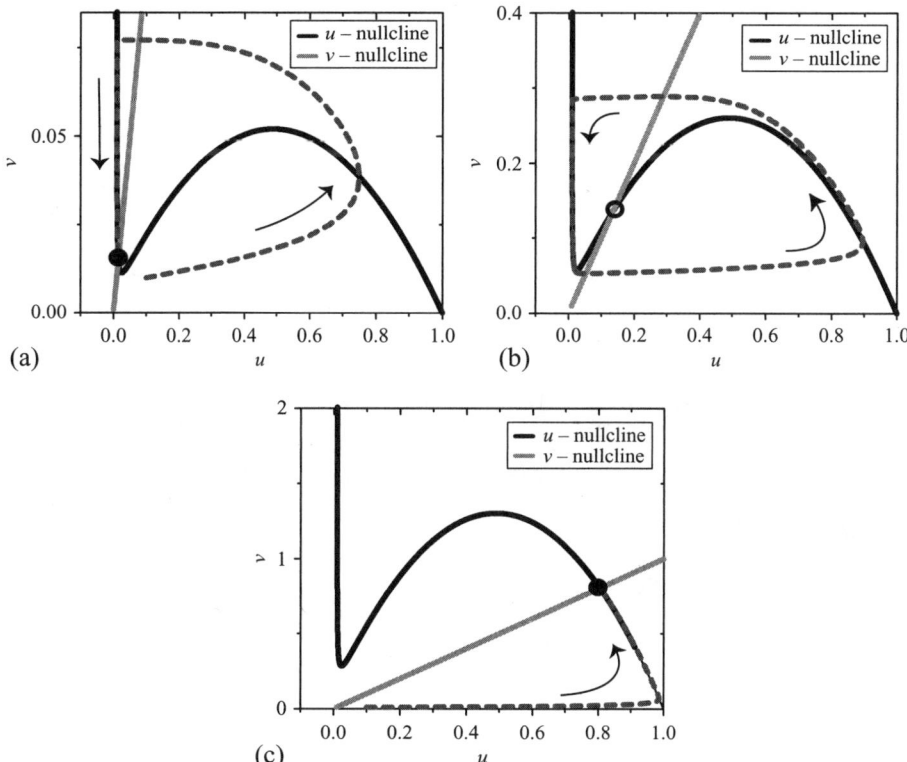

Figure 2.10 Trajectories of the system governed by the Oregonator model with several values of f. (A) The system approaches the intersection of u and v-nullclines denoted by the filled circles at $f=5$. The intersection corresponds to the reduced state of the BZ solution. (B) The system exhibits limit cycle oscillation at $f=1$. In this case, the intersection denoted by the open circle is unstable. (C) The system approaching the intersection corresponds to the oxidized state when $f=0.2$.

reaction. The governing equations are usually given by adding the diffusion terms to Eqs. (2-7) and (2-8) as

$$\varepsilon \frac{du(\boldsymbol{x},t)}{d\tau} = u(1-u) - \frac{fv(u-q)}{u+q} + D_u \nabla^2 u, \qquad (2\text{-}11)$$

$$\frac{dv(\boldsymbol{x},t)}{d\tau} = u - v + D_v \nabla^2 v, \qquad (2\text{-}12)$$

where D_u and D_v are the diffusion coefficients of u and v, respectively. Equations (2-11) and (2-12) are often called reaction–diffusion equations, and their solutions display a variety of behaviors. Figure 2.11 shows the results of the numerical simulation of Eqs. (2-11) and (2-12) in one dimension. When a pacemaker is set at the left end, the wave propagates from left to right in such a way that the excited region (large value of u)

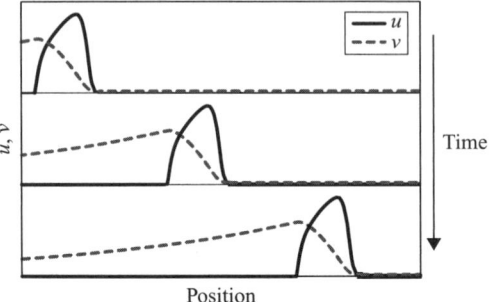

Figure 2.11 Wave propagation.

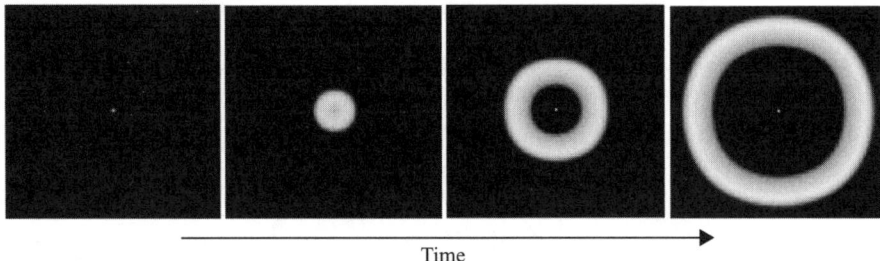

Figure 2.12 Time course of a circular wave. The bright region corresponds to a large value of u. A pacemaker is set at the center, from which circular waves are initiated.

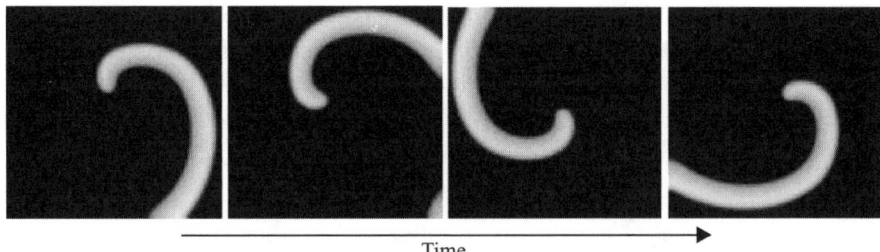

Figure 2.13 Time course of a rotating spiral wave. The bright region corresponds to a large value of u.

activates the neighboring region through diffusion. The increase in u leads to a slow increase in v. After v increases, u decreases, and then v begins to decrease. The wave is again initiated when v becomes sufficiently low. Figures 2-12 and 2-13 show the results of the numerical simulation in two dimensions using Eqs. (2-11) and (2-12). In Figure 2.12, a pacemaker is set at the center, from which circular waves are initiated. Since the waves are initiated periodically, a target pattern corresponding to that in Figure 2.5 is formed with time. Figure 2.13 shows a rotating spiral wave produced when appropriate initial conditions are given. The trajectory of the tip of the spiral is a circle. In this way, the Oregonator model qualitatively reproduces the experimental results, and it has been successfully employed to explain and even predict a variety of features observed in the BZ reaction (Miyazaki and Kinoshita, 2007).

2.4 Synchronization

2.4.1 Synchronization in Chemical Oscillation

A system of sustained oscillatory units generates a collective rhythm through mutual interactions. Such systems are commonly observed and often serve as significant functional units in nature and technology. These systems can be regarded as an ensemble of oscillators having slightly different natural oscillation frequencies, which then generate a collective rhythm through mutual interactions. A typical example of such a system is synchronously flashing fireflies (Buck, 1988; Strogatz, 2003). A swarm of male fireflies in south Asia generates synchronous flashing on a tree to attract female attention. Although each firefly blinks repeatedly with its own natural frequency when isolated from its environment, it is sensitive to light emitted by the other fireflies and it adjusts its flashing time. As a result, when a large number of fireflies swarm, their flashing timings are spontaneously synchronized to the same frequency through the mutual interaction.

Recently, synchronization phenomenon have been studied especially from a biological point of view, because a coupled biological oscillator, such as circadian rhythms (Reppert and Weaver, 2002), neurons in the brain (Uhlhaas and Singer, 2006; Hammond et al., 2007), and cardiac cells in the heart, are known to play an essential functional role in living organs. Furthermore, such collective behaviors have attracted the interests of physical and mathematical scientists, since the cooperative dynamical ordering have an analogy with conventional second-order phase transition in magnetic and dielectric substances.

As models of coupled oscillators, many investigations have been performed on coupled chemical oscillators, including the BZ reaction since its mechanism is well understood and a numerical simulation can be performed based on the model. The coupled BZ reactors display a wide range of behaviors. Marek and Stuchl (1975) observed various types of synchronization between two reactors separated by a plate perforated with a variable number of holes. Crowley and Epstein (1989) actively controlled the mass flow between two reactors and observed three types of behaviors depending on the coupling strength: entrainments at in-phase and out-of-phase, and oscillation death. Yoshimoto and colleagues (1993) performed the same experiment on three coupled reactors with symmetric and asymmetric mass exchanges controlled by peristaltic pumps. They observed biperiodic, all death, and two types of synchronized modes for the symmetric exchange with increasing coupling strength, while two additional types of synchronized modes were observed for the asymmetric exchange.

2.4.2 Analytical Method: Phase Model

Several theoretical approaches have been proposed so far to deal with such coupled oscillators. The most straightforward and commonly employed method is to deal directly with the differential equations describing coupled oscillators, that is, the

Oregonator model for the BZ reaction, including coupling terms on the basis of a detailed knowledge of the reaction kinetics and coupling scheme. However, in nature, it is generally difficult to determine the differential equations governing the dynamics of a system. Further, it is necessary to determine exact parameter values to correctly describe the system, because even a small change in the parameter set would lead to a completely different collective behavior of the system.

A complementary and the most promising way to deal with such coupled oscillators is a phase model (Kuramoto, 1984; Pikovsky et al., 2001) in which the dynamical behavior of each oscillator is described solely by a single variable of phase under the assumption that the coupling strength and dispersion of natural frequencies are sufficiently small. The time evolution of the phase is generally expressed as

$$\frac{d\phi_i}{dt} = \omega_i + \sum_{j \neq i}^{N} \epsilon_{ij} q(\phi_i - \phi_j), \tag{2-13}$$

where ϕ_i and ω_i represent the phase and natural frequency of the ith oscillator, respectively, N the number of oscillators, and ϵ_{ij} the coupling strength between the ith and jth oscillators. The quantity $q(\psi)$ is the coupling function that determines the collective behavior of the coupled oscillators. Kuramoto (1984) showed the transition from disorder to a macroscopic entrained state by using a tractable form of the coupling function $q(\psi) = \sin\psi$. Since then, many studies using this form of $q(\psi)$ have been reported to clarify the unique features of the transition. As for the BZ reaction, there have been several studies in which the behavior of coupled BZ reactors is analytically treated by the phase model by assuming $q(\psi) = \sin\psi$ (Yoshimoto et al., 1993; Fujii and Sawada, 1978).

However, it is necessary to exactly determine the shape of the coupling function to elucidate the synchronous behavior of the system. In particular, the BZ reaction behaves like a relaxation-type oscillator that is characterized by a different time scale during one period within a limit cycle, and the form of $q(\psi)$ is far from a simple sine function. Typical examples of such cases are oscillatory nerves and pacemaker cells in the heart, where the coupling function is expected to include higher-harmonic terms and the synchronization feature would differ considerably from that of a sinusoidal form of coupling function. In fact, Daido (1994) explained macroscopic entrainment in a large population of globally coupled phase oscillators with $q(\psi)$ including higher-order harmonic terms, and showed that the critical exponent of the order parameter at the onset of entrainment differs from that given by $q(\psi) = \sin\psi$. Another interesting phenomenon is slow switching (Hansel et al., 1993; Kori and Kuramoto, 2001), where the coupling function includes higher-order harmonic terms with a time-delay effect.

2.4.3 Method for Determining Coupling Function

Here, a simple method is presented to determine the coupling function for two given coupled oscillators (Miyazaki and Kinoshita, 2006a, 2006b). By writing the phase equations for coupled oscillators as

$$\frac{d\phi_1}{dt} = \omega_1 + \epsilon_{12}q(\phi_1 - \phi_2),$$

$$\frac{d\phi_2}{dt} = \omega_2 + \epsilon_{21}q(\phi_2 - \phi_1), \qquad (2\text{-}14)$$

and assuming that it takes $T_i + \Delta T_i$ ($i = 1, 2$) during one oscillation under the influence of mutual interactions, where ΔT_i is the deviation from the natural oscillation period, we obtain

$$2\pi = \oint d\phi_i = \int_0^{T_i + \Delta T_i} dt \frac{d\phi_i}{dt}. \qquad (2\text{-}15)$$

By substituting Eq. (2-14) into Eq. (2-15) and assuming that the phase difference $\psi \equiv \phi_i - \phi_j$ evolves slowly and changes very slightly during one oscillation, we obtain the relation

$$2\pi = \int_0^{T_i + \Delta T_i} dt \left[\omega_i + \epsilon_{ij}q(\psi)\right] \cong (T_i + \Delta T_i)\left[\omega_i + \epsilon_{ij}q(\psi)\right]. \qquad (2\text{-}16)$$

Thus, the coupling function is obtained for each ψ up to the first order of $\Delta T_i(\psi)$ as

$$q(\psi) = -\frac{2\pi \Delta T_i(\psi)}{\epsilon_{ij}T_i^2}. \qquad (2\text{-}17)$$

It is clear that $q(\psi)$ can be obtained by two simple ways: (1) specifying the phase difference between the two oscillators, or (2) measuring the time interval between the marked events, that is, ΔT, at various ψ. The determination of $q(\psi)$ does not require comprehensive knowledge of either the oscillation mechanism or the interaction between the oscillators.

2.4.4 Application to the Coupled BZ Oscillators

When two BZ reactors of CSTR are mutually connected through the mass flow, they exhibit in-phase and out-of-phase synchronization depending on the flow rate that is proportional to the coupling strength, as shown in Figure 2.14. In this experiment, the two reactors were immersed in separate water baths with the different temperatures and the natural oscillation periods were maintained at $T_1 = 107$ s and $T_2 = 107$ s, respectively. It is found that only the in-phase synchronization is observed at a high flow rate, while for flow rates between 0.035 ml/s and 0.044 ml/s, both out-of-phase and in-phase synchronizations are found to be present. The oscillation periods of the in-phase synchronization T_{in} and the out-of phase synchronization T_{out} are listed in Table 2.3 for several flow rates. It is found that $T_1 < T_{in} < T_2$, while $T_{out} > T_1, T_2$, and T_{out} tends to become longer as the flow rate is increased, while T_{in} is rather insensitive to the flow rate.

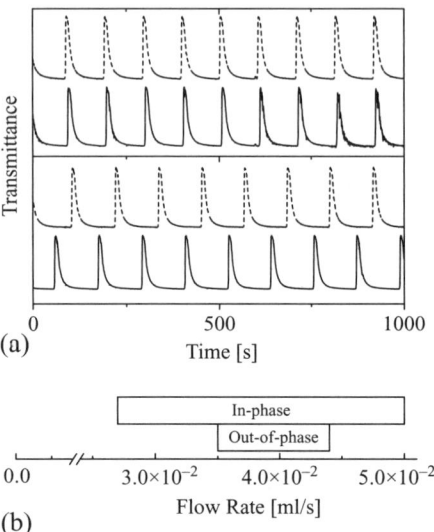

Figure 2.14 (a) Time traces of in-phase (top) and out-of-phase (bottom) synchronizations. (b) Phase diagram of the synchronization mode is expressed as a function of the flow rate (coupling strength).

Table 2.3 Oscillation Periods of In-phase and Out-of-phase Synchronizations for Various Flow Rates

	Flow Rate (ml/s) (Coupling Strength)	Period (s)
In-phase	0.029	102
	0.041	102
Out-of-phase	0.035	113
	0.039	115
	0.041	117

The preceding behavior of coupled BZ reactors can be explained in terms of the phase model through the direct measurement of $q(\psi)$. For this purpose, it is necessary to specify the phase of an oscillator. In the experiment, one can employ the phase defined by

$$\phi(t) = 2\pi \frac{t - t_k}{t_{k+1} - t_k} \text{ for } t_k < t < t_{k+1},$$

where t_k represents the time at which the kth-marked event, such as firing, occurs. This definition is essentially the same as that of the phase model with a zero-coupling limit and is applicable to weakly coupled systems.

Next, it is examined how ΔT can be measured at various ψ. Let us suppose that the two coupled oscillators have different natural oscillation periods and that they synchronize above a critical coupling strength ε_c. It is known that presynchronization occurs at a coupling strength just below ε_c, where the two oscillators repeatedly assume loosely synchronized and asynchronized states. In such a case, ψ evolves over 2π as shown in the left part of Figure 2.15. Therefore, ΔT is measureable at any value of ψ. On the other hand, when the coupling strength is above ε_c, the two oscillators are synchronized and the value of ψ is locked at a specific value of ψ_{sync}. However, if either oscillator is instantaneously perturbed or the coupling is transiently prevented, ψ will deviate from ψ_{sync}. Subsequently, ψ begins to recover to ψ_{sync} slowly as compared with the oscillation period, thereby providing an opportunity to measure $\Delta T(\psi)$. Thus, in principle, $q(\psi)$ can be determined in the presence of coupling, irrespectively of whether the coupling strength is below or above ε_c.

The right part of Figure 2.15 shows $q(\psi)$ obtained in the presynchronization region. $q(\psi)$ is most accurately determined at the coupling strength just below the synchronization threshold, because $\Delta T(\psi)$ due to coupling effect becomes larger. It is obvious that $q(\psi)$ is different from that of a sinusoidal function: they are characterized by a curve that gradually decreases in the region of small values of ψ while it abruptly increases at a large value of ψ. It is confirmed that almost identical $q(\psi)$'s are obtained as well in the synchronization region, and for reactors 1 and 2.

Here we will describe the analysis of the synchronization behavior by using the obtained $q(\psi)$. The time evolution of ψ is derived from Eq. (2-14) as

$$\frac{d\psi}{dt} = -\Delta\omega + \varepsilon Q(\psi) \tag{2-18}$$

where $Q(\psi) \equiv q(\psi) - q(-\psi)$ and $\Delta\omega = \omega_2 - \omega_1$. Figure 2.16 shows $Q(\psi)$, which is estimated from $q(\psi)$ in the bottom right of Figure 2.15. One can then easily predict the

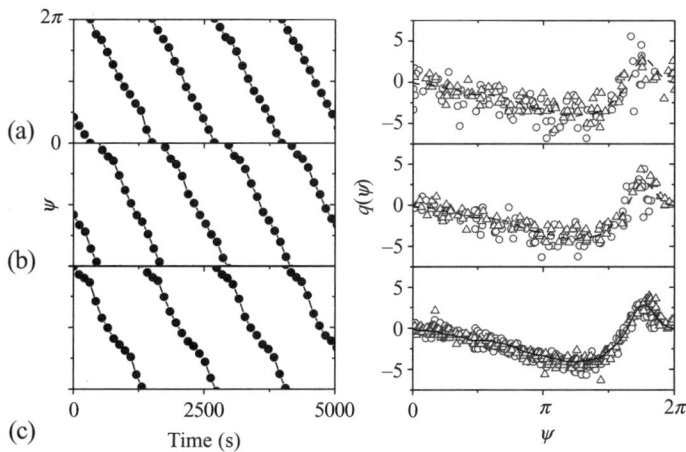

Figure 2.15 Time evolution of the phase difference (left) and coupling functions (right) determined in the presynchronization region. The flow rates are (a) $\rho = 0.011$ (ml/s), (b) $\rho = 0.016$ (ml/s), and (c) $\rho = 0.022$ (ml/s). Points obtained from reactors 1 and 2 are plotted with open circles and triangles, respectively.

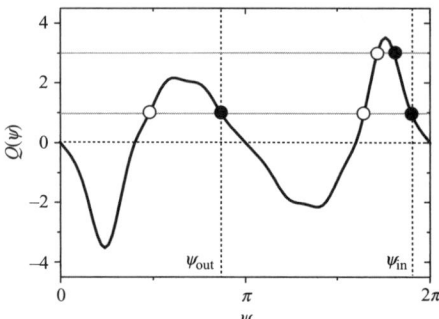

Figure 2.16 $Q(\psi)$ estimated from $q(\psi)$ in Figure 2.15(c) is shown as a solid line. Stable and unstable solutions of Eq. (2-18) are expressed as solid and open circles, respectively, at two values of $\Delta\omega/\varepsilon$. The phase differences designated as ψ_{out} and ψ_{in} correspond to those of the out-of-phase synchronization and in-phase synchronization.

result of changing the coupling strength ε. For this purpose, it is convenient to plot $\Delta\omega/\varepsilon$ for two typical values of ϵ. With increasing coupling strength, a pair of stationary solutions of Eq. (2-18) are obtained as intersection points of $Q(\psi)$ and $\Delta\omega/\varepsilon$. Since the stability of the entrainment requires $dQ(\psi)/d\psi < 0$, only one of the two solutions can actually be realized. The phase difference of the stable solution designated as ψ_{in} asymptotically approaches 2π with increasing ϵ. Thus, the in-phase synchronization is realized. Since the signs of $q(\psi_{in})$ and $q(-\psi_{in})$ are positive and negative, respectively, the oscillation period in the in-phase synchronization mode is intermediate between the two natural oscillation periods.

At high value of ε, a new stable solution appears slightly below π in addition to that of the in-phase synchronization. It is apparent that this stable solution will approach π as ε is increased. This corresponds to the out-of-phase synchronization at the phase difference designated as ψ_{out}. Thus, bistability between the out-of-phase and in-phase synchronization can be reproduced. In the out-of-phase synchronization, the signs of $q(\psi_{out})$ and $q(-\psi_{out})$ are negative, and therefore the two oscillators synchronize at a longer period as compared to their natural frequency. At a considerably high value of ε, it is natural to expect that only the in-phase synchronization should be realized as the concentrations of the two reactors approach homogeneity. However, the analysis using Eq. (2-18) always shows the coexistence of two stable synchronizations. In this case, the fast oscillating terms that are neglected in the derivation of the phase equation of Eq. (2-13) become crucial.

The coupling function determined using two coupled oscillators is generally applicable to the analysis of multicoupled oscillatory systems composed of nearly identical oscillators. There are several studies on multicoupled BZ oscillators such as three BZ reactors coupled through mass exchange (Yoshimoto et al., 1993), and beads with the catalyst arranged in an array of a lattice structure in the BZ solution (Fukuda et al., 2002). In these cases, it may be possible to deal with the oscillators by using a set of phase equations having an identical coupling function. As an example, an analysis for three coupled oscillators by using the obtained $q(\psi)$ predicts several synchronous modes—three phase, partial in-phase, and all in-phase—that were previously reported in the experiment of three coupled oscillators. In this way, the present method has potential applicability in the analysis of collective dynamical behavior of many oscillators without full knowledge of either the detailed oscillation mechanism or the interactions.

References

Agladze, K., Aliev, R.R., Yamaguchi, T., Yoshikawa, K., 1996. Chemical diode. J. Phys. Chem. 100, 13895–13897.
Ali, F., Menzinger, M., 1991. Inhomogeneity-induced isola formation in the chlorite/iodide reaction. J. Phys. Chem. 95, 6408–6411.
Belousov, B.P., 1958. A periodic reaction and its mechanism. Sbornik Referatov po Radiatsionni Meditsine, Medgiz, Moscow 145 (in Russia).
Belousov, B.P., 1985. Autowave Processes in Systems with Diffusion. In: Grecova, M.T. (Ed.), USSR Academy of Sciences, Gorky.
Buck, J., 1988. Synchronous rhythmic flashing of fireflies. II. Q. Rev. Biol. 63, 265–289.
Castets, V., Dulos, E., Boissonade, J., De Kepper, P., 1990. Experimental evidence of a sustained standing Turing-type nonequilibrium chemical pattern. Phys. Rev. Lett. 64, 2953–2956.
Crowley, M.F., Epstein, I.R., 1989. Experimental and theoretical studies of a coupled chemical oscillator: Phase death, multistability and in-phase and out-of-phase entrainment. J. Phys. Chem. 93, 2496–2502.
Daido, H., 1994. Generic scaling at the onset of macroscopic mutual entrainment in limit-cycle oscillators with uniform all-to-all coupling. Phys. Rev. Lett. 73, 760–763.
Davidenko, J.M., Pertsov, A.M., Salomonsz, R., Baxter, W.T., Jalife, J., 1991. Stationary and drifting spiral waves of excitation in isolated cardiac muscle. Nature 355, 349–351.
Epstein, I.R., Pojman, J.A., 1998. An Introduction to Nonlinear Chemical Dynamics: Oscillations, Waves, Patterns, and Chaos. Oxford University Press, New York.
Field, R.J., Körös, E., Noyes, R., 1972. Oscillations in chemical systems. II. Thorough analysis of temporal oscillation in the bromate–cerium–malonic acid system. J. Am. Chem. Soc. 94, 8649–8664.
Field, R.J., Noyes, R.M., 1974. Oscillations in chemical systems. IV. Limit cycle behavior in a model of a real chemical reaction. J. Chem. Phys. 60, 1877–1884.
Fujii, H., Sawada, Y., 1978. Phase-difference locking of coupled oscillating chemical systems. J. Chem. Phys. 69, 3830–3832.
Fukuda, H., Nagano, H., Kai, S., 2002. Stochastic synchronization in two-dimensional coupled lattice oscillators in the Belousov–Zhabotinsky reaction. J. Phys. Soc. Jpn. 72, 487–490.
Gorelova, N.A., Bures, J., 1983. Spiral waves of spreading depression in the isolated chicken retina. J. Neurobiol 14, 353–363.
Gyorgyi, L., Field, R.J., 1992. A three-variable model of deterministic chaos in the Belousov–Zhabotinsky reaction. Nature 355, 808–810.
Hammond, C., Bergman, H., Brown, P., 2007. Pathological synchronization in Parkinson's disease: Networks, models and treatments. Trends Neurosci. 30, 357–364.
Hansel, D., Mato, G., Meunier, C., 1993. Clustering and slow switching in globally coupled phase oscillators. Phys. Rev. E 48, 3470–3477.
Kitahata, H., Aihara, R., Magome, N., Yoshikawa, K., 2002. Convective and periodic motion driven by a chemical wave. J. Chem. Phys. 116, 5666–5676.

Kori, H., Kuramoto, Y., 2001. Slow switching in globally coupled oscillators: Robustness and occurrence through delayed coupling. Phys. Rev. E 63, 046214.
Kuhnert, L., Agladze, K.I., Krinsky, V.I., 1989. Image processing using light-sensitive chemical waves. Nature 337, 244–247.
Kuramoto, Y., 1984. Chemical Oscillation, Wave, and Turbulence. Springer-Verlag, Berlin.
Lechleiterand, J.D., Clapham, D.E., 1992. Molecular mechanisms of intracellular calcium excitability in X. laevis oocytes. Cell 69, 283.
Marek, M., Stuchl, I., 1975. Synchronization in two interacting oscillatory systems. Biophys. Chem. 3, 241–248.
Miike, H., Mori, Y., Yamaguchi, T., 1997. Hiheikou-kei no Kagaku III: Dynamics of Reaction–Diffusion Systems. Kodansha, Tokyo (in Japanese).
Miyazaki, J., Kinoshita, S., 2006a. Determination of a coupling function in multicoupled oscillators. Phys. Rev. Lett. 96, 194101.
Miyazaki, J., Kinoshita, S., 2006b. Method for determining a coupling function in coupled oscillators with application to Belousov–Zhabotinsky oscillators. Phys. Rev. E 74, 056209.
Miyazaki, J., Kinoshita, S., 2007. Stopping and initiation of a chemical pulse at the interface of excitable media with different diffusivity. Phys. Rev. E 76, 066201.
Motoike, I., Yoshikawa, K., 1999. Information operations with an excitable field. Phys. Rev. E 59, 5354–5360.
Müller, C., Plesser, T., Hess, B., 1985. The structure of the core of the spiral wave in the Belousov–Zhabotinskii reaction. Science 230, 661–663.
Noszticzius, Z., Bodnar, Z., Garamszegi, L., Wittmann, M., 1991. Hydrodynamic turbulence and diffusion-controlled reactions: Simulation of the effect of stirring on the oscillating Belousov–Zhabotinskii reaction with the Radicalator model. J. Phys. Chem. 95, 6575–6580.
Noyes, R., Field, R.J., Körös, E., 1972. Oscillations in chemical systems. I. Detailed mechanism in a system showing temporal oscillations. J. Am. Chem. Soc. 94, 1394–1395.
Pertsov, A.M., Davidenko, J.M., Salomonsz, R., Baxter, W.T., Jalife, J., 1993. Spiral waves of excitation underlie reentrant activity in isolated cardiac muscle. Circ. Res. 72, 631–650.
Pikovsky, A., Rosenblum, M., Kurths, J., 2001. Synchronization: A Universal Concept in Nonlinear Sciences. Cambridge University Press, Cambridge.
Pojman, J.A., Dedeaux, H., Fortenberry, D., 1992. Surface-induced stirring effects in the manganese-catalyzed Belousov–Zhabotinskii reaction with a mixed hypophosphite/acetone substrate in a batch reactor. J. Phys. Chem. 96, 7331–7333.
Reppert, S.M., Weaver, D.R., 2002. Coordination of circadian timing in mammals. Nature 418, 935–941.
Ruoff, P., 1993. Excitations induced by fluctuations: An explanation of stirring effects and chaos in closed anaerobic classical Belousov–Zhabotinskii systems. J. Phys. Chem. 97, 6405–6411.
Sekiguchi, T., Mori, Y., Okazaki, N., Hanazaki, I., 1994. Photoinduction and photoinhibition of chemical oscillations in the tris(2,2'-bipyridine) ruthenium (II)-catalyzed minimal bromate oscillator. Chem. Phys. Lett. 219, 81–85.
Srivastava, P.K., Mori, Y., Hanazaki, I., 1992. Photo-inhibition of chemical oscillation in the $Ru(bpy)_2^{+3}$-catalyzed Belousov–Zhabotinskii reaction. Chem. Phys. Lett. 190, 279–284.
Strogatz, S.H., 2003. Sync: The Emerging Science of Spontaneous Order. Hyperion, New York.
Toth, A., Showalter, K., 1995. Logic gates in excitable media. J. Chem. Phys. 103, 2058–2066.
Turing, A.M., 1952. The chemical basis of morphogenesis. Phil. Trans. R. Soc. B 237, 37–72.
Uhlhaas, P.J., Singer, W., 2006. Neural synchrony in brain disorders: Relevance for cognitive dysfunctions and pathophysiology. Neuron 52, 155–168.

Vanag, V.K., Melikhov, D.P., 1995. Asymmetrical concentration fluctuations in the autocatalytic bromate–bromide catalyst reaction and in the oscillatory Belousov–Zhabotinsky reaction in closed reactor: Stirring effects. J. Phys. Chem. 99, 17372–17379.

Winfree, A.T., 1980. The Geometry of Biological Time. Springer, New York.

Yoshimoto, M., Yoshikawa, K., Mori, Y., 1993. Coupling among three chemical oscillators: Synchronization, phase death, and frustration. Phys. Rev. E 47, 864–874.

Zhabotinsky, A.M., 1964a. Periodical process of oxidation of malonic acid solution (studies on the kinetics of Belousov's reaction). Biofizika 9, 306–311 (in Russia).

Zhabotinsky, A.M., 1964b. Periodic oxidation reactions in liquid phase. Dokl. Akad. Nauk USSR 157, 392–395 (in Russian).

Zhabotinsky, A.M., 1974. Spontaneously Oscillation Concentrations. Science Publishers, Moscow.

Zaikin, A.N., Zhabotinsky, A.M., 1970. Concentration wave propagation in two-dimensional liquid-phase self-oscillating system. Nature 225, 535–537.

3 Dynamics of Droplets

Hiroyuki Kitahata, Natsuhiko Yoshinaga, Ken H. Nagai, Yutaka Sumino

---

Chapter Contents
3.1 Introduction   85
    3.1.1 Active Matter   85
    3.1.2 Surface Tension   86
3.2 Surface Tension–Driven Spontaneous Motion   88
    3.2.1 Droplet Gliding on a Glass Surface   88
    3.2.2 Droplet Drifting on an Aqueous Surface   92
    3.2.3 Suspended Droplet Swimming in an Aqueous Phase   95
3.3 Hydrodynamics for Spontaneous Motion   97
    3.3.1 Basic Knowledge   97
    3.3.2 Stokes Flow   99
    3.3.3 Surface Tension in the Frame of Hydrodynamics   101
    3.3.4 Spherical Droplet Moving Under a Concentration Gradient   102
3.4 Motion Coupled with Pattern Formation   106
    3.4.1 Motion of a BZ Droplet   106
    3.4.2 Numerical Results   107
3.5 Concluding Remarks   111
Appendix   116

3.1 Introduction

3.1.1 Active Matter

The idea of pattern formation can be extended to the formation of spatiotemporal structures that are not at a free-energy minimum in the sense of equilibrium states. Active elements, which are elements that generate spontaneous motion, is an interesting topic related to the study of such pattern formation. Our natural curiosity leads us to investigate the situation in which active elements form a group similar to the matter consisting of passive elements. A group of such actively moving elements has recently been given the name *active matter* (Ramaswamy, 2010), and it has also attracted great interest as an example of a research field concerned with far-from-equilibrium systems. Several systems exhibiting spontaneous motion have been proposed, including vibrating granular materials (Kudrolli et al., 2008; Narayan et al., 2007), a droplet on a vibrating liquid interface (Couder et al., 2005; Couder and Fort, 2006), Leidenfrost droplets (Linke et al., 2006; Quéré and Ajdari, 2006; Snezhko et al., 2008), and vibrating pastes (Merkt et al., 2004).

We should note that spontaneous motion of each element in active matter can also be recognized as a spatiotemporal structure. In fact, spontaneous motion is recognized as motion in the absence of a macroscopic external force, such as of gravity or an electric field, acting as the driving force. To fulfill this condition, the system must have a mechanism that converts chemical or vibrational energy into a directional force. The manner of symmetry breaking for motion under far-from-equilibrium conditions is also an important physical question (Haken, 1983; Nicolis and Prigogine, 1977).

One type of spontaneous motion is that of a droplet. In a droplet system that shows spontaneous motion, some type of surfactant is often used in addition to the bulk phase and the droplet phase. The driving force of this type of system is the surface tension inhomogeneity, which is explained in the next section. This type of system usually consists of a small number of components; therefore, it is possible that experiments in a broad area of parameter space can be performed with such a system. This implies that such systems are ideal experimental systems for examining the general aspects of active matter.

3.1.2 Surface Tension

When two immiscible fluids are mixed, a boundary is formed between the two phases (de Gennes et al., 2004). The boundary has a free-energy cost that is proportional to the area of the interface, which is referred to as surface energy; this energy minimizes the area of the surface under the given restrictions. In the absence of other free-energy costs, this surface-energy cost causes a liquid droplet to take a spherical shape. On the other hand, under a gravitational field, the shape of a droplet is determined through the minimization of the total energy consisting of the surface energy and gravitational energy. For example, on the Earth, a static water droplet on the centimeter scale forms a pancakelike shape, which has a flat top with a rounded periphery, rather than a spherical shape since the gravitational energy is comparable to the surface energy. Surface energy also exists at a solid–liquid interface or a solid–air interface. Since solid surfaces cannot deform, the surface energy acts only if more than one fluid is placed on a substrate. When three phases coexist, in an equilibrium state, the three angles between the phases are fixed at values that minimize the total surface energy under volume conservation. This rule is referred to as Young's equation (Young, 1805).

In practical cases, surface energy is the same as surface tension. When we stretch a soap film with a bar and increase the film area by dS, we need to perform an amount of work at least γ dS, where γ denotes the surface tension. It is noted that the term *surface tension* is sometimes used only in the case of air–liquid interfaces; however, we use this term in the case of all types of interfaces discussed in this chapter. In the case of Figure 3.1, d$S = 2L$ dx when the bar is moved by dx. Here, the necessary work is $2\gamma L$ dx and the force exerted on the bar is $2\gamma L$; therefore, surface tension is also the force per unit length. From the viewpoint of surface tension, Young's equation expresses the force balance at the triple line, where the three phases coexist.

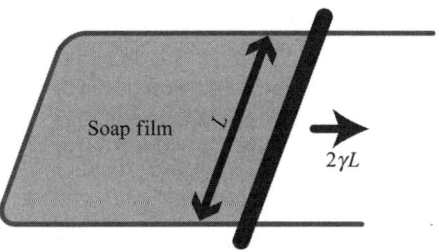

Figure 3.1 Surface tension exerted to the boundary. When a soap film is pulled with a bar, force proportional to its length L needs to be exerted.

Box 3.1 Measurement of Surface Tension

Here, two basic approaches to measure the surface tension are introduced. Both ways utilize the fact that surface tension is the force per unit length.

A conventional approach to measure a surface tension is the Wilhelmy method. The schematic diagram of this method is illustrated in Figure B3.1(a). A thin platinum plate with a width of l is used. First, the plate is attached to the interface and pulled up. During the pulling of the plate, the force exerted on the plate is simultaneously measured. At the moment when the plate is detached from the interface, the force suddenly decreases since the interface pulls the plate with a force proportional to l. Since the force is $2\gamma l\cos\theta$ immediately before the detachment of the plate, where γ is the surface tension and θ is the contact angle, we can measure the surface tension through the measurement of the force before it decreases. We note that the force is not γl but $2\gamma l$ since the plate has two surfaces.

The other simple approach is the pendant drop method. As shown in Figure B3.1(b), a tube filled with the target liquid is used. The observer increases the droplet volume gradually and measures the critical volume where the droplet falls down, V. The gravity force exerted on the droplet with V is balanced against the maximum of the force originating from the surface tension; therefore, we can measure the surface tension γ to be $\gamma = \rho g V/(2\pi d)$, where ρ is the density of the liquid, g is the gravitational acceleration, and d is the radius of the tube.

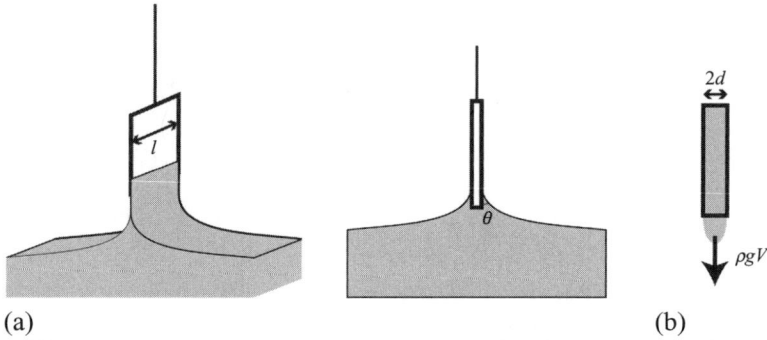

Figure B3.1 Schematic diagram of the measurement of surface tension: (a) Wilhelmy method and (b) pendant drop method.

Surface tension depends on thermodynamical variables, such as temperature and the concentration of the solute (Davis, 1987). In general, surface tension at the water–air and water–oil interfaces is a decreasing function of temperature and the bulk concentration of a surfactant in the dilute solution. When the temperature or the concentration is inhomogeneous, a stress field emerges due to the inhomogeneity of the surface tension. As a result, this stress field induces flow, which is called the Marangoni effect (Scriven and Sternling, 1960). A well-known case of the Marangoni effect is the flow in a thin film induced by a thermal gradient. A thin liquid film can slide up along a wall when the wall is cooled from the top (Schneemilch, 2000; Schneemilch and Cazabat, 2000). A similar phenomenon is also induced by a concentration gradient. When ethanol aqueous solution is put in a glass vessel, a thin water film slides up the glass wall driven by the ethanol concentration gradient created by the evaporation of ethanol (Thomson, 1855; Fournier and Cazabat, 1992; Vuilleumier et al., 1995). A detailed explanation of this phenomenon is in Box 3.2.

A surface tension gradient also induces three-dimensional (3D) convection in the bulk phase, such as Marangoni convection (Schatz and Neitzel, 2001). For example, when a source of a surfactant exists, the convection expanding from the source is induced by the Marangoni effect (Kovalchuk and Vollhardt, 2000). Using this type of convection induced by a surfactant source, spontaneous motion of a small droplet that supplies a surfactant can be realized without any external force (Nagai et al., 2005; Hanczyc et al., 2007; Thutupalli et al., 2011). Motion of a droplet is induced not only by the gradient of fluid–fluid surface tension but also by the gradient of fluid–solid surface tension. For example, Chaudhury and Whitesides (1992) reported motion of a droplet at the boundary between two substrates with different surface tensions (see also (Ichimura et al., 2000)). In this experiment, the droplet moved in the direction with the smaller contact angle. In the case that a droplet contains a chemical that modifies the substrate surface, the droplet keeps moving even when the substrate is uniform. That is, it can gain momentum without any external force. (Domingues Dos Santos and Ondarçuhu, 1995; Sumino et al., 2005). In this chapter, we present recent works on the spontaneous motion of a small droplet caused by surface tension inhomogeneity. In Section 3.2, recently reported experimental systems are introduced. In Sections 3.3 and 3.4, a theoretical framework based on hydrodynamics is introduced and explained.

3.2 Surface Tension–Driven Spontaneous Motion

In this section, we introduce several experimental systems that show spontaneous motion of a droplet driven by wettability difference and/or the Marangoni effect. More intuitively, we introduce three typical types of droplet motions: gliding (Section 3.2.1), drifting (Section 3.2.2), and swimming (Section 3.2.3). In Box 3.3, we also describe the procedures for conducting these experiments.

3.2.1 Droplet Gliding on a Glass Surface

A droplet on an inclined substrate surface slides down the substrate due to the effect of gravity. In this case, the droplet moves under the influence of an external force. Under appropriate conditions, a droplet can slide around spontaneously on a solid substrate.

Dynamics of Droplets

Box 3.2 Tears of Wine

It is quite easy to observe the thin film that slides up the wall of a glass spontaneously when liquor is poured into the glass. This phenomenon is the so-called *tears of wine* (Thomson, 1855; Fournier and Cazabat, 1992; Vuilleumier et al., 1995; Figure B3.2(a)), or *wine legs*, and is a typical example of the Marangoni effect. You can observe it with a simple experiment. Instead of using wine, prepare a 30% ethanol aqueous solution (Fournier and Cazabat, 1992; Figure B3.2 (b)) and add a small amount of edible dye for visualization. When the liquid is poured into a glass watch dish or an evaporation dish, there appears a thin film sliding up the slope. A key aspect of the mechanism is the difference of the evaporation rate with respect to the surface/bulk ratio, depending on the position inside the container (Figure B3.2(c)). The evaporation of ethanol is almost the same at all the surfaces; thus, there exists a gradient in the ethanol concentration since the volume of the liquid per unit surface is smaller in the thinner film region. Ethanol is a surface-active chemical; therefore, surface tension of the mixture is higher at the top of the film. This unbalanced surface tension results in Marangoni flow. As a result, the liquid spontaneously slides up along the glass surface, and the mixture forms droplets to minimize the sum of the surface and gravitational energies.

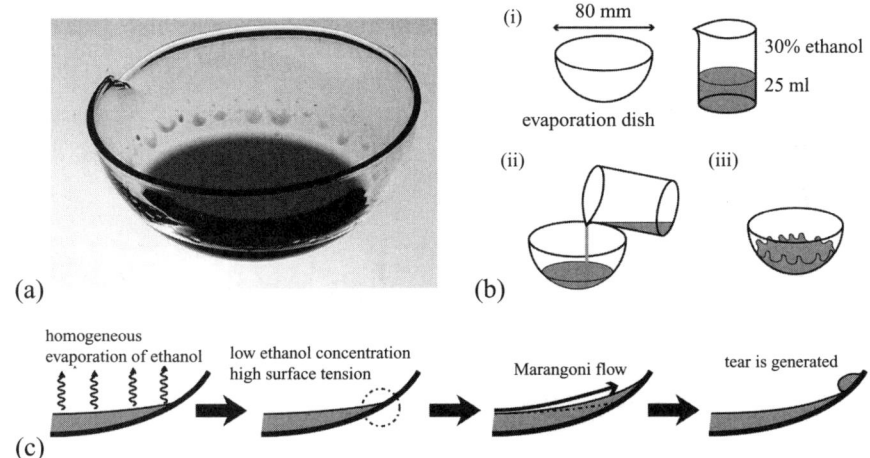

Figure B3.2 (a) Appearance of tears of wine, (b) experimental procedure, and (c) schematic representation of mechanism of tear formation. (For color version of this figure, the reader is referred to the online version of this chapter.)

It might sound incredible, but the droplet can even slide up slopes, ascend stairs, and move in loops inside a vertical ring, against gravity. The setup used for demonstrating this is as follows: an oil droplet is immersed in an aqueous phase, and a glass substrate is situated beneath the droplet (Sumino et al., 2005; Sumino and Yoshikawa, 2008). The oil droplet consists of nitrobenzene containing both iodine and potassium iodide. This oil droplet has higher density than the aqueous phase. The aqueous phase contains a cationic surfactant, stearyltrimethylammonium chloride (STAC, which is the salt of STA^+ and Cl^- ion), which modifies the surface wettability of the glass substrate.

Figure 3.2(a) shows a typical droplet motion, in which a droplet freely slides on a glass substrate. It is observed that the droplet moves spontaneously without the application of any external force. The glass surface plays an important role in droplet motion. A glass surface is known to be negatively charged at the natural pH. Since STA^+ is a cationic surfactant and the glass is negatively charged, STA^+ tends to be adsorbed at the aqueous–glass interface. Thus, the glass surface is covered with STA^+ when an oil droplet is absent. In the presence of an oil droplet, the adsorbed STA^+ is desorbed into the oil droplet. If this were the only process at work, the oil droplet would become saturated quickly. In fact, when the oil droplet does not contain iodine, it cannot move spontaneously. However, the iodine and potassium iodide in the oil droplet tend to form $STAI_3$ in the organic phase by the following chemical reaction:

$$STA^+ + I^- + I_2 \rightarrow STAI_3. \tag{3-1}$$

Thus, the chemical reaction prevents the saturation of STA^+ in the oil droplet. The glass surface under the droplet has a lower density of STA^+, resulting in a less wettable surface for the oil under the oil droplet. Therefore, even with an infinitesimal perturbation, the oil droplet starts to move toward the more wettable region around it provided the driving force from the wettability difference is larger than the viscous damping (Sumino and Yoshikawa, 2008). We should note that the region of the glass surface with lower STA^+ density recovers in terms of wettability before the droplet returns to it since STA^+ is dissolved in the aqueous phase (Figure 3.2(b)).

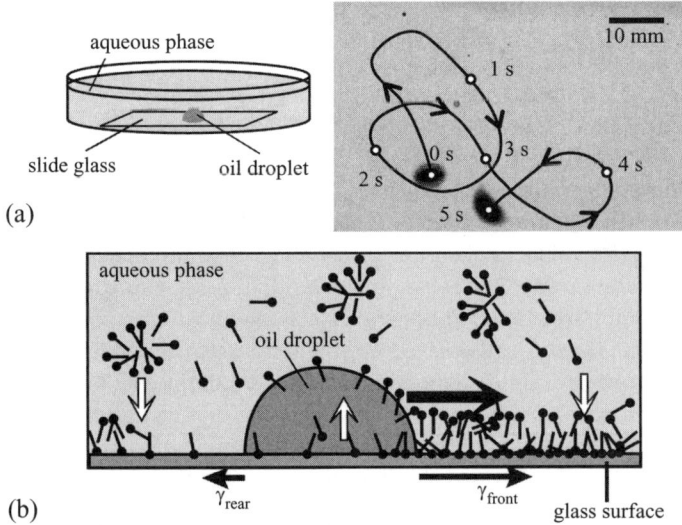

Figure 3.2 (a) Setup of experimental results, and the actual trajectory of an oil droplet. (b) Schematic representation of the mechanism of spontaneous crawling (Sumino and Yoshikawa, 2008).

In this system, the energy source is the flux of STA^+ from the aqueous phase into the oil droplet via the glass surface. The chemical reaction (Eq. 3-1) causes STA^+ to be stored in the oil droplet and maintains its flux. As sufficient potassium iodide is present in the oil droplet, the amount of iodine inside the oil droplet determines the lifetime of the oil droplet motion, which is around 1 min for a 50 μl oil droplet with 5 mM iodine.

The oil droplet is therefore propelled to move on the glass substrate. As mentioned earlier, the driving force is sufficiently strong to have the droplet roll inside of loops and move upstairs, as shown in Figure 3.3, even though the density of the oil droplet is larger than that of the aqueous phase. A more interesting result can be seen in the situation when the droplet is placed in an evaporation dish. Although gravity attracts the droplet to the center (bottom) of the evaporation dish, the droplet continues to move along a circular path around the center because it enters an unstable condition when it stops at the center.

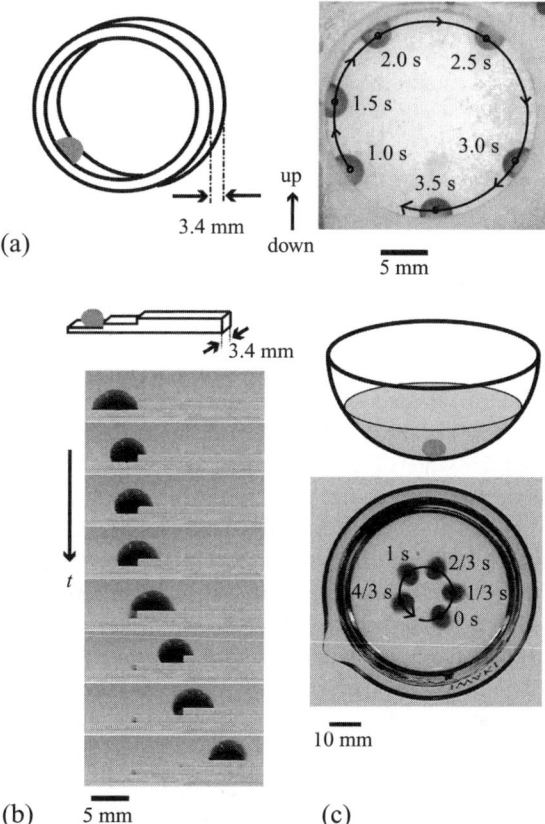

Figure 3.3 (a) Snapshots of a droplet rolling inside a ring. (b) Snapshots (every 0.2 s) of a droplet ascending upstairs. (c) Snapshots of a droplet rotating in an evaporation dish (Sumino et al., 2005; Sumino and Yoshikawa, 2008).

This type of droplet motion is so-called reactive wetting, and a theoretical prediction of such motion was first made by Greenspan (1978); further theoretical analysis has been performed by a French group (Brochard-Wyart and de Gennes, 1985). Experimental realization of reactive wetting was reported around 1990 by several groups (Bain and Whitesides, 1989; Chaudhury and Whitesides, 1992; Domingues Dos Santos and Ondarçuhu, 1995). The theoretical analyses required a rather delicate treatment of the contact line as there is a singularity in viscous dissipation (Huh and Scriven, 1971; de Gennes et al., 2004). Recently, the concept of disjoining pressure has been used to successfully describe spontaneous droplet motion in a continuous field (Thiele et al., 2004; John et al., 2005).

3.2.2 Droplet Drifting on an Aqueous Surface

When ethanol is poured into a puddle, the surface of the puddle ripples extensively until the ethanol completely dissolves into the aqueous phase or evaporates from the surface. This rippling appears due to the flow generated by the Marangoni effect. As ethanol lowers the surface tension of the water, the ethanol-rich surface has lower surface tension than the surrounding water surface. As a result, a flow toward the higher surface tension is observed, which is called Marangoni flow, and this effect is called the Marangoni effect. When more ethanol is poured into the puddle, the rippling becomes weaker, as the proportion of ethanol in the puddle increases. However, the ethanol does not form droplets on the aqueous surface, as it is perfectly miscible with water. When alcohol with a longer alkyl chain than ethanol is poured into a puddle, the alcohol forms droplets on the surface. These droplets, with an appropriate length of the alkyl chain, move spontaneously by generating a flow inside the aqueous phase. Here, we refer to this droplet as a drifting droplet as it moves by generating a flow inside the aqueous phase, and consequently resembles a sailing object from underneath flow.

The experimental setup of the spontaneous motion of a floating alcohol droplet is as follows (Figure 3.4(a)) (Nagai et al., 2005, 2006, 2007). The droplet consists of 1-pentanol. The aqueous phase is an aqueous solution of pentanol, of which the concentration is lower than the saturation concentration of 2.7 g/100 ml. Since pentanol has a lower density than water, the pentanol droplet floats on the aqueous surface.

When a droplet is set on the aqueous phase, pentanol flux occurs until the droplet disappears. Pentanol molecules are transferred from the droplet and diffuse into the aqueous surface. Then, pentanol can evaporate into the air or dissolve into the aqueous phase; consequently, the concentration of pentanol at the aqueous surface decreases with the distance from the droplet. This spatial gradient in the pentanol concentration at the aqueous surface induces Marangoni flow. Once the droplet starts to move, the concentration gradient of pentanol is larger in the front than in the rear. Therefore, the Marangoni flow in front of the oil droplet becomes stronger, and it sustains the spontaneous motion of the pentanol droplet (Figure 3.4(b)) (Nagai et al., 2005, 2006, 2007). This type of spontaneous motion can be realized with various chemicals that have finite miscibility with water and are surface active (Bekki et al., 1990, 1992; Santiago-Rosanne et al., 1997, 2001). A famous analogous example is a system with the so-called camphor boat (Nakata et al., 1997; Nagayama et al., 2004), in which a floating camphor grain spontaneously drifts across on the water surface.

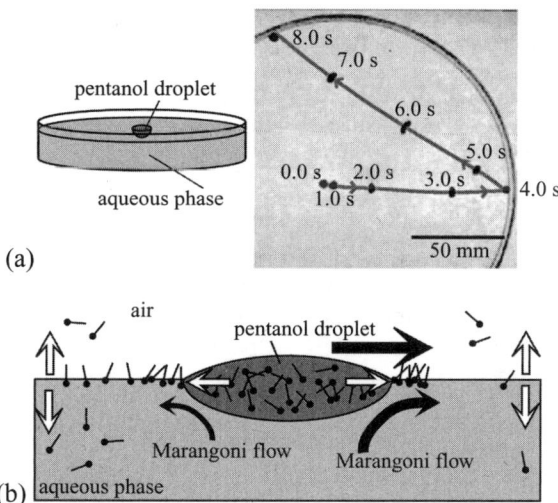

Figure 3.4 (a) Experimental setup, and the actual pentanol droplet motion behavior. (b) Schematic representation of the mechanism of a spontaneous motion of a pentanol droplet (Nagai et al., 2005, 2006, 2007).

Interestingly, this type of droplet motion shows strong coupling with the deformation of the droplet (Nagai et al., 2005). For a small droplet, the droplet maintains an almost circular cross-section as seen from above, and its trajectory of motion is irregular. If the size of a droplet is larger than a certain critical size, the droplet deforms into a croissant shape, and the trajectory becomes straight. Such strong coupling between deformation and motion is also seen in motile cellular fragments (Verkhovsky et al., 1999; Yam et al., 2007; Kozlov and Mogilner, 2007). It is expected that understanding the behavior of droplets will provide insights into the general features of the coupling between droplet deformation and motion (Ohta and Ohkuma, 2009).

Box 3.3 Observation of a Spontaneously Moving Pentanol Droplet

The experimental setup required for the observation of pentanol droplet motion is simple and easy to organize. Here, we give a brief description of the setup, along with some comments that need to be noted. Please ensure proper ventilation at the location of the experiment because pentanol is a slightly volatile liquid.

The experiment is very simple; you need to prepare pure water in a petri dish. The amount of water can be arbitrarily selected. If you cannot prepare pure

Continued

Box 3.3 Observation of a Spontaneously Moving Pentanol Droplet—cont'd

water, tap water can be used. In this case, the motion of the droplet may be less vivid. Place a pentanol droplet gently at the water surface. The water surface should fluctuate extensively as the droplet dissolves into the water (Figure B3.3). Upon adding further pentanol, the fluctuation becomes progressively weaker, and the droplet remains undissolved for a longer time. At some point, the dissolution of the droplet should become almost negligible, and the droplet starts to exhibit translational motion. In this situation, droplets can be observed to turn, bounce, and cling to the dish, while fusing and splitting.

To observe single droplet motion, prepare an aqueous mixture of pentanol with a weight fraction of about 2%. Pour this aqueous mixture into a petri dish and deposit on it a pentanol droplet with a volume of more than 0.1 μl and less than 200 μl. Deformation of the droplet and its motion along a straight trajectory should now be possible to observe.

It is difficult to observe the motion of a pentanol droplet because pentanol is a transparent colorless liquid. One solution for this problem is to dissolve pigmented ink used for stamping into the pentanol. A drop of the ink is enough for 10 ml of pentanol. However, as pentanol dissolves into the water or evaporates into the air, the ink remains at the surface of the aqueous phase and contaminates the system. Another solution is to illuminate the system with a strong light from above, and then observe the shadow of the droplet. The edge of the droplet makes a shadow underneath it. In this case, the system is not contaminated.

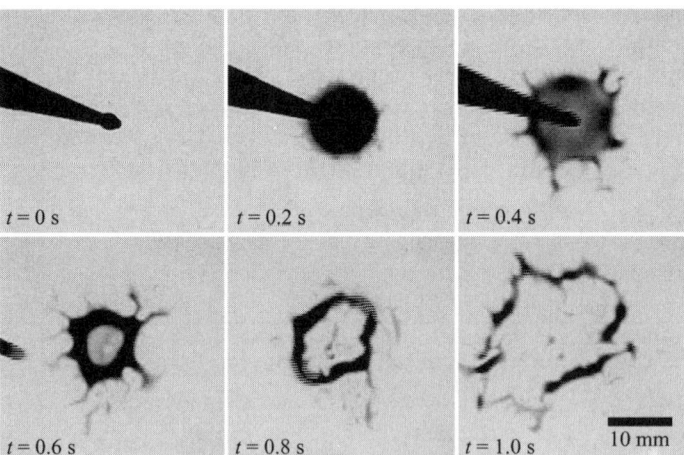

Figure B3.3 Strong fluctuation in the shape of a pentanol droplet in the case that the aqueous phase was almost pure water. Please compare this with the result in Figure 3.4(a), for which the aqueous phase is close to saturation.

3.2.3 Suspended Droplet Swimming in an Aqueous Phase

A droplet can also swim inside a bulk phase. In this case, the Marangoni flow generated by the interface of the droplet propels it. To realize this swimming motion, a gradient of surface tension must be generated and maintained with the aid of convection and/or a chemical reaction. One famous example is an oil droplet producing surfactant when the oil is in contact with the outer aqueous phase (Hanczyc et al., 2007). Another example is a system involving a droplet containing a surfactant that dissociates when it is in contact with the surrounding bulk phase (Thutupalli et al., 2011). In both cases, the close coupling between the convection generated by the Marangoni effect, the adsorption/desorption of the surfactant, and the formation/degradation of the surfactant are essential (Yoshinaga et al., 2012; Yabunaka et al., 2012).

In addition to the mechanism in the preceding examples, we propose in this chapter an active pattern of chemical reaction inside the droplet that can couple with Marangoni flow and cause the suspended droplet to swim. Here, a suspended droplet swims inside an aqueous phase by the coupling between Marangoni flow and the active pattern generated by the Belousov–Zhabotinsky (BZ) reaction (Kitahata et al., 2002, 2011, 2012).

The BZ reaction is well known for generating spatiotemporal structures such as oscillations and traveling waves, which were already mentioned in the previous chapters. Due to a variety of patterns in a BZ reaction and their unsteady nature, the motion is not restricted in a simple translational motion. This implies that a few components of chemical species—that is, scalar concentration fields—may lead to various types of motion through the patterns depending on initial and/or external stimulus.

The key factor for connecting the spatial pattern of chemical concentrations with Marangoni flow is the catalyst of iron complex catalyst, that is, ferroin. The surface tension of the BZ solution in the oxidized state (consisting of ferriin, which is blue in color) is higher than that in the reduced state (consisting of ferroin, red in color). Therefore, Marangoni flow occurs from the interface in the oxidized state to that in the reduced state. As a result of this Marangoni flow, the aqueous droplet exhibits spontaneous motion that is coupled with the propagation of a chemical wave. The BZ reaction generates different types of spatiotemporal patterns inside a droplet: a target pattern (Figure 3.5(b)), a spiral wave (Figure 3.6(a)), and a scroll ring (Figure 3.6(b)). Coupled with these chemical patterns, the droplets showed different patterns of swimming behavior, as shown in Figures 3.5(c) and 3.6.

The system is composed of an aqueous droplet and a surrounding organic phase placed in a petri dish. The aqueous droplet contains the chemicals necessary for the BZ reaction. The surrounding organic phase is oleic acid, having a relative density of about 0.9. Thus, the droplet has larger density than the surrounding organic phase and remains at the bottom of the petri dish, moving in a horizontal direction. The experimental setup is shown in Figure 3.5(a). We inserted a 1 µl droplet of the BZ medium into a layer of oleic acid in a petri dish. We used a petri dish made of polytetrafluoroethylene to prevent the droplet from coming into contact with the bottom of the dish.

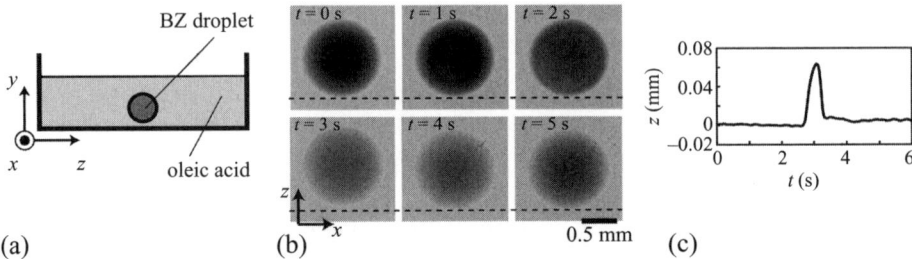

Figure 3.5 (a) Experimental setup for the observation of the spontaneous motion of a BZ droplet. (b) Snapshots of a spontaneously moving BZ droplet induced by the target pattern inside it. (c) Plot of the position of the droplet center as a function of time. (The movies are available at *http://link.aps.org/supplemental/10.1103/PhysRevE.84.015101* in the supplementary information of (Kitahata et al., 2011).) (For color version of this figure, the reader is referred to the online version of this chapter.)

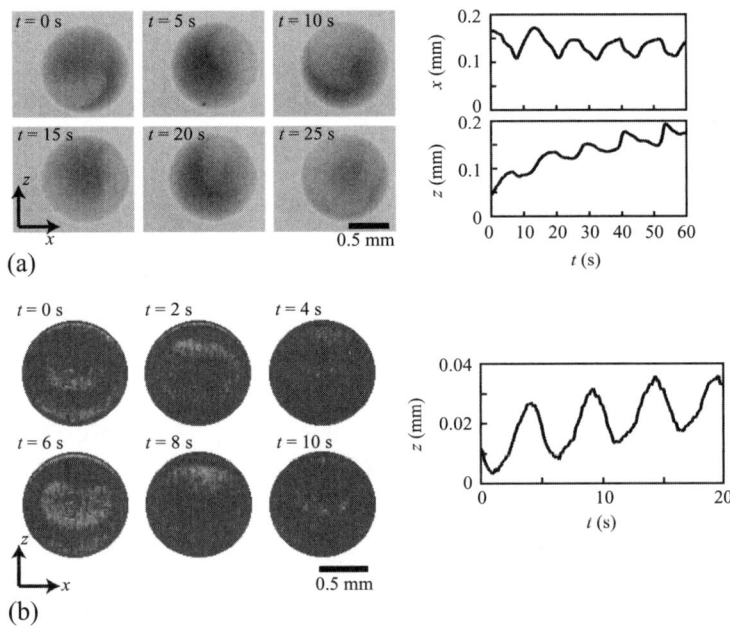

Figure 3.6 Snapshots of a spontaneously moving BZ droplet induced by (a) a spiral wave and (b) a scroll ring inside it. The right plots represent the position of the droplet center as a function of time. (The movies are available at *https://www.jstage.jst.go.jp/article/cl/41/10/41_CL-120495/_article* in the supplementary information of (Kitahata et al., 2012).) (For color version of this figure, the reader is referred to the online version of this chapter.)

The aqueous droplet was obtained using the following procedure. The aqueous solution of ferroin, tris 1,10-phenanthroline iron(II) sulfate, was prepared by mixing stoichiometric amounts of 1,10-phenanthroline and ferrous sulfate in pure water. The water was purified with a Millipore-Q system. The composition of the BZ solution was $[NaBrO_3] = 0.3$ M, $[H_2SO_4] = 0.6$ M, $[CH_2(COOH)_2] = 0.1$ M,

[NaBr] = 0.03 M, and [Fe(phen)$_3$SO$_4$] = 5 mM. In the process of preparing the BZ reaction medium, we mixed all the chemicals other than Fe(phen)$_3$SO$_4$, waited until the solution color turned from yellow to transparent, and then added the Fe(phen)$_3$SO$_4$. The reaction medium was in the oscillatory state. To obtain a droplet with a spiral wave or a scroll ring, as shown in Figures 3.6(a) and (b), we set a longer waiting time before mixing the Fe(phen)$_3$SO$_4$ solution ($\sim$6 min) than in the case for the target pattern ($\sim$4 min waiting time).

From these experiments, it is shown that a BZ droplet shows various motions such as a single shuttling motion, rotational motion, and back-and-forth motion, accompanied with chemical patterns inside it. In addition to these qualitative aspects, the relation of the chemical pattern and the droplet motion should be quantitatively verified. For example, the velocity, the net translational displacement, should be connected with the dynamics and details of the chemical patterns. To understand the droplet motion quantitatively, the aid of mathematical modeling is necessary for the difficulty of precise control of the chemical pattern in the experiment. At a first glance, the mathematical analysis of the BZ droplet motion seems to be equally complicated because the model should be described by a reaction-diffusion equation with hydrodynamics. Furthermore, the model necessarily includes a Marangoni effect and surfactant adsorption/desorption process. However, the situation can be drastically simplified by treating the problem in a quasi-analytical way for a droplet with small size—that is, when the system has a low Reynolds number. In this situation, hydrodynamics can be written with the Stokes equation, which can be solved with appropriate boundary conditions. Assuming a system has axisymmetry, the flow field can be described with sets of special functions. Based on the obtained analytical solution of the flow field, the motion of a BZ droplet is calculated and the chemical pattern is computed numerically. In the following section, we describe the basic of hydrodynamics at a low Reynolds number, and apply it to analyze the swimming of a BZ droplet.

3.3 Hydrodynamics for Spontaneous Motion

3.3.1 Basic Knowledge

When a droplet moves in a fluid, flow is created by mechanical force acting on a fluid inside and outside the droplet. The flow is thus governed by the equation of force balance, which is equivalent to the law of momentum conservation. A velocity field $v(r,t)$ satisfies the following Navier–Stokes equation (Bird et al., 1987):

$$\rho \frac{\partial v}{\partial t} + (v \cdot \nabla)v = \nabla \cdot \boldsymbol{\sigma}, \tag{3-2}$$

where $\boldsymbol{\sigma}$ is a stress tensor and the left side of the equation represents the inertia. The stress tensor is a tensor $\boldsymbol{\sigma} = \sigma_{ij}$ ($i,j = (x,y,z)$) in three dimensions); for instance,

σ_{xy} is the force per unit area acting in the positive y direction on a surface perpendicular to the x direction. The force acting on a unit volume is expressed as

$$f = \sigma \cdot n \mathrm{d}S, \tag{3-3}$$

where n is a unit normal vector pointing outward and $\mathrm{d}S$ is a differential element of the surface S covering the unit volume. For a simple fluid (which is typically called Newtonian fluid), the force originates from pressure and the product of the shear viscosity and a velocity gradient:

$$\sigma_{ij} = -p\delta_{ij} + \frac{\eta}{2}\left(\frac{\partial v_i}{\partial x_j} + \frac{\partial v_j}{\partial x_i}\right), \tag{3-4}$$

where p is pressure and η is the viscosity of a fluid and corresponds to the necessary force required to shear the fluid. Water is a good example of such a fluid. Assuming incompressibility, we have neglected the contribution of $\sim \zeta(\partial v_i/\partial x_i)$, where ζ is the bulk viscosity coefficient. Examples of the viscosity are as follows: water ($\simeq 1$ mPa·s $= 1$ cP), oil ($\simeq 80$ mPa·s for olive oil), honey ($\simeq 400$ mPa·s), and air ($\simeq 0.01$ mPa·s) at room temperature. Note that, in general, the viscosity is dependent on temperature and composition if the material is composite. The viscosity of honey is strongly temperature dependent; it increases more than 30 times when the temperature decreases from 70°C to 30°C. In this chapter, we neglect such dependence and assume that the viscosity remains constant and uniform. This is a good approximation for water. The compressibility of water at room temperature is $\simeq 10^{-10}$ (Pa^{-1}), which implies a volume change of $\Delta V/V \simeq 0.001\%$ under a change in pressure within the order of atomistic pressure (Rana and Frank, 1973). Therefore, we may assume that the fluid is incompressible:

$$\nabla \cdot v = 0. \tag{3-5}$$

When flow velocity is steady and slow enough, the inertia in Eq. (3-2) is negligible and Eq. (3-2) is simplified to

$$\eta \nabla^2 v - \nabla p = 0. \tag{3-6}$$

This equation is called the Stokes equation, which can be solved with appropriate boundary conditions together with Eq. (3-5) (see Section 3.3.2 and Section A3.1 in the Appendix at the end of this chapter). The boundary conditions depend on a system. For our present purpose, it is reasonable to impose the condition that the velocity inside and outside the droplet is continuously connected:

$$v^{(i)}(r_s) = v^{(o)}(r_s), \tag{3-7}$$

where $r = r_s$ is the position at the interface. Let us consider a spherical liquid droplet with a radius of R in another liquid as shown in Figure 3.7. The superscripts (i) and (o) denote the fluids inside and outside of the droplet, respectively. We consider an

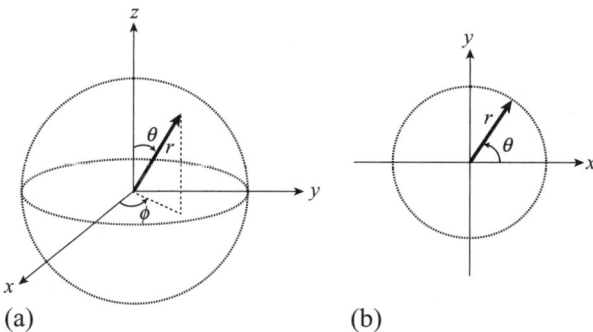

Figure 3.7 Schematic pictures of (a) spherical and (b) polar coordinates.

axisymmetric system; the z axis and x axis are selected as the symmetry axis in three dimensions and in two dimensions, respectively. In three dimensions, the droplet moves at a constant velocity, $u = ue_z$, where e_z is the unit vector in the z direction. Since the boundary condition is given at the interface of the droplet, it is convenient to use spherical coordinates co-moving with the droplet. Since the droplet is moving at a velocity u, the flow velocity far away from the droplet in the co-moving frame is $-u$, taking the relative velocity into consideration; therefore, $v \to -u$ as $r \to \infty$. If a spherical or a circular droplet is undeformable, the velocity perpendicular to the interface vanishes. For these coordinates, the condition of no deformation becomes

$$v_r^{(i)}(r_s) = v_r^{(o)}(r_s) = 0. \tag{3-8}$$

We may also have to determine the boundary conditions for stress tensors, which are dependent on a situation and are discussed later.

3.3.2 Stokes Flow

We start by considering the elementary problem of Stokes law in three dimensions. When a droplet in a fluid is moved by an external force such as gravity or buoyancy due to density contrast, the droplet experiences a frictional force proportional to its velocity,

$$F = -\xi u, \tag{3-9}$$

where F is an external mechanical force acting on the droplet. The frictional coefficient ξ is obtained by using the Hadamard–Rybczynski solution (Happel and Brenner, 1965):

$$\xi = 2\pi \eta^{(o)} R \frac{3\eta^{(i)} + 2\eta^{(o)}}{\eta^{(i)} + \eta^{(o)}}. \tag{3-10}$$

This is known as Stokes law, and as we will see later, this is the consequence of viscous dissipation due to the distortion of flow. In particular, when the droplet is a solid particle—that is, $\eta^{(i)} \to \infty$—we recover the famous Stokes–Einstein relation $\xi = 6\pi\eta^{(o)}R$. The opposite limit $\eta^{(i)} \to 0$ corresponds to a bubble in a fluid where there is no force (stress) acting on its interface. In this case, $\xi = 4\pi\eta^{(o)}R$.

The frictional coefficient is calculated by solving the Stokes equation. To do so, it is convenient to introduce the stream function ψ such that

$$v_r = -\frac{1}{r^2 \sin\theta} \frac{\partial \psi}{\partial \theta}, \tag{3-11}$$

$$v_\theta = \frac{1}{r \sin\theta} \frac{\partial \psi}{\partial r}, \tag{3-12}$$

because the incompressible condition is always satisfied with this form. The Stokes equation is rewritten as

$$\mathcal{E}^2 \mathcal{E}^2 \psi = 0, \tag{3-13}$$

where the operator is

$$\mathcal{E}^2 \equiv \frac{\partial^2}{\partial r^2} + \frac{\sin\theta}{r^2} \frac{\partial}{\partial \theta} \frac{1}{\sin\theta} \frac{\partial}{\partial \theta}. \tag{3-14}$$

Since this equation is linear, the general solution of ψ can be written using eigenfunction expansion (Young et al., 1959; Levan, 1981; Kitahata et al., 2011):

$$\psi = \sum_{n=2}^{\infty} (a_n r^{n+2} - b_n r^{-n+3} + c_n r^n + d_n r^{-n+1}) G_n^{-1/2}(\cos\theta), \tag{3-15}$$

where a_n, b_n, c_n, and d_n are constants to be determined by the boundary conditions and $G_n^{-1/2}$ is the Gegenbauer polynominal of order n and degree $-1/2$. We are seeking the solution of translational motion, which corresponds to the $n=2$ mode. Using the boundary conditions discussed in the previous section, the stream functions inside and outside the droplet can be written as

$$\psi^{(i)} = \frac{1}{2}\left(\frac{A_2}{R} - \frac{3u}{2}\right) r^2 \left(1 - \frac{r^2}{R^2}\right) \sin^2\theta, \tag{3-16}$$

$$\psi^{(o)} = \frac{1}{2}\left[A_2 r \left(\frac{R^2}{r^2} - 1\right) - ur^2\left(\frac{R^3}{r^3} - 1\right)\right] \sin^2\theta. \tag{3-17}$$

The constants A_2 and u are determined by the boundary conditions of the stress tensor. In the absence of inhomogeneity in the surface tension, the boundary condition is simply that the shear stress is continuous across the interface:

$$\sigma_{r\theta}^{(i)}\Big|_{r=R} = \sigma_{r\theta}^{(o)}\Big|_{r=R}, \tag{3-18}$$

where the shear stress is explicitly written in spherical coordinates under axisymmetric flow as

$$\sigma_{r\theta} = \eta \left(\frac{1}{r}\frac{\partial v_r}{\partial \theta} + \frac{\partial v_\theta}{\partial r} - \frac{v_\theta}{r} \right). \tag{3-19}$$

The force acting on the droplet is

$$\boldsymbol{F} = \int \mathrm{d}S\, \boldsymbol{n} \cdot \sigma^{(o)}|_{r=R} = -4\pi\eta^{(o)} A_2 \boldsymbol{e}_z. \tag{3-20}$$

where one could find the constant A_2 is essentially the magnitude of an external force. We may also obtain u using the condition in Eq. (3-18) and finally determine the translational velocity in Eqs. (3-9) and (3-10) as a function of an external mechanical force.

3.3.3 Surface Tension in the Frame of Hydrodynamics

The crucial difference between the motion under gravity discussed in the previous section and the motion driven by the Marangoni effect is that there is no external mechanical force acting on the droplet in the latter system. Instead, the force is generated by the gradient of the surface tension, which is caused by chemical concentration and/or temperature gradients. The force is localized at the interface between two fluids (e.g., water/oil or water/air) although the net force is zero. The mechanical aspect of surface tension is discussed in detail by Davis and Scriven (1982). Here we first discuss the simplest example of Laplace pressure under a quiescent state. We may not be able to directly perceive it when the pressure inside a bubble is higher than the atomistic pressure, but we can see that liquid can be poured into a glass slightly above its brim without spilling. In the latter case, the gravitational force is balanced by the surface tension. Interestingly, the force due to the surface tension is sensitive to the geometry of the interface, or, more precisely, to its curvature. Therefore, we may predict excess pressure inside a spherical bubble of radius R—it is $p^{(i)} - p^{(o)} = 2\gamma/R$. This is known as the Laplace pressure (de Gennes et al., 2004). For example, the surface tension at an air–water interface is $\gamma = 74$ mN/m. The excess pressure is not significantly large compared with the atomistic pressure $p^{(o)} \simeq 100$ kPa when the size of a droplet is large ($R \gtrsim 1$ mm). However, for a micrometer-scale droplet, the pressure created by the surface tension becomes comparable to the atomistic pressure.

The force creating excess pressure is normal to the interface and one might think this is strange because we have seen that the surface tension is the force (or stress, to be precise) that acts *along* the interface and tends to reduce the interface area (see also Section 3.1.2). How does the force tangential to the interface create the force in a perpendicular direction? To explain this, one could imagine a rubber band. In this case, the surface tension is replaced by a tension of the rubber, and the normal force is the one that is acting on the wrist when the band is worn around it. The normal force acting

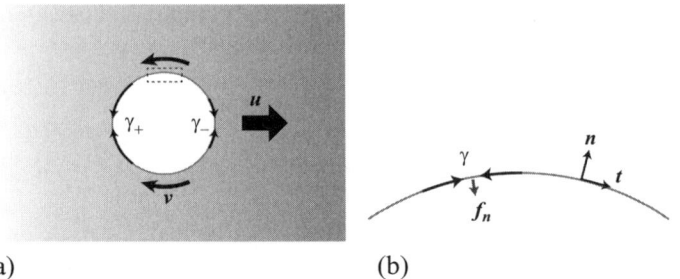

Figure 3.8 (a) Inhomogeneous surface tension along the interface and the surrounding flow. (b) Inside the box of a broken line in (a). The normal and tangential directions at the interface are shown as *n* and *t*, respectively. The surface tension is contractile force along the interface and generates the normal force pointing in the inward direction of the droplet as shown denoted with f_n along the gray line.

on the wrist can certainly be felt when the band is worn around it. This is because the wrist is rounded as shown in a simplified geometric manner in Figure 3.8(b). The force in the normal direction is larger for a more curved interface and is proportional to the curvature κ.

Inhomogeneous surface tension along the interface results in an additional force; because one side (γ_+) is more compressive and the other side (γ_-) is less compressive (less tense), a force directed from γ_- to γ_+ exists (see Figure 3.8(a)). This force is tangential to the interface and is proportional to the gradient of the surface tension. When there is a velocity gradient around the interface, we may also have to consider the force due to shear viscosity. We finally arrive at the following boundary condition for the stress tensors (Antanovskii, 1992):

$$\boldsymbol{n} \cdot \boldsymbol{\sigma}^{(o)} - \boldsymbol{n} \cdot \boldsymbol{\sigma}^{(i)} = \partial_s(\gamma \boldsymbol{t}), \tag{3-21}$$

where *t* is a unit tangent vector along the interface. Note that $\partial_s \boldsymbol{t} = \kappa \boldsymbol{n}$. By construction, the total force acting on the interface—that is, the integral of the right side of Eq. (3-21) over the contour of the interface—is zero. This proves that there is no external mechanical force acting on the droplet. Nevertheless, flow is induced by the force acting at the interface and leads to the net motion of the droplet as we will see in the next section.

3.3.4 Spherical Droplet Moving Under a Concentration Gradient

In this section, we discuss a droplet placed at a gradient of chemical concentration in three dimensions. Under these circumstances, the droplet moves in the direction of the gradient. This has been discussed both theoretically and experimentally in Young et al. (1959) when the gradient is temperature. We follow their work, but instead of a temperature field, we consider a concentration field $c(\boldsymbol{r}, t)$ of surfactants that reduce surface tension. Theoretically, the two systems are essentially the same.

We again assume axisymmetry here; the (inhomogeneous) surface tension depends only on θ, and therefore $\gamma = \gamma(\theta)$. When the concentration is not high, the surface tension is decomposed into homogeneous and excess parts, where the excess part is proportional to the concentration:

$$\gamma(\theta) = \gamma_0 + \gamma_c c(R, \theta, t). \tag{3-22}$$

Surface tension decreases in the presence of surfactants, and therefore $\gamma_c < 0$. The concentration field is, in general, time-dependent but in this section we focus on the steady distribution. In Sections 3.4.1 and 3.4.2 we consider a time-dependent concentration field and discuss how it results in complex motion of a droplet.

The steady-state distribution of concentration is described with the diffusion equation

$$D\nabla^2 c = 0. \tag{3-23}$$

We assume the diffusion constants inside and outside the droplet are the same, $D^{(i)} = D^{(o)} = D$, although, in general, they are different. Under the condition that the concentration field at both sides of the system is externally maintained—that is, the gradient is fixed as c_1—the concentration field is given as

$$c(\mathbf{r}) = c_0 + c_1 z, \tag{3-24}$$

where c_0 is the mean concentration.

For the given surface tension $\gamma(\theta)$ at the droplet interface, we can solve the hydrodynamic equations. First, we expand the surface tension using Legendre polynomials as

$$\gamma(\theta) = \sum_{n=0}^{\infty} \Gamma_n P_n(\cos\theta). \tag{3-25}$$

We note that Γ_n ($n = 0, 1, 2, \ldots$) can be calculated using $\gamma(\theta)$ as

$$\Gamma_n = \frac{2n+1}{2} \int_0^{2\pi} \gamma(\theta) P_n(\cos\theta) \sin\theta \, d\theta. \tag{3-26}$$

The boundary condition for a shear stress was discussed in the previous section and is expressed as

$$\sigma_{r\theta}^{(i)}|_{r=R} = \sigma_{r\theta}^{(o)}|_{r=R} + \frac{1}{R}\frac{\partial\gamma}{\partial\theta}. \tag{3-27}$$

We also have to consider the condition on the forces exerted on the droplet. In the present case, since the droplet is moving at a constant velocity, the net force exerted on the droplet should be zero. This is called a force-free condition, which is expressed as $\boldsymbol{F}=0$, and according to Eq. (3-20) we obtain

$$A_2 = 0. \tag{3-28}$$

With these results, we finally obtain the velocity of the droplet:

$$u = -\frac{2\gamma_c c_1 R}{9\eta^{(i)} + 6\eta^{(o)}}. \tag{3-29}$$

For completeness, we list the explicit forms of the velocity fields:

$$v_r^{(i)} = \left(\frac{A_2}{R} - \frac{3u}{2}\right)\left(\frac{r^2}{R^2} - 1\right)\cos\theta, \tag{3-30}$$

$$v_r^{(o)} = \left[\frac{A_2}{r}\left(1 - \frac{R^2}{r^2}\right) - u\left(1 - \frac{R^3}{r^3}\right)\right]\cos\theta, \tag{3-31}$$

$$v_\theta^{(i)} = -\frac{1}{2}\left(\frac{A_2}{R} - \frac{3u}{2}\right)\left(\frac{4r^2}{R^2} - 2\right)\sin\theta, \tag{3-32}$$

$$v_\theta^{(o)} = \frac{1}{2}\left[\frac{A_2}{r}\left(-1 - \frac{R^2}{r^2}\right) + u\left(2 + \frac{R^3}{r^3}\right)\right]\sin\theta \tag{3-33}$$

Figure 3.9 shows the velocity field inside and outside the droplet. The inhomogeneous surface tension creates flow around the droplet, and thus, we may say that the droplet is *swimming* in the outer fluid. The important difference between the flow under a surface tension gradient and that under an external mechanical force as discussed in Section 3.3.2 is that the flow velocity of the former decays much faster than that of the latter. It can also be seen in Eqs. (3-31) and (3-33) that the decay of the flow velocity is proportional to $1/r^3$, under the force-free condition (Eq. (3-28)). This is in contrast with the flow proportional to $1/r$ under $A_2 \neq 0$ as given by Eq. (3-9), which is

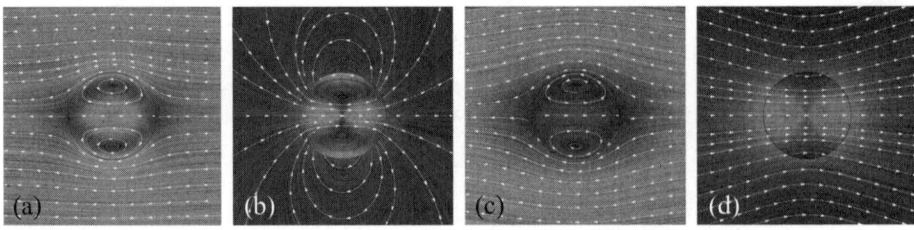

Figure 3.9 Streaming flow around the droplet driven by a linear gradient of chemical concentration. The flow under a gradient of a chemical concentration is shown in a co-moving frame with the droplet (a) and in a lab frame (b). The flow under an external mechanical force is shown in a co-moving frame with the droplet (c) and in a lab frame (d). (For color version of this figure, the reader is referred to the online version of this chapter.)

consistent with the typical range of hydrodynamics interactions. The short-ranged velocity gradient is found in the flow in a co-moving frame with the droplet (see Figure 3.9(a) and compare it with Figure 3.9(c)). In a lab frame, the flow around the droplet is accompanied by the motion of the droplet itself when it is driven by an external force (Figure 3.9(d)), while the surrounding flow of the motion under the surface tension gradient is accumulated near the surface of the droplet.

A similar calculation applies to the two-dimensional (2D) system (see Section A3.1 in the Appendix), but the surface tension in this case is expanded as

$$\gamma = \sum_{m=0}^{\infty} \gamma_m \cos(m\theta). \tag{3-34}$$

Without axisymmetry, terms similar to those containing $\sin(m\theta)$ are required in addition to the terms containing $\cos(m\theta)$. The force acting in the x direction is

$$F_x = R \int_0^{2\pi} d\theta (\sigma_{rr}^{(o)} \cos\theta - \sigma_{r\theta}^{(o)} \sin\theta) = 4\pi \eta^{(o)} \frac{B_1}{R}. \tag{3-35}$$

Here, from orthogonality, the modes for $m \geq 2$ do not contribute to the integral. This force must vanish because no external mechanical force is being exerted. The velocity of the droplet is finally obtained similarly to Eq. (3-29) as

$$u = -\frac{\gamma_c c_1 R}{4(\eta^{(i)} + \eta^{(o)})} \tag{3-36}$$

It should be stressed that the Stokes paradox does not apply to the present problem. In the case of the motion under mechanical force, as discussed in Section 3.3.2, the Stokes equation in two dimensions does not have a solution satisfying the boundary condition at infinity. This occurs due to the presence of a logarithmic term in the solution, which is originated from the external mechanical force. This is in contrast with the case of the motion under a surface tension gradient where there is no net force acting on the droplet, and, consequently, no such logarithmic term in the solution.

The motion under a gradient of a concentration discussed in this section inevitably occurs in the direction of the gradient. This is because there is no nonlinear term in the equations, which is necessary for breaking the symmetry and causing one direction of motion to be selected. Nevertheless, the theoretical approach shown in this section is extended to the spontaneous motion, the direction of which is not necessarily imposed by an external condition. One way to accomplish this is to include the coupling between the flow and concentration fields in Eq. (3-23) through the advection term $v \cdot \nabla c$ (Yoshinaga et al., 2012). Another way is to include nonlinear effects through cooperative behavior of reaction–diffusion systems leading to pattern formation. In the next section, we demonstrate the coupling of pattern formation with hydrodynamics that leads to various motions of a droplet.

3.4 Motion Coupled with Pattern Formation

3.4.1 Motion of a BZ Droplet

In Section 3.2.3 we presented the experimental results on the spontaneous motion of a BZ droplet coupled with chemical wave propagation inside it. The Reynolds number in the experimental system is calculated to be around 0.1. Thus, we may adopt the analysis based on the Stokes equation explained in Section 3.3.4.

To apply the theoretical analysis to the spontaneous motion of a BZ droplet, we adopt the reaction–diffusion–advection equation with the Oregonator model (see Section A3.2 in the Appendix) for the reaction kinetics of a BZ reaction inside the droplet:

$$\frac{\partial U}{\partial t} + \boldsymbol{v} \cdot \nabla U = \frac{1}{\epsilon}\left[U(1-U) - fV\frac{U-q}{U+q}\right] + D\nabla^2 U, \tag{3-37}$$

$$\frac{\partial V}{\partial t} + \boldsymbol{v} \cdot \nabla V = U - V + D\nabla^2 V, \tag{3-38}$$

where U and V correspond to the nondimensionalized concentrations of $HBrO_2$ and oxidized catalyst (ferriin), respectively. ϵ, q, and f are the parameters that determine the characterisitcs of the BZ reaction. D is the diffusion coefficients of the chemicals. The Oregonator model is nondimensionalized, and therefore, the time unit in the calculation is set as T.

Let us consider the experimental system shown in Figure 3.5(a). Since the BZ reaction occurs only inside the droplet, the reaction–diffusion–advection equations (Eqs. (3-37) and (3-38)) should be numerically solved in a spherical region. To reduce calculation time, we calculated on the 2D field by taking the axisymmetry into consideration. We assumed that the change in the flow velocity both inside and outside the droplet is much faster than the diffusive flow of chemicals, and we use the flow field analytically derived in Section 3.3.4. The surface tension at the oil–water interface depends on the chemical concentrations at the interface, considering that the BZ medium in the oxidized state is reported to have a higher surface tension than that in the reduced state (Yoshikawa et al., 1993; Inomoto et al., 2000). This means that the surface tension is an increasing function of V. Here we adopt the following linear function:

$$\gamma(V) = \gamma_0 + kV. \tag{3-39}$$

We note that γ_c in Eq. (3-22) is set to be equal to k ($k > 0$).

Under this assumption, the flow velocity can be written as a function of the profiles of V. The parameters for the Oregonator model were chosen from the concentration in the actual experiments except for f, as $q = 0.0000952$, $\epsilon = 0.033$, and $f = 1.2$. By adjusting the value of f, the system is set in an excitable condition for easier control of the timing of chemical wave propagation; that is, a chemical wave could propagate only from the point where it was initiated. In this condition, the time unit T corresponds to ~ 2 s in the actual system. The diffusion constant D is set to be $D = 125$, which corresponds to the condition that the diameter of the droplet is 1 mm. In the numerical calculation, the velocity field is calculated by summing the modes from $n = 1$ to 8 in the expansion of $\gamma(\theta)$. We confirmed that the cutoff modes did not critically affect the numerical results.

By integrating the velocity with respect to time, the position of the droplet was calculated as a function of time:

$$z(t) = \int_0^t u(t')dt'. \tag{3-40}$$

For the target pattern and the scroll ring corresponding to Figures 3.5(b) and 3.6(b), respectively, the pattern in the chemical concentration is axisymmetric and the framework mentioned previously can be used. However, for the spiral wave corresponding to Figure 3.6(a), the pattern is not axisymmetric, and the same method cannot be used. Thus, we used the results for the 2D circular droplet obtained using Eq. (3-36), which may result in an inconsistency from the actual system.

3.4.2 Numerical Results

This section describes numerical calculations based on the method explained in Section 3.3.4. We introduce the three results related to the corresponding experimental results in Figures 3.5(b), 3.6(a), and 3.6(b).

Figure 3.10 shows the results of numerical calculation in the case of a target pattern, which corresponds to Figure 3.5(b). Figure 3.10(a) shows the snapshots of V on a section, in which we can observe a circular wave propagating outward from one point. In Figure 3.10(b), both the spatiotemporal plot of the chemical wave inside the droplet and the position of the droplet are shown. The numerical results are consistent with the coupling between the chemical wave propagation and the droplet motion in the experiments. The chemical wave propagated radially outward from the point where it was initiated. When the wave reached the lower boundary of the droplet, the droplet began to move upward. Then when the chemical wave reached the upper boundary, the droplet moved back in the opposite direction. In this manner, a net motion was observed in each cycle.

Let us examine the detailed mechanism. In Figure 3.11(a), the profile of V along the interface against the angle from the symmetry axis, θ, is shown for each snapshot in Figure 3.10(a). The velocity of motion is proportional to the 1-mode component, when the profile $V(\theta)$ is expanded with Legendre polynomials (see Eqs. (3-25) and (3-26)). Since $P_1(\cos \theta) = \cos \theta$, the 1-mode component is positive when $V(\theta)$ has a larger value near $\theta = 0$ and a smaller value near $\theta = \pi$. From Figure 3.11(a), we can guess that the 1-mode component changes its sign, which results in a change in the direction of the motion at around $t = 1.4T$. In Figures 3.11(b)–(d), the time series in the components of n-modes ($n = 1, 2, 3, 4$), velocity, and position are shown. We can also confirm that the direction of the motion changes at around $t = 1.4T$.

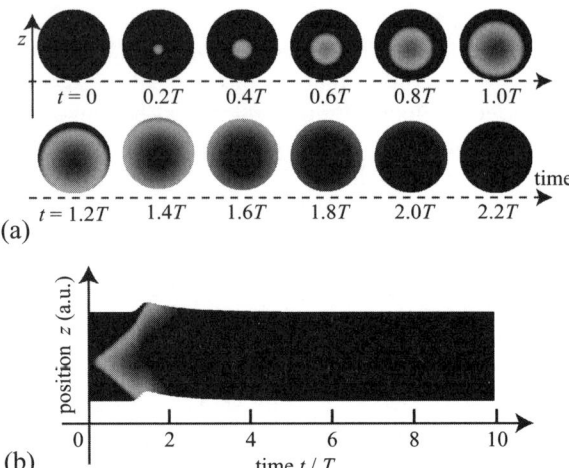

Figure 3.10 Results of numerical calculation for the target pattern inside the droplet. (a) Snapshots of V inside a droplet. At the time corresponding to $t=0$, a chemical wave was initiated by setting $U=1$ in the region corresponding to the experimental result shown in Figure 3.5(b). Bright and dark regions correspond to the area of the higher and lower V, respectively, which correspond to the blue- and red-colored regions in the actual BZ reaction. (b) Spatiotemporal plot of the droplet motion and the chemical wave inside it; the images along the z axis are arranged chronologically.

The flow profiles at $t = 1.0T$, $1.2T$, $1.4T$, and $1.6T$ are also shown in Figure 3.12(a), in which the direction of the roll structure is inverted at around $t = 1.4T$. We can clearly see a pair of rolls inside the droplet in the co-moving frame. In the lab frame, the roll structure cannot be clearly seen, as evident from Figure 3.12(b). The flow in the same direction as that of the droplet motion is mainly seen inside the droplet, and the flow outside is the same direction but weaker than that inside the droplet.

As for the motion of a BZ droplet in which a scroll ring (Winfree, 1973) occurs, as experimentally demonstrated and as shown in Figure 3.6(b), we can calculate this phenomenon using almost the same procedure. The differences are the point where the chemical wave is initiated and an additional operation of erasing chemical waves within a certain region. Through this operation, tips of the chemical waves are generated, and a scroll ring can be developed. The numerical results shown in Figure 3.13 reveal the occurrence of a scroll ring in the droplet. The droplet exhibits a back-and-forth motion coupled with the rotation of the scroll ring. In Figure 3.14, the flow profiles in the co-moving frame and in the lab frames are shown. The back-and-forth motion can be understood as a result from the fact that the droplet tends to move away from the point of contact of the scroll ring with the interface, at which the surface tension is the highest.

We cannot apply the same logic as that used earlier to the experimentally observed spiral wave shown in Figure 3.6(a), because in this case the droplet motion is not restricted in one direction but occurs in a 2D plane. Thus, we assume that the flow field is the same as that calculated in the 2D space (see Section 3.3.4). To create

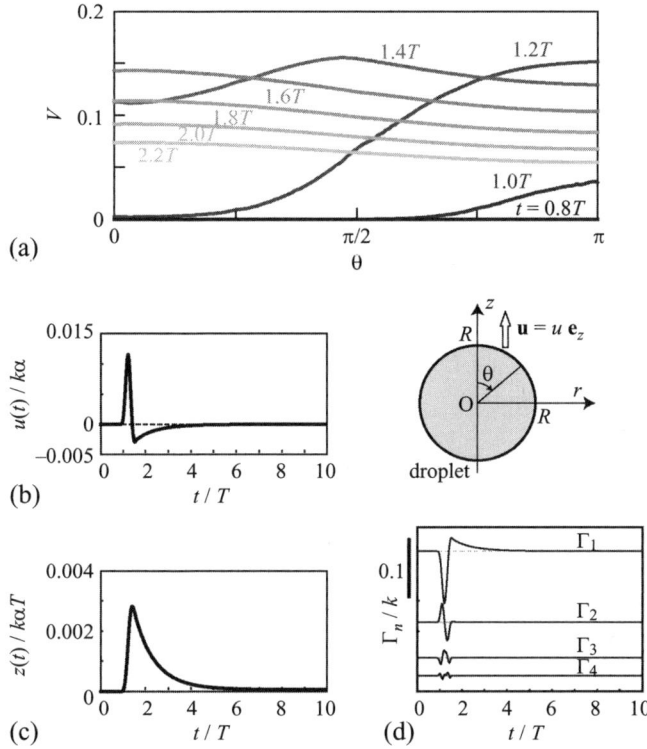

Figure 3.11 Numerical results for a BZ droplet with a target pattern. (a) Profiles of V against θ corresponding to the snapshots in Figure 3.10(a). The definition of θ is shown below. (b) Time course of the droplet velocity, $u(t)$. The velocity is normalized by $k\alpha = 5k/(3\eta^{(i)} + 2\eta^{(o)})$. (c) Time course of the droplet position, $z(t)$. The position is normalized by $k\alpha T$. (d) Time courses of Γ_n ($n = 1, 2, 3, 4$), which are normalized with k. The gray broken lines correspond to 0 for each Γ_n. (For interpretation of the references to color in this figure legend, the reader is referred to the online version of this chapter.)

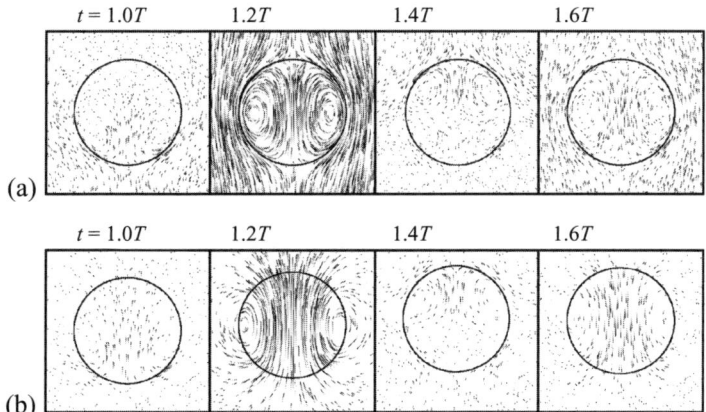

Figure 3.12 Numerical results of the flow inside and outside the droplet when a BZ droplet is moving coupled with a target pattern, corresponding to the results shown in Figure 3.10. Flow profiles at $t = 1.0T$, $1.2T$, $1.4T$, and $1.6T$ are shown. (a) Flow profiles in the co-moving frame. (b) Flow profiles in the lab frame.

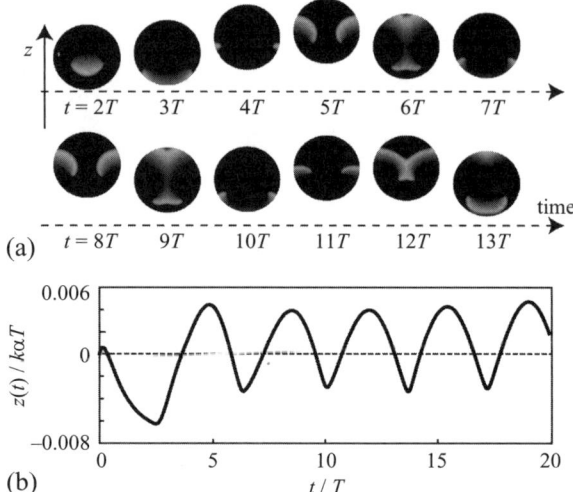

Figure 3.13 Numerical results for a BZ droplet showing a scroll ring. Inside the droplet, a chemical wave is initiated at the bottom, and at a time $1.3T$ later, the center of a spherical wave was artificially erased to form a tip. The time at which the tip is generated is set as $t=0$. Snapshots of the chemical pattern are shown, in which the value of V is shown. Brighter and darker regions correspond to the areas of higher and lower V, respectively. (b) Time course of the droplet position, $z(t)$. The position is normalized by $k\alpha T = 5kT/(3\eta^{(i)} + 2\eta^{(o)})$.

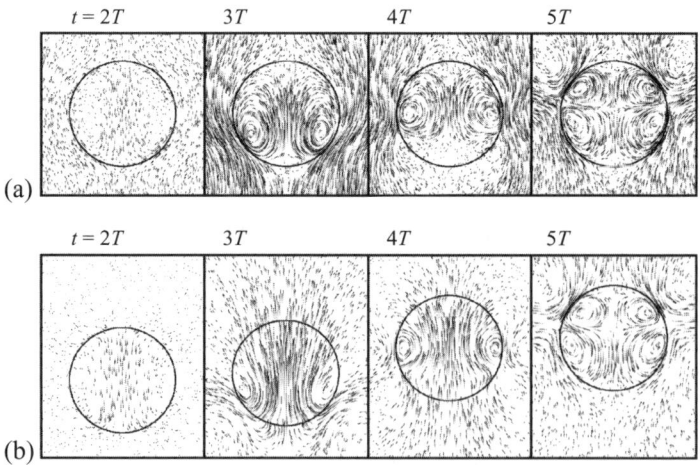

Figure 3.14 Numerical results of the flow inside and outside of the droplet when a BZ droplet moves coupled with a scroll ring. Flow profiles at $t=2T$, $3T$, $4T$, and $5T$ are shown. (a) Flow profiles in the co-moving frame. (b) Flow profiles in the lab frame. The time value corresponds to that in Figure 3.13.

Dynamics of Droplets

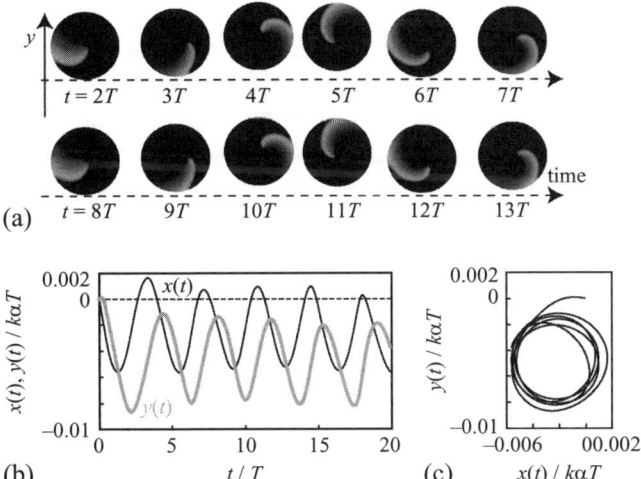

Figure 3.15 Numerical results for a BZ droplet showing a spiral wave. We performed this calculation in a 2D space. Inside the droplet, a chemical wave is initiated at the bottom, and at a time $2T$ later, a half of a circular wave was artificially erased to make a spiral core. The time at which a spiral core was generated was set to $t=0$. (a) Snapshots of the chemical pattern showing the value of V. Brighter and darker regions correspond to the areas of higher and lower V, respectively. (b) Time course of the droplet position coordinates, $x(t)$ and $y(t)$. The position is normalized by $k\alpha T = 2kT/(\eta^{(i)} + \eta^{(o)})$. (c) Trajectory of the center of mass of the droplet from $t=0$ to $20T$. (For color version of this figure, the reader is referred to the online version of this chapter.)

a spiral wave, we erase a part of the chemical wave and form a tip. The numerical results shown in Figure 3.15 reveal that a spiral wave occurs in the droplet and that the droplet itself exhibits rotational motion. The time courses and the trajectory of the center of mass of the droplet are shown in Figures 3.15(b) and (c), respectively. We can see that the BZ droplet moves in a circular orbit coupled with the spiral wave rotation inside it. The flow profiles at $t = 2T, 3T, 4T,$ and $5T$ are also shown in Figure 3.16(a), in which the asymmetric roll structure can be observed. The flow profiles from the lab frame are also shown in Figure 3.16(b). The rotational motion can also be understood as resulting from the fact that the droplet tends to move away from the point of contact of the spiral wave with the interface, at which the surface tension is the highest.

3.5 Concluding Remarks

In this chapter, we focused on the spontaneous motion of a droplet driven by the surface tension inhomogeneity. These systems have attracted interest of many researchers working on nonlinear and nonequilibrium phenomena, which still remain unexplored, in the expectation of bringing a novel experimental and theoretical approach to tackle the problems. They may also have potential application to

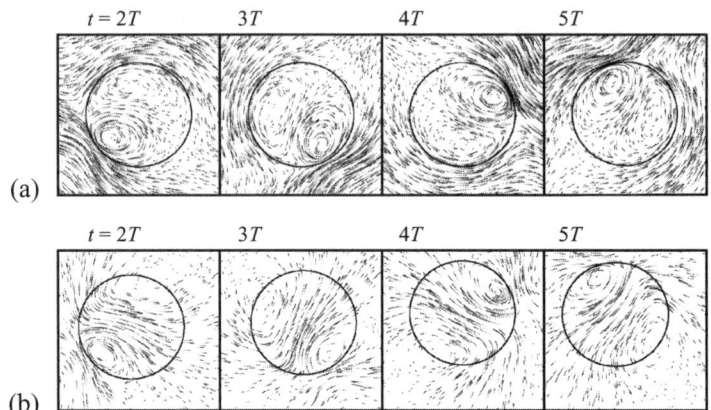

Figure 3.16 Numerical results of the flow inside and outside of the droplet when a BZ droplet moves coupled with a spiral wave. Flow profiles at $t=2T$, $3T$, $4T$, and $5T$ are shown. (a) Flow profiles in the co-moving frame. (b) Flow profiles in the lab frame.

chemical, biological, and industrial systems especially related to the motion of living organisms.

We introduced three typical types of droplet motion: a gliding droplet, a drifting droplet, and a swimming droplet. We especially focused on the swimming droplet, and explained the approach to the mechanism using the Stokes approximation in hydrodynamics. Using such an approach, we demonstrated the analytical results on the surface tension–driven motion of a BZ droplet.

In our framework, we do not directly calculate the hydrodynamics but use the analytical results on the translational motion of a spherical droplet. We believe the approach is transparent and helpful to understand more about the essential mechanism of self-propulsion. We may also know the universality and difference compared with other self-propelled systems.

Several remarks for future works are in order. We did not consider the deformation of the droplet. We have to confess that our system is too simple to compare with much more complicated systems as biological phenomena. In this respect, some extension may be interesting; for instance, introducing molecules that have internal degrees of freedom as polymers and liquid crystals may result in non-Newtonian behavior where not only viscosity but also elasticity appears in a fluid including the inside of a droplet and also in boundary conditions (Boukellal, et al., 2004; Sumino, et al., 2007, 2011). In the system of a BZ droplet, the chemical pattern generated asymmetry and this lead to a spontaneous motion of a droplet. Here, the feedback from motion to chemical pattern is rather minor, despite that the pattern is distorted by convection inside the droplet. Introducing chemical reaction that is sensitive to shear flow would make a direct pathway from motion to the chemical pattern. It might also be relevant to make a reconstitutive system consisting of biological systems. In any cases, controllable systems should be appreciated to stimulate further development driven by collaboration between experimental and theoretical studies.

In our theoretical treatment, we did not consider the deformation of the droplet. Several experiments have exemplified the relevance of deformation (Nagai et al., 2005) and indeed there have been several theoretical efforts (Ohta and Ohkuma, 2009). The effects of inertia and acceleration in the hydrodynamic equation (Eq. 3.2) may also be relevant in other self-propelled systems, although in the current context we may show they slightly modify our quantitative results (see Kitahata et al., 2011). We leave these issues for future works and hope that many researchers in various backgrounds would come to this field and clarify them.

Acknowledgments

We are grateful to Takao Ohta and Kenichi Yoshikawa for helpful discussions. The present studies are partly supported by PRESTO, JST (Alliance for Breakthrough between Mathematics and Sciences) to H. K., Grants-in-Aid for young scientists to H. K. (nos. 21740282 and 24740256), to N. Y. (no. 23740317), to Y. S. (no. 24740287) from MEXT, Japan, and JSPS fellowship for young scientists to K. H. N. (no. 23-1819).

References

Antanovskii, L., 1992. Creeping thermocapillary motion of a two-dimensional deformable bubble—Existence theorem and numerical simulation. Euro. J. Mech. B Fluids 11, 741–758.
Bain, C.D., Whitesides, G.M., 1989. A study by contact angle of the acid-base behavior of monolayers containing w-mercaptocarboxylic acids adsorbed on gold: An example of reactive spreading. Langmuir 5, 1370–1378.
Bekki, S., Vignes-Adler, M., Nakache, E., Adler, P.M., 1990. Solutal Marangoni effect: I. Pure interfacial transfer. J. Colloid Interface Sci. 140, 492–505.
Bekki, S., Vignes-Adler, M., Nakache, E., 1992. Solutal Marangoni effect: II Discussion. J. Colloid Interface Sci. 152, 314–324.
Bird, R., Armstrong, R., Hassager, O., 1987. Dynamics of Polymeric Liquids. Vol. 1: Fluid Mechanics. John Wiley and Sons, New York.
Boukellal, H., Campás, O., Joanny, J.F., Prost, J., Sykes, C., 2004. Soft Listeria: Actin-based propulsion of liquid drops Listeria: Actin-based propulsion of liquid drops. Phys. Rev. E 69, 061906.
Brochard-Wyart, F., de Gennes, P.G., 1985. Spontaneous motion of a reactive droplet. C. R. Acad. Sci. Paris Series II b 321, 285–288.
Chaudhury, M.K., Whitesides, G.M., 1992. How to make water run uphill. Science 256, 1539–1541.
Couder, Y., Fort, E., 2006. Single-particle diffraction and interference at a macroscopic scale. Phys. Rev. Lett. 97, 154101.
Couder, Y., Protière, S., Fort, E., Boudaoud, A., 2005. Dynamical phenomena: Walking and orbiting droplets. Nature 437, 208.
Davis, S.H., 1987. Thermocapillary instabilities. Annu. Rev. Fluid Mech. 19, 403–435.
Davis, H., Scriven, L., 1982. Stress and structure in fluid interfaces. Adv. Chem. Phys. 49, 357–454.

de Gennes, P.G., Brochard-Wyart, F., Quere, D., 2004. Capillarity and Wetting Phenomena: Bubbles Pearls Waves: Drops, Bubbles, Pearls, Waves. Springer, New York.

Domingues Dos Santos, F., Ondarçuhu, T., 1995. Free-running droplets. Phys. Rev. Lett. 75, 2972–2975.

Fournier, J.B., Cazabat, A.M., 1992. Tears of wine. Europhys. Lett. 20, 517–522.

Greenspan, H.P., 1978. On the motion of a small viscous droplet that wets a surface. J. Fluid Mech. 84, 125–143.

Haken, H., 1983. Synergetics: An Introduction: Nonequilibrium Phase Transitions and Self-Organization in Physics, Chemistry, and Biology. Springer-Verlag, Berlin (Springer Series in Synergetics).

Hanczyc, M.M., Toyota, T., Ikegami, T., Packard, N., Sugawara, T., 2007. Fatty acid chemistry at the oil–water interface: Self-propelled oil droplets. J. Am. Chem. Soc. 129, 9386–9391.

Happel, J., Brenner, H., 1965. Low Reynolds Number Hydrodynamics: With Special Applications to Particulate Media. Prentice-Hall, Englewood Cliffs, NJ.

Huh, C., Scriven, L., 1971. Hydrodynamic model of steady movement of a solid/liquid/fluid contact line. J. Colloid Interface Sci. 35, 85–101.

Ichimura, K., Oh, S.K., Nakagawa, M., 2000. Light-driven motion of liquids on a photoresponsive surface. Science 288, 1624–1626.

Inomoto, O., Abe, K., Amemiya, T., Yamaguchi, T., Kai, S., 2000. Bromomalonic acid–induced transition from trigger wave to big wave in the Belousov–Zhabotinsky reaction. Phys. Rev. E 61, 5326–5329.

John, K., Bär, M., Thiele, U., 2005. Self-propelled running droplets on solid substrates driven by chemical reactions. Euro. Phys. J. E 18, 183–199.

Kapral, R., Showalter, K., 1995. Chemical Waves and Patterns. Kluwer Academic, Dordrecht.

Keener, J.P., Tyson, J.J., 1986. Spiral waves in the Belousov–Zhabotinskii reaction. Physica D 21, 307–324.

Kitahata, H., Aihara, R., Magome, N., Yoshikawa, K., 2002. Convective and periodic motion driven by a chemical wave. J. Chem. Phys. 116, 5666–5672.

Kitahata, H., Yoshinaga, N., Nagai, K.H., Sumino, Y., 2011. Spontaneous motion of a droplet coupled with a chemical wave. Phys. Rev. E 84, 015101.

Kitahata, H., Yoshinaga, N., Nagai, K.H., Sumino, Y., 2012. Spontaneous motion of a Belousov–Zhabotinsky reaction droplet coupled with a spiral wave. Chem. Lett. 41, 1052–1054.

Kovalchuk, N.M., Vollhardt, D., 2000. Auto-oscillations of surface tension in water–alcohol systems. J. Phys. Chem. B 104, 7987–7992.

Kozlov, M.M., Mogilner, A., 2007. Model of polarization and bistability of cell fragments. Biophys. J. 93, 3811–3819.

Kudrolli, A., Lumay, G., Volfson, D., Tsimring, L.S., 2008. Swarming and swirling in self-propelled polar granular rods. Phys. Rev. Lett. 100, 058001.

Levan, M.D., 1981. Motion of a droplet with a Newtonian interface. J. Colloid Interface Sci. 83, 11–17.

Linke, H., Aleman, B.J., Melling, L.D., Taormina, M.J., Francis, M.J., Dow–Hygelund, C.C., et al., 2006. Self-propelled Leidenfrost droplets. Phys. Rev. Lett. 96, 154502.

Merkt, F.S., Deegan, R.D., Goldman, D.I., Rericha, E.C., Swinney, H.L., 2004. Persistent holes in a fluid. Phys. Rev. Lett. 92, 184501.

Nagai, K., Sumino, Y., Kitahata, H., Yoshikawa, K., 2005. Mode selection in the spontaneous motion of an alcohol droplet. Phys. Rev. E 71, 065301.

Nagai, K., Sumino, Y., Kitahata, H., Yoshikawa, K., 2006. Change in the mode of spontaneous motion of an alcohol droplet caused by a temperature change. Prog. Theor. Phys. Suppl. 161, 286–289.

Nagai, K., Sumino, Y., Yoshikawa, K., 2007. Regular self-motion of a liquid droplet powered by the chemical Marangoni effect. Colloid Surf. B. 56, 197.

Nagayama, M., Nakata, S., Doi, Y., Hayashima, Y., 2004. A theoretical and experimental study on the unidirectional motion of a camphor disk. Physica D 94, 151–165.

Nakata, S., Iguchi, Y., Ose, S., Kuboyama, M., Ishii, T., Yoshikawa, K., 1997. Self-rotation of a camphor scraping on water: New insight into the old problem. Langmuir 13, 4454–4458.

Narayan, V., Ramaswamy, S., Menon, N., 2007. Long-lived giant number fluctuations in a swarming granular nematic. Science 317, 105–108.

Nicolis, G., Prigogine, I., 1977. Self-organization in Nonequilibrium Systems: From Dissipative Structures to Order through Fluctuations. John Wiley and Sons, New York.

Ohta, T., Ohkuma, T., 2009. Deformable self-propelled particles. Phys. Rev. Lett. 102, 154101.

Quéré, D., Ajdari, A., 2006. Liquid drops: Surfing the hot spot. Nat. Mater. 5, 429–430.

Ramaswamy, S., 2010. The mechanics and statistics of active matter. Matter Annu. Rev. Cond. Matt. Phys. 1, 323–345.

Rana, A.F., Frank, J.M., 1973. Compressibility of water as a function of temperature and pressure. J. Chem. Phys. 59, 5529–5536.

Santiago-Rosanne, M., Vignes-Adler, M., Velarde, M.G., 1997. Dissolution of a drop on a liquid surface leading to surface waves and interfacial turbulence. J. Colloid Interface Sci. 191, 65–80.

Santiago-Rosanne, M., Vignes-Adler, M., Velarde, M.G., 2001. On the spreading of partially miscible liquids. J. Colloid Interface Sci. 234, 375–383.

Schatz, M.F., Neitzel, G.P., 2001. Experiments on thermocapillary instabilities. Annu. Rev. Fluid Mech. 33, 93–127.

Schneemilch, M., 2000. Wetting films in thermal gradients. Langmuir 16, 8796–8801.

Schneemilch, M., Cazabat, A.M., 2000. Shock separation in wetting films driven by thermal gradients. Langmuir 16, 9850–9856.

Scriven, L.E., Sternling, C.V., 1960. The Marangoni effects. Nature 187, 186–188.

Snezhko, A., Jacob, E.B., Aranson, I.S., 2008. Pulsating-gliding transition in the dynamics of levitating liquid nitrogen droplets. New J. Phys. 10, 043034.

Sumino, Y., Yoshikawa, K., 2008. Self-motion of an oil droplet: A simple physicochemical model of active Brownian motion. Chaos 18, 026106.

Sumino, Y., Magome, N., Hamada, T., Yoshikawa, K., 2005. Self-running droplet: Emergence of regular motion from nonequilibrium noise. Phys. Rev. Lett. 94, 068301.

Sumino, Y., Kitahata, H., Seto, H., Yoshikawa, K., 2007. Blebbing dynamics in an oil–water-surfactant system through the generation and destruction of a gel-like structure. Phys. Rev. E 76, 055202.

Sumino, Y., Kitahata, H., Seto, H., Yoshikawa, K., 2011. Dynamical blebbing at a droplet interface driven by instability in elastic stress: A novel self-motile system. Soft Matter 7, 3204–3212.

Thiele, U., John, K., Bär, M., 2004. Dynamical model for chemically driven running droplets. Phys. Rev. Lett. 93, 027802.

Thomson, J., 1855. On certain curious motion observable at surfaces of wine and other alcoholic liquors. Phil. Mag. 10, 330–333.

Thutupalli, S., Seemann, R., Herminghaus, S., 2011. Swarming behavior of simple model squirmers. New J. Phys. 13, 073021.

Verkhovsky, A.B., Svitkina, T.M., Borisy, G.G., 1999. Self-polarization and directional motility of cytoplasm. Curr. Biol. 11–20.

Vuilleumier, R., Ego, V., Neltner, L., Cazabat, A.M., 1995. Tears of wine: The stationary state. Langmuir 11, 4117–4121.

Winfree, A.T., 1973. Scroll-shaped waves of chemical activity in 3 dimensions. Science 181, 937–939.
Yabunaka, S., Ohta, T., Yoshinaga, N., 2012. Self-propelled motion of a fluid droplet under chemical reaction. J. Chem. Phys. 136, 074904.
Yam, P.T., Wilson, C.A., Ji, L., Hebert, B., Barnhart, E.L., Dye, N.A., et al., 2007. Actin myosin network reorganization breaks symmetry at the cell rear to spontaneously initiate polarized cell motility. J. Cell Biol. 178, 1207–1221.
Yoshikawa, K., Kusumi, T., Ukitsu, M., Nakata, S., 1993. Generation of periodic force with oscillating chemical reaction. Chem. Phys. Lett. 211, 211–213.
Yoshinaga, N., Nagai, K.H., Sumino, Y., Kitahata, H., 2012. Drift instability in the motion of a fluid droplet with a chemically reactive surface driven by Marangoni flow. Phys. Rev. E 86, 016108.
Young, T., 1805. An essay on the cohesion of fluids. Phil. Trans. R. Soc. Lond. 95, 65–87.
Young, N.O., Goldstein, J.S., Block, M.J., 1959. The motion of bubbles in a vertical temperature gradient. J. Fluid Mech. 6, 350–356.
Zaikin, A.N., Zhabotinsky, A.M., 1970. Concentration wave propagation in two-dimensional liquid-phase self-oscillating system. Nature 225, 535–537.

Appendix

A3.1 Solution of the Stokes Equation in Two Dimensions

The Stokes equation can be rewritten using the streaming function $\Psi(r)$ taking the curl ($\nabla \times$) of Eq. (3-6):

$$\nabla^2 \nabla^2 \Psi = 0, \tag{A3-1}$$

that is, $\nabla^2 \Phi = 0$ and $\nabla^2 \Psi = \Phi$. The fluid velocity is directed in parallel with contour lines of the streaming function, namely

$$v = -e_z \times \nabla \Psi = \nabla \times \Psi e_z. \tag{A3-2}$$

In a Cartesian coordinate system, this can be written as $v_x = \partial \Psi/\partial y$ and $v_y = -\partial \Psi/\partial x$. In a polar coordinate system, it can be written as $v_r = (1/r)\partial \Psi/\partial \theta$ and $v_\theta = -\partial \Psi/\partial r$. The pressure is given as

$$\nabla p = -e_z \times \eta \nabla \Phi. \tag{A3-3}$$

The general solution for Φ in polar coordinates is

$$\Phi(r, \theta) = \left(\frac{8A_1}{R^2}\frac{r}{R} + \frac{2B_1}{R^2}\frac{R}{r}\right) \sin\theta + \sum_{m=2}^{\infty} \left(\frac{A_m}{R^2}\left(\frac{r}{R}\right)^m + \frac{B_m}{R^2}\left(\frac{R}{r}\right)^m\right) \sin(m\theta), \tag{A3-4}$$

and that for Ψ is

$$\Psi(r,\theta) = \left(C_1\frac{r}{R} + D_1\frac{R}{r} + B_1\frac{r}{R}\ln\frac{r}{R} + A_1\left(\frac{r}{R}\right)^3\right)\sin\theta$$

$$+ \sum_{m=2}^{\infty}\left(\frac{A_m}{4(m+1)}\left(\frac{r}{R}\right)^{m+2} - \frac{B_m}{4(m-1)}\left(\frac{r}{R}\right)^{-m+2}\right.$$

$$\left. + C_m\left(\frac{r}{R}\right)^m + D_m\left(\frac{r}{R}\right)^{-m}\right)\sin(m\theta). \quad \text{(A3-5)}$$

The velocity field for a translational motion is calculated neglecting higher modes ($m > 1$ in two dimensions) as

$$v_r^{(o)} = -u\left(1 - \frac{R^2}{r^2}\right)\cos\theta, \quad \text{(A3-6)}$$

$$v_\theta^{(o)} = u\left(1 + \frac{R^2}{r^2}\right)\sin\theta, \quad \text{(A3-7)}$$

$$v_r^{(i)} = u\left(1 - \frac{r^2}{R^2}\right)\cos\theta, \quad \text{(A3-8)}$$

$$v_\theta^{(i)} = -u\left(1 - 3\frac{r^2}{R^2}\right)\sin\theta. \quad \text{(A3-9)}$$

A3.2 Mathematical Model for the BZ Reaction

To construct a mathematical model for the pattern dynamics seen in the BZ reaction (Zaikin and Zhabotinsky, 1970; Kapral and Showalter, 1995), we have to consider the three aspects of such patterns: reaction, diffusion, and advection. In other words, the time evolution of the pattern can be represented by the local concentration of several kinds of chemicals, which can be described as functions of time and space, $c_i(\mathbf{r},t)$, where i represents the kinds of chemicals, $\mathbf{r}$ represents position in space, and t represents time.

Of course, the chemicals change due to the chemical reaction. This change is described by a mass-action law. For example, if the reaction

$$A + B \rightleftharpoons C \quad \text{(A3-10)}$$

occurs, the evolution of the concentration of the chemical C, [C], is written as

$$\frac{d[C]}{dt} = k_+[A][B] - k_-[C] \quad \text{(A3-11)}$$

In a BZ reaction, various kinds of chemical reactions occur; however, Keener and Tyson (1986) investigated such complicated reactions and suggest that the rather simple dynamics for the reactions can be written using only two variables, as follows:

$$\frac{dU}{dt} = \frac{1}{\epsilon}\left[U(1-U) - fV\frac{U-q}{U+q}\right], \tag{A3-12}$$

$$\frac{dV}{dt} = U - V, \tag{A3-13}$$

where U and V are the nondimensionalized concentrations of $HBrO_2$ and the oxidized catalyst, respectively, which are often referred to as an activator and an inhibitor, respectively. ϵ, q, and f are the parameters that determine the characterisitcs of the BZ reaction. We note that the oxidized catalyst corresponds to the ferriin, which is indicated in bright, implying that the higher V corresponds to the blue-colored state in the actual BZ reaction.

Diffusion and advection are the processes that are not related to the chemical reaction. If there exists a concentration gradient, the diffusion tends to homogenize the concentration. The advection term corresponds to the transport by the bulk flow in the solution. Generally, the dynamics of the concentration can be expressed by the reaction–diffusion–advection equation,

$$\frac{\partial c_i}{\partial t} + \mathbf{v} \cdot \nabla c_i = f_i(c_i) + D_i \nabla^2 c_i, \tag{A3-14}$$

where $\mathbf{v}$ is the bulk velocity, and D_i is the diffusion constant for the ith chemical. In the absence of flow, the second term in the left side disappears, and the resulting simplified form like Eq. (A3-14) is called the reaction–diffusion equation, which is often used in the study of pattern formation.

4 Density Oscillators

Takeshi Kano

Chapter Contents
4.1 Introduction to Density Oscillators   119
    4.1.1 Self-Oscillatory Phenomena   119
    4.1.2 Relaxation Oscillations   121
    4.1.3 Density Oscillators   123
4.2 Phenomenological Description   127
    4.2.1 Experimental Procedure and General Oscillation Trend   127
    4.2.2 Hydrodynamic Analysis of Each Upflow and Downflow Branch   131
    4.2.3 Phenomenological Model   137
4.3 Fundamental Mechanism of Oscillation   140
    4.3.1 Hydrodynamic Analysis of Flow Reversal   141
    4.3.2 Viscosity-Dependent Flow Reversal   143
    4.3.3 Model Including Flow-Reversal Process   149
4.4 Concluding Remarks   158
Appendix   162

4.1 Introduction to Density Oscillators

4.1.1 Self-Oscillatory Phenomena

Systems exhibiting spontaneous regular rhythms abound in nature (Field and Noyes, 1974; Botros and Bruce, 1990; DiFrancesco, 1993; Landa, 1996; Blasius et al., 1997; Stern and McClintock, 1998; Pikovsky et al., 2001; Takamatsu et al., 2001; Lagomarsino et al., 2003; Bennett and Zukin, 2004; Miyazaki and Kinoshita, 2006). These systems, known as self-oscillatory systems (Andronov and Chaikin, 1949; Minorsky, 1974; Landa, 1996; Pikovsky et al., 2001), have been extensively studied in various fields. One of the most well-known examples is a heart that beats regularly even without conscious external control (DiFrancesco, 1993). This regular rhythm originates from the periodic activity of the pacemaker cells in the sinoatrial node. Another example is the menstrual cycle of several female mammals including human beings (Stern and McClintock, 1998), and physiological studies have shown that the cycle is caused by rhythmic changes in the secretions of hormones. Even a plasmodial slime mold, which is an amoeboid multinucleated unicellular organism, exhibits various oscillatory phenomena such as oscillations in the adenosine triphosphate and Ca^{2+} concentrations, the thickness of the plasmodium, and protoplasmic streaming (Takamatsu et al., 2001). These oscillations are thought to be generated by complicated mechanochemical reactions among chemicals, actin, intracellular organelles, and so on. Self-oscillatory systems are also found in nonliving systems, for example, in the Belousov–Zhavotinski (BZ) reaction, a complicated process of chemical reactions that causes rhythmic changes in the color of the reactants (see Chapter 2) (Field and Noyes, 1974; Miyazaki and Kinoshita, 2006).

From physical and mathematical viewpoints, a number of studies have been devoted to capturing the common characteristics of these self-oscillatory systems (Andronov and Chaikin, 1949; Minorsky, 1974; Landa, 1996; Pikovsky et al., 2001). In these studies, the self-oscillatory systems are considered as dynamical systems of which the behaviors are predetermined by a set of rules (algorithms), and the oscillatory phenomena are described by simple differential equations with several degrees of freedom.

Such an approach began more than a century ago. Near the end of the nineteenth century, Lord Rayleigh, known for his fluid dynamics and optics research, devoted his studies to oscillatory phenomena in acoustic systems. In his famous treatise *The Theory of Sound* (Rayleigh, 1877), he noted that vibrations of several acoustic systems are maintained in connection with a constant energy source, and described their behaviors through a simple nonlinear equation. Although the significance of his discovery was not realized immediately, it was developed in the studies of van der Pol in the 1920s (1920, 1926, 1927; van der Pol and van der Mark, 1927, 1928). His intensive studies of electric generators found that their behaviors were described by an equation similar to that derived by Rayleigh.

Around the same time as Rayleigh, Poincaré (1892) developed another approach from a mathematical viewpoint. He found that a dynamical system described by a pair of differential equations would exhibit a closed orbit in the phase plane, toward which neighboring paths were attracted (Figure 4.1). He called this orbit a *(stable) limit*

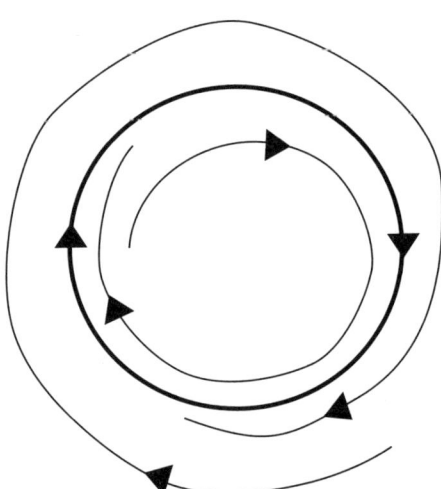

Figure 4.1 Limit cycle.

cycle. Later, Andronov and Chaikin (1949) noticed the similarity between Poincaré's limit cycles and the periodic oscillations of the electronic generators studied by van der Pol and named this type of oscillation a *self-oscillation*.

Self-oscillatory systems have some properties that differ from those of harmonic oscillators. First, they exhibit undamped oscillations by taking and dissipating energy from various sources; thus, the systems have the typical characteristics of nonequilibrium open systems. Second, the dynamical behavior of a self-oscillatory system does not depend on the initial condition but on the properties of the system itself, because the orbit of the self-oscillation is attracted to a stable limit cycle in the phase plane.

Third, self-oscillatory systems often exhibit "synchronization" (Kuramoto, 1984; Pikovsky et al., 2001; Manrubia et al., 2004). Synchronization is a phenomenon in which several oscillators adjust their own frequencies and behave cooperatively when the oscillators are coupled to each other or subjected to a periodical external field. Indeed, synchronization is a well-known phenomenon in nature. For example, thousands of male fireflies common in Southeast Asia emit light pulses synchronously to attract females (Kaempfer, 1727); in a concert hall, rhythmic applause is often synchronized (Pikovsky et al., 2001); and snowy tree crickets are able to synchronize their chirps by responding to the preceding chirps of their neighbors (Pikovsky et al., 2001). Quite a few scientists have studied such synchronization phenomena from a physical viewpoint. In particular, the emergence of cooperative behavior in coupled oscillators has been described by an analogy to the second-order phase transition (Kuramoto, 1984). Recently, coupled oscillator systems are not only studied to understand their underlying principles but also used in engineering (Wiesenfeld et al., 1998; Tass, 1999; Hong and Scaglione, 2005; Kiss et al., 2007; Ijspeert, 2008; Umedachi et al., 2010; Sato et al., 2011). Coupled oscillators, for example, are widely used for autonomous decentralized control of robots that behave like living organisms (Ijspeert, 2008; Umedachi et al., 2010; Sato et al., 2011).

4.1.2 Relaxation Oscillations

Relaxation oscillations are a particular type of self-oscillations and were first proposed by van der Pol (1926). Van der Pol devoted his study to electronic generators (van der Pol, 1920, 1926, 1927; van der Pol and van der Mark, 1927, 1928), and described their behaviors by the following equation, which is the now-famous van der Pol equation:

$$\ddot{v} - \varepsilon(1 - v^2)\dot{v} + v = 0. \tag{4-1}$$

This equation has the form of a harmonic oscillator when $\varepsilon = 0$. When $\varepsilon > 0$, the second term on the left side results in a damping effect, which works positively for $|v| > 1$ and negatively for $|v| < 1$. Owing to this, the system exhibits stable oscillations with a finite amplitude. While a nearly sinusoidal solution is obtained in the case of $0 < \varepsilon \ll 1$, van der Pol found that a periodic solution is obtained even in the case of $\varepsilon \gg 1$, where the system is highly nonlinear and dissipative. He pointed out that this condition is actually satisfied in an electrical system called a *multivibrator* (van der Pol, 1926).

Van der Pol noticed that the oscillatory behavior in the case of $\varepsilon \gg 1$ is distinctly different from that in the case of $0 < \varepsilon \ll 1$. First, the waveform deviates considerably from a sinusoidal function and contains many higher harmonics. Indeed, in the case of $\varepsilon \gg 1$, the value of v evolves in the following way: it initially increases slowly but jumps abruptly to a larger value when it reaches a certain threshold. Then, v decreases slowly and jumps abruptly to a smaller value when it reaches another threshold. Thus, the oscillation is characterized by slow and fast processes. Second, the oscillation period is characterized by the relaxation time of the system. In fact, a simple analysis shows that the period of the multivibrator can be estimated as the product of the capacitance and resistance, which corresponds to the relaxation time of the system. Van der Pol (1926) named oscillations showing these characteristics *relaxation oscillations*.

Following from this, van der Pol and van der Mark (1927, 1928) presented an electric circuit that exhibits relaxation oscillations (Figure 4.2). The circuit consists of a battery, capacitor, resister, and a neon tube that conducts electric current only when the voltage reaches a certain critical level. The capacitor is initially slowly charged, and when the voltage reaches the threshold, the neon tube begins to conduct electric current. As a result, the capacitor quickly discharges, the voltage drops, and the neon tube becomes nonconductive again. This process then repeats ad infinitum. In this case also, the period of the oscillation is characterized by the relaxation time of the capacitor charge—that is, the product of the capacitance and resistance.

In general, relaxation oscillations are intuitively understood in the following way. Let us consider a system of which the behavior is determined by a switch. When the switch is on, the system approaches a certain equilibrium state. Conversely, when the switch is off, the system approaches another equilibrium state. Thus, the state the system approaches changes depending on the state of the switch. A relaxation oscillator is considered to be a system of which the switch is inherent (Figure 4.3). When the system approaches one of the equilibrium states, the switch is flicked automatically, and the system in turn approaches the other equilibrium state. The examples shown by van der Pol can be understood within this framework. While the durations of the on and off states are equivalent in the system described by Eq. (4-1), for the system in Figure 4.2, the duration of the on state (conductive neon tube) is shorter than the off state (nonconductive neon tube).

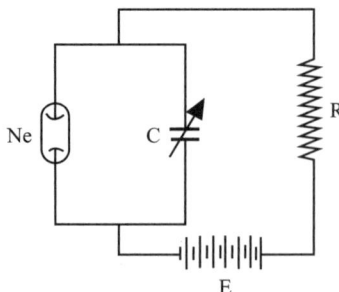

Figure 4.2 Electrical circuit that exhibits relaxation oscillations. It consists of a neon lamp (Ne), a variable capacitor (C), a resistance (R), and a battery (E).

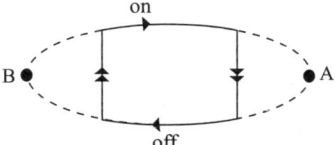

Figure 4.3 Intuitive interpretation of relaxation oscillation. When the switch is on, the system approaches equilibrium state A. However, the switch is turned off before it reaches state A, which makes the system approach equilibrium state B. The switch is then turned on again before the system reaches state B. In this way, the oscillation continues.

Van der Pol and van der Mark (1928) predicted in their paper that relaxation oscillations could be common in nature. Relaxation oscillations have indeed become well-known phenomena, observed in systems such as the endogenous circadian rhythms in plants (Blasius et al., 1997), heartbeats (DiFrancesco, 1993), respiratory rhythms (Botros and Bruce, 1990), and neuronal spiking in animals (Bennett and Zukin, 2004). Furthermore, relaxation oscillators have also been applied to technology; those of electric circuits (Chua et al., 1987) have been applied to produce square or sawtooth waves, which are used for triggering logic circuits or raster scanning in televisions.

4.1.3 Density Oscillators

In 1970, Martin (1970) discovered a curious phenomenon. He took a saltwater-filled straight tube, the cross-section of which shrinks drastically at its bottom (e.g., like that of a funnel, hypodermic syringe, pipette, or a tin can with a pin hole in the bottom), and held it within a beaker filled with fresh water. He found that this system exhibited oscillations of finite amplitude, where a downward jet of salt water and an upward jet of fresh water would repeatedly appear (Figure 4.4). He named this system a *salt oscillator*.

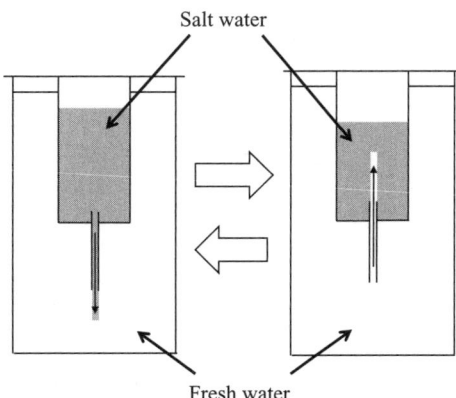

Figure 4.4 The salt oscillator discovered by Martin (1970).

The salt oscillator exhibits fluid oscillations even when the salt and fresh water are replaced with two other miscible fluids of different densities, because the density difference is the most crucial factor for generating oscillations (Martin, 1970; Alfredsson and Lagerstedt, 1981; Yoshikawa et al., 1988; Noyes, 1989; Yoshikawa et al., 1990; Yoshikawa and Nakata, 1990; Yoshikawa et al., 1991; Steinbock et al., 1998; Nakata et al., 1998; Miyakawa and Yamada, 1999, 2001; Aoki, 2000; Okamura and Yoshikawa, 2000; Ueno et al., 2006; Kano and Kinoshita, 2007, 2008, 2009; Kano, 2008; González et al., 2008). Thus, this system is also called a *density oscillator* (Alfredsson and Lagerstedt, 1981; Steinbock et al., 1998; Ueno et al., 2006; Kano and Kinoshita, 2007, 2008, 2009; Kano, 2008). Figure 4.5 shows schematically the oscillatory process of the density oscillator in which an inner container with a thin pipe in its bottom is filled with a heavy fluid, and this inner container is held within an outer container filled with a light fluid. During upflow, the system approaches a hydrostatic equilibrium in which the pipe is filled with the light fluid. However, the flow reverses before it reaches the hydrostatic equilibrium and downflow begins. During downflow, the system approaches another hydrostatic equilibrium, and this time the pipe is filled with the heavy fluid. However, the flow once again reverses before it reaches the hydrostatic equilibrium. From similarities between Figures 4.3 and 4.5, the density oscillator can be classified as a relaxation oscillator; thus, this oscillator has been used as a simple model system to investigate the fundamental mechanisms of relaxation oscillations.

When several inner containers are held within one outer container, synchronization occurs due to the mutual interaction of the oscillators through changes to the hydrostatic

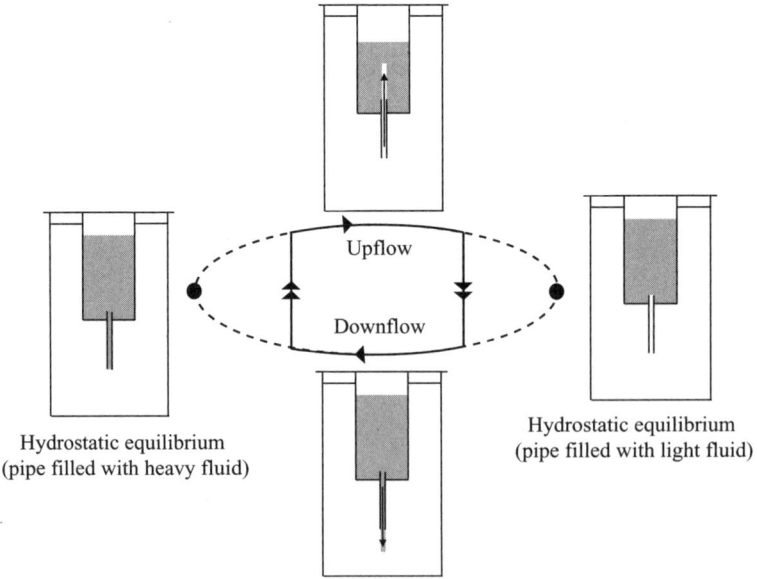

Figure 4.5 Overview of the oscillation process.

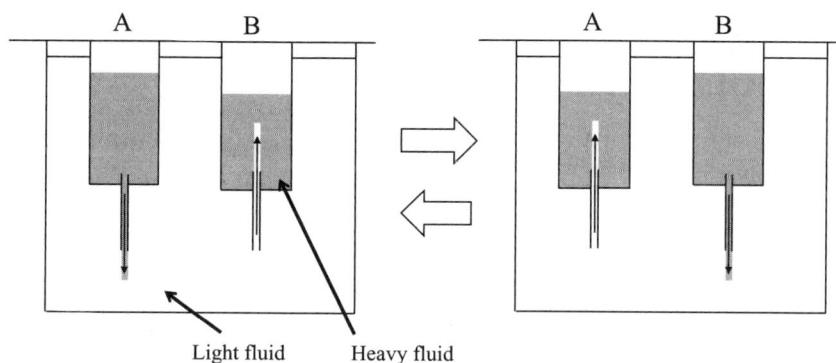

Figure 4.6 Coupled density oscillators.

pressure in the fluid in the outer container (Yoshikawa et al., 1990; Yoshikawa and Nakata, 1990; Yoshikawa et al., 1991; Nakata et al., 1998; Miyakawa and Yamada, 1999, 2001). For example, consider the case shown in Figure 4.6; containers A and B are held within a large outer container, and when container A is in the downflow state, the fluid surface of the outer container increases due to the inflow of fluid from container A. The hydrostatic pressure in the outer container then increases, which works so as to force container B into the upflow state. Thus, the rhythms of the two containers are synchronized in anti-phase. Similarly, when three inner containers are held within an outer container, their rhythms are synchronized with equidistant phases (Yoshikawa et al., 1990).

We can observe other types of synchronization by devising different experimental setups. For example, in-phase synchronization can occur in a system consisting of two oscillators coupled to each other through a window in the partition wall (Miyakawa and Yamada, 1999) and in a system in which one inner container has two orifices (Yoshikawa et al., 1991; Nakata et al., 1998) (Figure 4.7). Cluster states and more

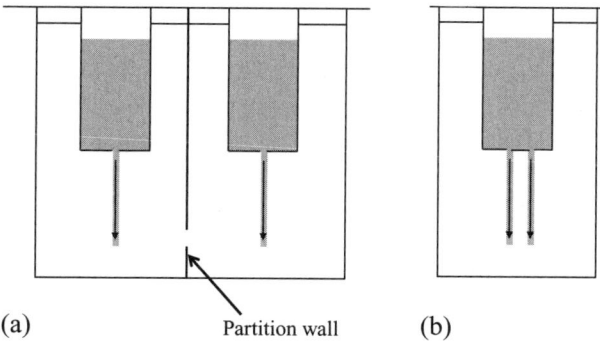

Figure 4.7 Two experimental conditions: (a) two oscillators coupled with each other through a gap in the partition wall, and (b) one inner container with two orifices.

complex patterns are observed when more than three inner containers are held within one outer container (Miyakawa and Yamada, 2001). Furthermore, when the fluid in the outer container is periodically infused and drained, the flow of the inner container is synchronized with the external perturbation in various ways depending on the perturbation period (González et al., 2008).

The oscillation of the electrical potential, which occurs when a pair of electrodes are placed in the light and heavy fluids, has also been extensively studied (Yoshikawa et al., 1989; Adamčiková, 1992; Upadhyay et al., 1992; Das and Srivastava, 1993; Srivastava, 1994; Srivastava et al., 1994; Villarreyes et al., 2000; Cervellati and Solda, 2001; Rastogi et al., 2005). The cause of the oscillations has long been a source of controversy and has not yet been completely clarified. However, research in this field has a wide applicability. For example, it has been suggested that this system could be used as an alternating voltage battery (Cervellati and Solda, 2001) and as a model of biological membranes (Yoshikawa and Nakata, 1990; Cervellati and Solda, 2001) and taste-sensing mechanisms (Srivastava et al., 1994).

Thus, the density oscillator exhibits various interesting phenomena. However, when we look into the fundamental mechanism of the fluid oscillation, we notice that it is not trivial and the following question naturally arises: Why does a density oscillator oscillate? This is indeed an important question, because clarifying this mechanism will lead us to an understanding of the common mechanism underlying the relaxation oscillations widely found in nature. To answer this question, we need to clarify the flow-reversal mechanism, because without this process the system will not exhibit oscillations but continue to approach hydrostatic equilibrium. We recently clarified the essential mechanism of the flow-reversal process by constructing a simple model on the basis of detailed experiments (Kano and Kinoshita, 2007, 2008, 2009; Kano, 2008). In the following, we review studies on this topic and introduce the relevant work.

Density Oscillator Experiment

Since the experimental setup for the density oscillator is quite simple, it is widely used in undergraduate teaching classes as well as for academic research. Even outside the laboratory, we can construct a density oscillator from everyday items. One example is shown in Figure B4.1. The inner container is a plastic cup with a straw in the bottom. A pair of chopsticks are attached on the top of the inner container and fixed to an aquarium. Then, by filling the plastic cup and the aquarium with salt water and pure water, respectively, the fluid oscillations are observed. We can observe the oscillations more easily by blending a few drops of red food coloring into the salt water.

Continued

Density Oscillator Experiment—cont'd

The oscillation period and amplitude can be tuned by changing the diameter and length of the straw and the density of the salt water. Furthermore, synchronization phenomena will be observed when several inner containers are placed within one aquarium.

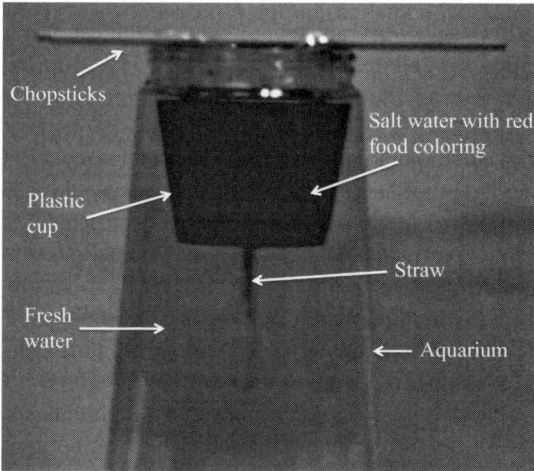

Figure B4.1 Density oscillator constructed of everyday items.

4.2 Phenomenological Description

4.2.1 Experimental Procedure and General Oscillation Trend

A typical experimental setup for a density oscillator is shown in Figure 4.8. The density oscillator consists of an outer container and an inner container with a pipe or orifice at the bottom. The heavy fluid is poured into the inner container with the exit plugged and the inner container is fixed to the outer container filled with the light fluid. The amount of fluid is adjusted so that the heights of the heavy and light fluids are nearly identical. We can then observe the oscillatory behavior by removing the plug; downflow occurs first.

When we perform the experiments, we need to pay attention to the following points. First, the inner container and pipe should be placed so that they are perpendicular to the ground. If they are considerably tilted, oscillations will not occur, and instead simultaneous up- and downflow within the pipe will be observed. Second, air bubbles should not be generated during the experiment because otherwise the oscillation stops when they adhere to the end of the pipe. To prevent this, any water used in the fluids should be boiled before use. Third, when we observe the long-time behavior of the oscillations, we need to prevent the fluid from evaporating. This can be achieved

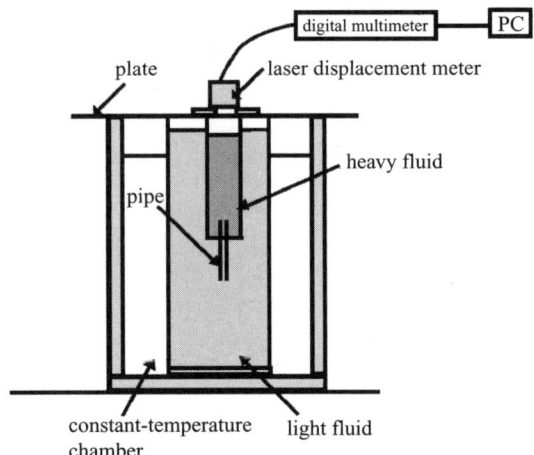

Figure 4.8 Overview of the experimental setup.

by covering the inner and outer containers with plates or adhesive tape. Fourth, it is better to control the fluid temperature to obtain accurate data, although it does not considerably affect the oscillation properties. In our experiment, we controlled the temperature by using a thermostat heater with a switch that was regulated according to the temperature measured by a thermocouple (Kano and Kinoshita, 2007; Kano, 2008).

We can observe oscillations under various experimental conditions. First, as described before, any fluid can be used in this system as long as the two fluids are miscible. Second, the oscillation occurs for a wide range of pipe (orifice) lengths and diameters and density differences between the two fluids. Although accurate parameter ranges that lead to oscillations have yet to be defined, previous studies (Alfredsson and Lagerstedt, 1981; Steinbock et al., 1998; Miyakawa and Yamada, 1999; Kano and Kinoshita, 2007; Kano, 2008) have shown that oscillation occurs at least for pipe (orifice) lengths in the range of 0.4–100 mm, pipe (orifice) diameters of 0.73–20.0 mm, and density differences of 0.006×10^3 to 0.230×10^3 kg•m^3. Third, the oscillation occurs even when the geometry of the containers or the pipe is varied. For example, oscillations can be seen even for pipes with a rectangular cross-section.

The oscillatory behavior can be measured in a number of ways. One well-known method is by measuring the electrical potential via a pair of electrodes placed in the light and heavy fluids (Yoshikawa et al., 1989; Adamčiková, 1992; Upadhyay et al., 1992; Das and Srivastava, 1993; Srivastava, 1994; Srivastava et al., 1994; Villarreyes et al., 2000; Cervellati and Solda, 2001; Rastogi et al., 2005). Another method, developed by Miyakawa and Yamada (2001) in a study of coupled density oscillators, involves illuminating the bottoms of the inner containers from below and detecting the light passing through orifices with a charge-coupled device (CCD) camera. The light intensity differs for up- and downflow, which makes this imaging possible. Although these two methods are useful for detecting up- and downflow, they are not suitable for quantitatively measuring the height of the fluid surface. To quantify the oscillatory behavior, a laser displacement meter can be used to measure the height

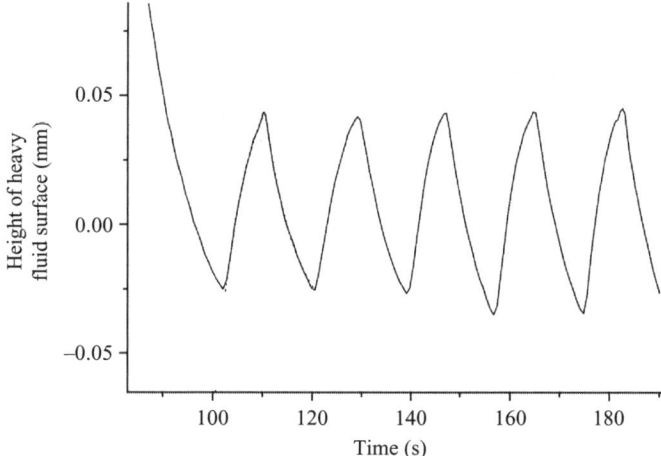

Figure 4.9 Temporal evolution of the height of the heavy fluid surface when the inner container has an orifice in the bottom. The thickness of the bottom of the inner container is 2 mm, while the diameter of the orifice is 1.12 mm. The viscosities of the heavy and light fluids are 1.77×10^{-3} Pa·s and 0.89×10^{-3} Pa·s, respectively.

of the fluid surface by detecting the light reflected from the surface (Steinbock et al., 1998; Okamura and Yoshikawa, 2000; Kano and Kinoshita, 2007; Kano, 2008). To reflect the light effectively, a water-repellent nontransparent plate needs to be floated on the surface.

Figure 4.9 shows a typical example of the temporal evolution of the height of the heavy fluid surface when the inner container has a small orifice in its bottom. The fluid surface regularly moves up and down in a range of ∼0.07 mm within a period of ∼20 s. If we instead attach a thin pipe to the bottom of the inner container (Figure 4.10), the amplitude and period differ considerably from those in the case shown in Figure 4.9: the amplitude and the period are ∼2.5 mm and ∼5000 s, respectively.

Thus, it becomes clear from Figures 4.9 and 4.10 that the amplitude and period vary significantly depending on the experimental conditions. In fact, they are affected by several experimental parameters: the amplitude increases as the pipe (orifice) length or density difference between the heavy and light fluids increases, while the period increases as the pipe (orifice) diameter decreases, the pipe (orifice) length increases, and the surface area of the heavy or light fluid increases. These tendencies are summarized in Table 4.1.

Turning our attention to each up- and downflow branch, in the case of a thin pipe the temporal evolution for each flow is well fitted by an exponential function (Figure 4.10a). Although the flow deviates slightly from the exponential function before the flow reversal, this deviation is extremely small compared to the oscillation amplitude (Figure 4.10b). As we will describe later, the difference between the asymptotic values of two adjacent exponential functions is in surprisingly good agreement with the difference between the heights of the heavy fluid surface at the

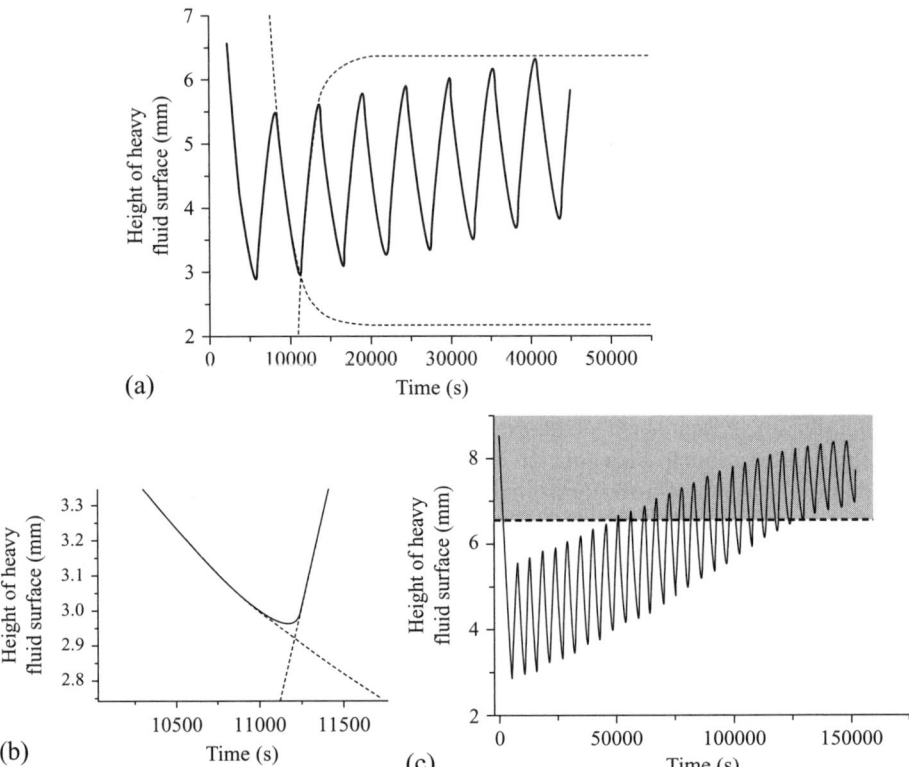

Figure 4.10 (a) Temporal evolution of the height of the heavy fluid surface when a pipe is attached to the bottom of the inner container. The pipe diameter and length are 0.73 and 70 mm, respectively. The viscosities of the heavy and light fluids are 2.64×10^{-3} Pa·s and 1.98×10^{-3} Pa·s, respectively, and the densities of the fluids are 1.059×10^3 kg·m^{-3} and 0.997×10^3 kg·m^{-3}, respectively. Each branch for the up- and downflow is well fitted by an exponential curve (dashed line). (b) Magnified view of (a) at the flow-reversal point from down- to upflow. (c) Long-time behavior of the oscillation ((a) is a magnified view of the first several cycles). In the gray region, the output of the laser displacement meter is out of the measurement range, and therefore the obtained data are somewhat inaccurate.

Table 4.1 Oscillation Amplitudes and Periods for Various Experimental Conditions

	Amplitude	Period
Increase the pipe length	Increase	Increase
Increase the pipe diameter	Not significantly affected	Decrease significantly
Increase the density difference between the heavy and light fluids	Increase	Not significantly affected
Increase the fluid viscosity	Not significantly affected	Increase
Increase the surface areas of containers	Not significantly affected	Increase

Density Oscillators 131

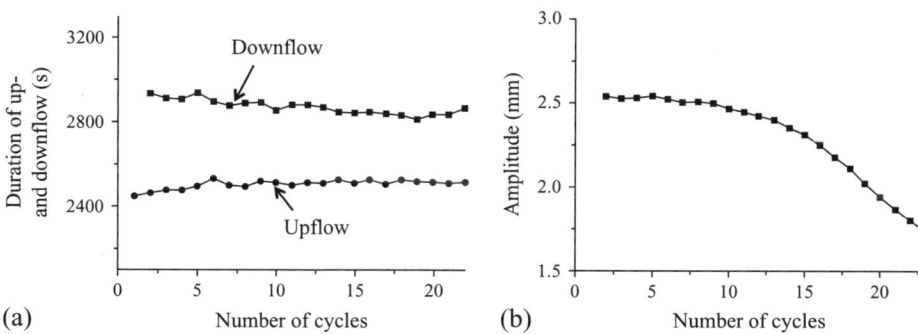

Figure 4.11 (a) Durations of the up- and downflow and (b) the oscillation amplitude plotted against the number of cycles. The experimental conditions are the same as those in Figure 4.10.

two hydrostatic equilibria shown in Figure 4.5. This fact suggests that each up- and downflow branch approaches the hydrostatic equilibrium in which the pipe is filled with light or heavy fluid, respectively. In contrast, in the case of an orifice (Figure 4.9), the difference between the asymptotic values of the exponential functions fitted to adjacent up- and downflow branches does not agree with the height difference between the two hydrostatic equilibria, indicating that each flow does not exhibit an accurate exponential response in the case of an orifice.

When we examine the long-time behavior of the oscillations, as in Figure 4.10(c), we find that the heavy fluid surface gradually elevates as the number of cycles increases. Figure 4.11 shows the durations of up- and downflow branches and the oscillation amplitude plotted against the number of cycles. The amplitude is seen to clearly decrease gradually, whereas the durations do not vary significantly. As the number of cycles increases further, the amplitude continues to decrease, and finally, the oscillation stops. These results can be understood in the following way. Since the light fluid accumulates above the heavy fluid in the inner container during upflow, the density difference between the two fluids decreases as the number of cycles increases. Therefore, the height difference between the two hydrostatic equilibria shown in Figure 4.5 decreases, and finally, the heights of the heavy- and light-fluid surfaces become identical since the density difference diminishes.

4.2.2 Hydrodynamic Analysis of Each Upflow and Downflow Branch

The oscillation process of the density oscillator is divided into two parts: one-way flow (up or down) and flow reversal. Compared with the flow-reversal process, one-way flow is easy to analyze theoretically. Martin (1970) performed a hydrodynamic analysis of each up- and downflow branch, and in this section, we review his theoretical analysis with a slight modification.

Consider the case where the pipe length d is much larger than the pipe radius a, and that the flow inside the pipe is parallel to the pipe axis (Figure 4.12). Cylindrical

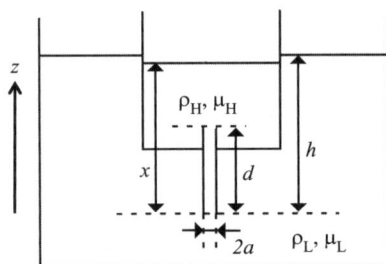

Figure 4.12 Definition of the parameters. The densities of the heavy and light fluids are ρ_H and ρ_L, while μ_H and μ_L are the viscosities of the heavy and light fluids, respectively.

coordinates are used, where the z axis is taken to coincide with the axis of the pipe and r denotes the radial coordinate. The origin of the z coordinate is set to coincide with the lower end of the pipe, and the upper direction is taken as positive. Then, the z component of the Navier-Stokes equation is given as follows:

$$\rho \frac{\partial u}{\partial t} = -\frac{\partial P(z)}{\partial z} - \rho g + \frac{\mu}{r} \frac{\partial}{\partial r}\left(r \frac{\partial u}{\partial r}\right), \tag{4-2}$$

where u is the z component of the flow velocity, $P(z)$ is the hydrostatic pressure, and ρ and μ are the fluid density and viscosity, respectively. The first, second, and third terms on the right side express the hydrostatic pressure gradient, gravitational force, and viscous drag force, respectively. When the space inside the pipe is completely occupied by the light fluid (i.e., in the case of upflow), the integration of Eq. (4-2) over this space yields

$$\rho_L \int_0^d dz \int_0^a 2\pi r dr \frac{\partial u}{\partial t} = -\int_0^d dz \int_0^a 2\pi r dr \frac{\partial P(z)}{\partial z} - \pi a^2 d \rho_L g \\ + \int_0^d dz \int_0^a 2\pi r dr \frac{\mu_L}{r} \frac{\partial}{\partial r}\left(r \frac{\partial u}{\partial r}\right), \tag{4-3}$$

where ρ_L and μ_L are the average density and viscosity of the fluid in the outer container, respectively (see Figure 4.12). Thus, we obtain

$$\rho_L \frac{\partial \bar{u}}{\partial t} = -\frac{P(d) - P(0)}{d} - \rho_L g + \frac{2\mu_L}{a} \left.\frac{\partial u}{\partial r}\right|_{r=a}, \tag{4-4}$$

where $\bar{u}$ is the average velocity given by

$$\bar{u} = \frac{2}{a^2} \int_0^a u(r,t) r dr. \tag{4-5}$$

Similarly, when the space inside the pipe is completely occupied by the heavy fluid (i.e., in the case of downflow), we obtain

$$\rho_H \frac{\partial \bar{u}}{\partial t} = -\frac{P(d) - P(0)}{d} - \rho_H g + \frac{2\mu_H}{a} \frac{\partial u}{\partial r}\bigg|_{r=a}, \qquad (4\text{-}6)$$

where ρ_H and μ_H are the average density and viscosity of the fluid in the inner container, respectively (see Figure 4.12). Note that although ρ_H, ρ_L, μ_H, and μ_L do depend on time due to inflow of the heavy fluid into the outer container and that of the light fluid into the inner container, only ρ_H is assumed to be a function of time since the time dependence of the other parameters is not essential. Because the flow inside the pipe is regarded as almost 3D Hagen–Poiseuille flow (i.e., steady flow of an incompressible fluid through a straight pipe of circular cross-section; Schlichting, 1960), the following relation should hold (see also Section A4.1 in the Appendix at the end of this chapter):

$$\frac{\partial u}{\partial r}\bigg|_{r=a} = -\frac{4\bar{u}}{a}. \qquad (4\text{-}7)$$

In addition, the hydrostatic pressure should satisfy

$$P(0) - P(d) = \rho_L g h - \rho_H(t) g (x - d) - \frac{3}{4} \rho_j \bar{u}|\bar{u}|, \qquad (4\text{-}8)$$

where x and h are the heights of the heavy- and light-fluid surfaces, respectively. The first and second terms on the right side of Eq. (4-8) give the difference in the hydrostatic pressure derived simply from the heights of the fluid surface, while the third term denotes the loss of pressure due to the fluid passage through the pipe, called *head loss*. Here, x and h have the following relation:

$$h - h_{de}^{(i)} = -R\left(x - x_{de}^{(i)}\right), \qquad (4\text{-}9)$$

where $R = S/S_{out}$ with S and S_{out} being the surface areas of the inner and outer containers, respectively, and $x_{de}^{(i)}$ and $h_{de}^{(i)}$ are the heights of the heavy- and light-fluid surface, respectively, at the hydrostatic equilibrium in which the pipe is filled with heavy fluid in the ith cycle (Figure 4.13) (hereafter, the superscript (i) is only included if the specific cycle is important). Here, the term *cycle* is defined as a sequence of down- and upflow branches. In addition, from the continuity condition,

$$\dot{x} = \frac{\pi a^2}{S} \bar{u}. \qquad (4\text{-}10)$$

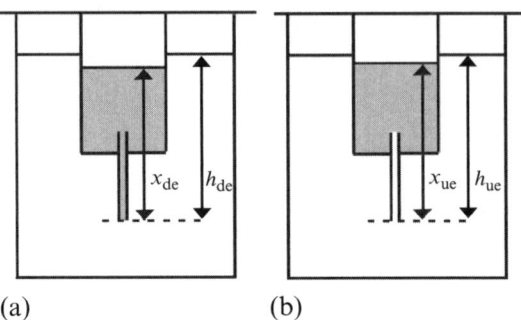

Figure 4.13 Two hydrostatic equilibria for a pipe filled with (a) heavy fluid and (b) light fluid.

By substituting Eqs. (4-7)–(4-10) into Eqs. (4-4) and (4-6), the following equations are obtained:

$$\frac{S}{a^2\pi}\ddot{x} + \frac{3S^2}{4da^4\pi^2}\dot{x}^2 + \frac{8v_L S}{a^4\pi}\dot{x} + \frac{g}{d}\left(\frac{\rho_H(t)}{\rho_L} + R\right)x$$
$$= \frac{g}{d}\left(h_{de}^{(i)} + Rx_{de}^{(i)}\right) + \frac{\rho_H(t) - \rho_L}{\rho_L}g, \qquad (4\text{-}11)$$

for upflow, and

$$\frac{S}{a^2\pi}\ddot{x} - \frac{3S^2}{4da^4\pi^2}\dot{x}^2 + \frac{8v_H(t)S}{a^4\pi}\dot{x} + \frac{\rho_L}{\rho_H(t)}\frac{g}{d}\left(\frac{\rho_H(t)}{\rho_L} + R\right)x$$
$$= \frac{\rho_L}{\rho_H(t)}\frac{g}{d}\left(h_{de}^{(i)} + Rx_{de}^{(i)}\right), \qquad (4\text{-}12)$$

for downflow, where $v_H(t) = \mu_H/\rho_H(t)$ and $v_L = \mu_L/\rho_L$ are the kinematic viscosities of the heavy and light fluids, respectively.

Equations (4-11) and (4-12) can be further simplified through the following considerations. First, $\rho_H(t)$, which decreases as the number of cycles increases owing to inflow of the light fluid, does not depend on the time during downflow, and so

$$\rho_H(t) = \rho_H^{(i)}, \qquad (4\text{-}13)$$

for the ith downflow. On the other hand, because the light fluid accumulates above the heavy fluid in the inner container during upflow (Figure 4.14), the average density $\rho_H(t)$ should satisfy the following relation for the ith upflow:

$$\rho_H(t)(x - d) = \rho_H^{(i)}\left(x_{de}^{(i)} - d\right) + \rho_L\left(x - x_{de}^{(i)}\right). \qquad (4\text{-}14)$$

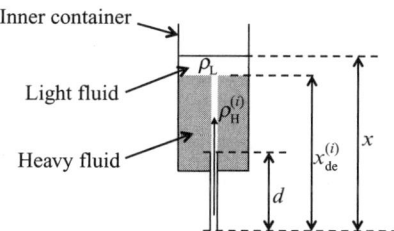

Figure 4.14 Accumulation of light fluid in the inner container during upflow.

Second, from the hydrostatic pressure balance at the hydrostatic equilibrium in which the pipe is filled with heavy fluid, the following relation should hold (see Figure 4.13a):

$$\rho_H^{(i)} x_{de}^{(i)} = \rho_L h_{de}^{(i)}. \tag{4-15}$$

Third, the nondimensionalized variables $\tilde{x}^{(i)}$ and $\tilde{t}$ are introduced as follows:

$$\tilde{x}^{(i)} = \frac{x - x_{de}^{(i)}}{(\delta\rho^{(i)}/\rho_L)d}, \tag{4-16}$$

$$\tilde{t} = \frac{4\nu_L \beta}{a^2 \sigma^{1/2}} t, \tag{4-17}$$

where $\beta^2 = ga^6\pi/(16S\nu_L d)$ and $\sigma = 3S\delta\rho^{(i)}/(2a^2\pi\rho_L)$ with $\delta\rho^{(i)} = \rho_H^{(i)} - \rho_L$. Using Eqs. (4-13)–(4-17), Eqs. (4-11) and (4-12) are written in nondimensional form as follows:

$$\frac{d^2\tilde{x}^{(i)}}{d\tilde{t}^2} + \frac{2\sigma^{1/2}}{\beta}\left[\frac{d\tilde{x}^{(i)}}{d\tilde{t}} + \frac{\beta\sigma^{1/2}}{4}\left(\frac{d\tilde{x}^{(i)}}{d\tilde{t}}\right)^2\right] + \sigma\left[(1+R)\tilde{x}^{(i)} - 1\right] = 0, \tag{4-18}$$

for upflow, and

$$\frac{d^2\tilde{x}^{(i)}}{d\tilde{t}^2} + \frac{2\sigma^{1/2}}{\beta}\left[\frac{\nu_H^{(i)}}{\nu_L}\frac{d\tilde{x}^{(i)}}{d\tilde{t}} - \frac{\beta\sigma^{1/2}}{4}\left(\frac{d\tilde{x}^{(i)}}{d\tilde{t}}\right)^2\right] + \sigma(1+DR)\tilde{x}^{(i)} = 0, \tag{4-19}$$

for downflow, where $D = \rho_L/\rho_H^{(i)}$ and $\nu_H^{(i)} = \mu_H/\rho_H^{(i)}$. Note that both $\beta < 1$ and $\sigma > 3/2$ should be satisfied so that oscillation occurs; these conditions were determined experimentally.

Thus, Eqs. (4-18) and (4-19) are the fundamental equations describing each up- and downflow branch. The solutions of Eqs. (4-18) and (4-19) approach $\tilde{x}^{(i)} = 1/(1+R)$ and $\tilde{x}^{(i)} = 0$ over time, which correspond to $x = x_{de}^{(i)} + \delta\rho^{(i)}d/[(1+R)\rho_L] \equiv x_{ue}^{(i)}$

and $x = x_{de}^{(i)}$ in dimensional form, respectively. These asymptotic values correspond to the two hydrostatic equilibria in the i th cycle (Figure 4.13).

We note that the difference between the asymptotic values for the $(i-1)$th upflow and the ith downflow, $x_{ue}^{(i-1)} - x_{de}^{(i)}$, and that for the ith up- and downflow, $x_{ue}^{(i)} - x_{de}^{(i)}$, are not equal because the height of the heavy-fluid surface increases as the number of cycles increases due to an accumulation of light fluid in the inner container, as previously described. The former difference, $x_{ue}^{(i-1)} - x_{de}^{(i)} (\equiv \delta x^{(i)})$, is derived in the following way. From Eq. (4-15), $\rho_H^{(i)} x_{de}^{(i)} = \rho_L h_{de}^{(i)}$ should be satisfied, and since the $(i-1)$th upflow approaches the hydrostatic equilibrium in which the pipe is filled with light fluid with $\rho_H(t)$ being approximately $\rho_H^{(i)}$, then $\rho_H^{(i)}(x_{ue}^{(i-1)} - d) = \rho_L(h_{ue}^{(i-1)} - d)$ with $h_{ue}^{(i)} \equiv h_{de}^{(i)} - R(x_{ue}^{(i)} - x_{de}^{(i)})$ should also be satisfied (Figure 4.13b). From these, we find that $\delta x^{(i)}$ is given as

$$\delta x^{(i)} = x_{ue}^{(i-1)} - x_{de}^{(i)} = \frac{\delta \rho^{(i)} d}{\rho_H^{(i)} + \rho_L R}. \quad (4\text{-}20)$$

We now consider the case of $\beta \sigma^{1/2} \ll 1$, which is satisfied when the pipe is sufficiently thin and long (like the case in Figure 4.10). In this case, Eqs. (4-18) and (4-19) can be solved analytically since inertia and the nonlinear terms are neglected (Kano, 2008). Thus, Eqs. (4-18) and (4-19) become

$$\frac{2\sigma^{1/2}}{\beta} \frac{d\tilde{x}^{(i)}}{d\tilde{t}} = -\sigma \left[(1+R)\tilde{x}^{(i)} - 1\right], \quad (4\text{-}21)$$

$$\frac{2\sigma^{1/2}}{\beta} \frac{v_H^{(i)}}{v_L} \frac{d\tilde{x}^{(i)}}{d\tilde{t}} = -\sigma(1+DR)\tilde{x}^{(i)}. \quad (4\text{-}22)$$

The solutions of Eqs. (4-21) and (4-22) are given in dimensional form as follows:

$$x = -C_u e^{-t/\tau_u} + x_{ue}^{(i)}, \quad (4\text{-}23)$$

for upflow, and

$$x = -C_d e^{-t/\tau_d} + x_{de}^{(i)}, \quad (4\text{-}24)$$

for downflow, where C_u and C_d are positive constants, and τ_u and τ_d are given as

$$\tau_u = \frac{8Sd\mu_L}{(1+R)ga^4 \pi \rho_L}, \quad (4\text{-}25)$$

$$\tau_d = \frac{8Sd\mu_H}{(1+DR)ga^4 \pi \rho_H^{(i)}}. \quad (4\text{-}26)$$

Thus, each up- and downflow branch exhibits an exponential response.

Equations (4-23) and (4-24) along with Eqs. (4-25) and (4-26) reproduce the experimental observations fairly well. Taking the case in Figure 4.10 as an example, the time constants of the exponential functions fitted to the up- and downflow branches, found to be 1.99×10^3 s and 1.58×10^3 s from the average of the second to the fifth cycles, are in good agreement with the values obtained from Eqs. (4-25) and (4-26) of 1.97×10^3 s and 1.57×10^3 s, respectively. Moreover, the difference between the asymptotic values of the two adjacent exponential functions is found to be 4.10 mm (averaged over the second to the fifth cycles), which is in good agreement with the $\delta x^{(i)}$ value calculated from Eq. (4-20) of 4.03 mm. Steinbock and colleagues (1998), Kano and Kinoshita (2007), and Kano (2008) also confirmed that the time constants and the difference between the asymptotic values of two adjacent exponential functions are generally in good agreement with the theoretical values even when the pipe length, density difference between the heavy and light fluids, and fluid viscosities are varied.

4.2.3 Phenomenological Model

In the previous section, the dynamical behavior during the up- and downflow branches was theoretically analyzed. However, the oscillatory behavior could not be fully described by this theory because the flow reversal—that is, switching between Eqs. (4.18) and (4.19)—was not considered. To describe the entire oscillatory behavior, the equations for the up- and downflow branches should be properly connected.

Consider the case in which $D \approx 1$ and $\tilde{v}_H = \tilde{v}_L \equiv \tilde{v}$, and assume that $\rho_H^{(i)}$ and $\tilde{x}^{(i)}$ do not depend on the number of cycles i. Then, Eqs. (4-18) and (4-19) are combined into one equation as follows:

$$\frac{d^2 \tilde{x}'}{d\tilde{t}^2} = -A_1 \left(\frac{d\tilde{x}'}{d\tilde{t}}\right) \left|\frac{d\tilde{x}'}{d\tilde{t}}\right| - A_2 \frac{d\tilde{x}'}{d\tilde{t}} - A_3(1+R)\tilde{x}' + \frac{A_3}{2}\mathrm{sgn}\left(\frac{d\tilde{x}'}{d\tilde{t}}\right), \quad (4\text{-}27)$$

where $\tilde{x}' = \tilde{x} - 1/[2(1+R)]$, $A_1 = \sigma/2$, $A_2 = 2\sigma^{1/2}/\beta$, and $A_3 = \sigma$. Because $d\tilde{x}'/d\tilde{t}$, $(d\tilde{x}'/d\tilde{t})|d\tilde{x}'/d\tilde{t}|$, and $\mathrm{sgn}(d\tilde{x}'/d\tilde{t})$ are odd functions of $d\tilde{x}'/d\tilde{t}$, the $d\tilde{x}'/d\tilde{t}$ terms on the right side of Eq. (4-27) are approximated to the third order, and the oscillatory behavior is then described by

$$\frac{d^2 \tilde{x}'}{d\tilde{t}^2} = B_1 \left(\frac{d\tilde{x}'}{d\tilde{t}}\right) - B_2 \left(\frac{d\tilde{x}'}{d\tilde{t}}\right)^3 - A'_3 \tilde{x}', \quad (4\text{-}28)$$

where B_1 and B_2 are positive constants, and $A'_3 = A_3(1+R)$. Equation (4-28) is known as the Rayleigh equation, which was used for describing vibrations in acoustic systems (Rayleigh, 1877).

Figure 4.15 shows a typical example of the temporal evolution of $\tilde{x}'$ obtained by solving Eq. (4-28) numerically. We find that the waveform is quite similar to that obtained from the real-world experiment (see Figures 4.9 and 4.10). The validity of this phenomenological model has also been investigated through a hydrodynamic simulation: Okamura and Yoshikawa (2000) performed a simulation based on the

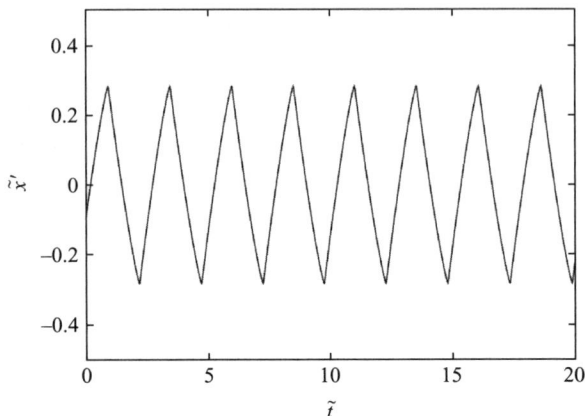

Figure 4.15 Numerical solution of Eq. (4-28). The parameter values are given as follows: $B_1 = 140$, $B_2 = 600$, and $A_3' = 100$.

fundamental equations of fluid dynamics and confirmed that the oscillatory behavior was well approximated by Eq. (4-28).

The reason why Eq. (4-28) describes the oscillatory behavior so well can be understood in the following way. We first rewrite Eqs. (4-27) and (4-28) as

$$\frac{d\tilde{x}'}{d\tilde{t}} = \tilde{y}', \tag{4-29}$$

$$\frac{d\tilde{y}'}{d\tilde{t}} = -A_1 \tilde{y}' |\tilde{y}'| - A_2 \tilde{y}' - A_3(1+R)\tilde{x}' + \frac{A_3}{2}\text{sgn}(\tilde{y}'), \tag{4-30}$$

and

$$\frac{d\tilde{x}'}{d\tilde{t}} = \tilde{y}', \tag{4-31}$$

$$\frac{d\tilde{y}'}{d\tilde{t}} = B_1 \tilde{y}' - B_2 \tilde{y}'^3 - A_3' \tilde{x}'. \tag{4-32}$$

Here, note that $\tilde{y}'$ is generally a fast variable compared to $\tilde{x}'$ because A_1, A_2, and A_3 are usually much larger than unity. Vector fields can be drawn in the $\tilde{y}' - \tilde{x}'$ plane, as shown by the examples in Figure 4.16. In the case of Eqs. (4-29) and (4-30), the system approaches the equilibrium points $(\tilde{y}', \tilde{x}') = (0, \pm 1/[2(1+R)])$ in the phase plane, and thus, oscillation does not occur. In the system described by Eqs. (4-31) and (4-32), however, limit-cycle oscillation occurs: $\tilde{x}'$ slowly increases along the curve $A_3'\tilde{x}' = B_1\tilde{y}' - B_2\tilde{y}'^3$, then at a threshold $\tilde{y}'$ jumps to a lower value, and $\tilde{x}'$ decreases slowly along the same curve. At the other threshold, $\tilde{y}'$ jumps to a larger value, and $\tilde{x}'$ slowly increases again along the same curve. In this way, the self-oscillatory behavior is well described by Eq. (4-28).

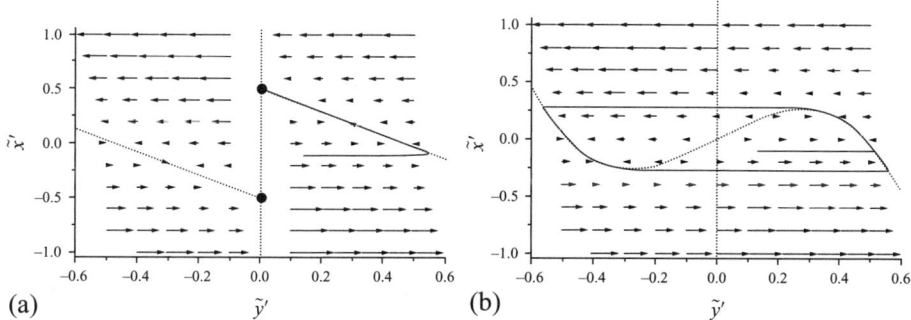

Figure 4.16 Vector fields described by (a) Eqs. (4-29) and (4-30) and (b) Eqs. (4-31) and (4-32). The trajectories with an initial condition of $\tilde{y}' = 0.1$ and $\tilde{x}' = -0.1$ are also shown (solid lines). Dotted lines denote the steady solutions of (a) Eqs. (4-29) and (4-30) and (b) Eqs. (4-31) and (4-32). Whereas the system is attracted to $(\tilde{y}', \tilde{x}') = (0, \pm 1/[2(1+R)])$ (black circles) for any initial condition in the case of (a), the trajectory draws a limit cycle in the case of (b). The parameter values are as follows: $A_1 = 10$, $A_2 = 100$, $A_3 = 100$, $B_1 = 140$, $B_2 = 600$, and $R = 0$.

Equation (4-28) can be easily extended to describe coupled oscillators. If we consider a system consisting of N identical inner containers held within one outer container, then, since the height of the light-fluid surface h is determined by those of the heavy-fluid surfaces in the inner containers, the following relation should be satisfied:

$$h - h_{\text{de}}^{(i)} = -R \sum_{j=1}^{N} \left(x_j^{(i)} - x_{\text{de}}^{(i)} \right), \quad j = 1, 2, \ldots, N \tag{4-33}$$

By using Eq. (4-33) instead of Eq. (4-9), Eqs. (4-11) and (4-12) change to

$$\frac{S}{a^2\pi}\ddot{x}_k + \frac{3S^2}{4da^4\pi^2}\dot{x}_k^2 + \frac{8v_L S}{a^4\pi}\dot{x}_k + \frac{g}{d}\frac{\rho_H(t)}{\rho_L}x_k + \frac{g}{d}R\sum_{j=1}^{N}x_j$$
$$= \frac{g}{d}\left(h_{\text{de}}^{(i)} + RNx_{\text{de}}^{(i)}\right) + \frac{\rho_H(t) - \rho_L}{\rho_L}g, \tag{4-34}$$

$$\frac{S}{a^2\pi}\ddot{x}_k - \frac{3S^2}{4da^4\pi^2}\dot{x}_k^2 + \frac{8v_H(t)S}{a^4\pi}\dot{x}_k + \frac{g}{d}x_k + \frac{\rho_L}{\rho_H(t)}\frac{g}{d}R\sum_{j=1}^{N}x_j$$
$$= \frac{\rho_L}{\rho_H(t)}\frac{g}{d}\left(h_{\text{de}}^{(i)} + RNx_{\text{de}}^{(i)}\right), \tag{4-35}$$

where k is the oscillator number. In nondimensional form, the following equations instead of Eqs. (4-18) and (4-19) are obtained:

$$\frac{d^2\tilde{x}_k^{(i)}}{d\tilde{t}^2} + \frac{2\sigma^{1/2}}{\beta}\left[\frac{d\tilde{x}_k^{(i)}}{d\tilde{t}} + \frac{\beta\sigma^{1/2}}{4}\left(\frac{d\tilde{x}_k^{(i)}}{d\tilde{t}}\right)^2\right] + \sigma\left(\tilde{x}_k^{(i)} - 1\right) + \sigma R\sum_{j=1}^{N}\tilde{x}_j^{(i)} = 0, \tag{4-36}$$

for upflow, and

$$\frac{d^2 \tilde{x}_k^{(i)}}{d\tilde{t}^2} + \frac{2\sigma^{1/2}}{\beta}\left[\frac{v_H^{(i)}}{v_L}\frac{d\tilde{x}_k^{(i)}}{d\tilde{t}} - \frac{\beta\sigma^{1/2}}{4}\left(\frac{d\tilde{x}_k^{(i)}}{d\tilde{t}}\right)^2\right] + \sigma\tilde{x}_k^{(i)} + \sigma DR\sum_{j=1}^{N}\tilde{x}_j^{(i)} = 0, \quad (4\text{-}37)$$

for downflow.

Therefore, by again assuming that $D \approx 1$ and $\tilde{v}_H = \tilde{v}_L \equiv \tilde{v}$, and that $\rho_H^{(i)}$ and $\tilde{x}^{(i)}$ do not depend on the number of cycles i, the following equation is obtained in the same way as Eq. (4-28):

$$\frac{d^2 \tilde{x}_k'}{d\tilde{t}^2} = B_1\left(\frac{d\tilde{x}_k'}{d\tilde{t}}\right) - B_2\left(\frac{d\tilde{x}_k'}{d\tilde{t}}\right)^3 - A_3\tilde{x}_k' - A_3R\sum_{j=1}^{N}\tilde{x}_j', \quad (4\text{-}38)$$

Here, $\tilde{x}_k'$ is defined as $\tilde{x}_k' = \tilde{x}_k - 1/2$. The last term on the right side of Eq. (4-38) originates from the change in the light-fluid surface, through which the oscillators are coupled.

Equation (4-38) has been widely used for describing the behavior of coupled density oscillators. In fact, various types of synchronization phenomena have been reproduced by simulating Eq. (4-38) (or its modifications) (Yoshikawa et al., 1988, 1990, 1991; Nakata et al., 1998; Miyakawa and Yamada, 1999, 2001). Thus, Eq. (4-38) is now the most common description of a coupled density oscillator.

4.3 Fundamental Mechanism of Oscillation

The model described in the previous section (Eq. 4-28) describes the oscillation behavior phenomenologically. However, when we examine it in more detail, we notice that this model does not answer a fundamental question: Why does the density oscillator oscillate? This is because the flow-reversal process is not correctly described but an approximation is made instead to derive Eq. (4-28) from Eq. (4-27); the system exhibits oscillations only because this approximation is employed. Indeed, if Eq. (4-27) is expanded by Tailor series expansion to infinite order rather than approximated by third-order function, then oscillatory behavior will not be observed.

In a general context, the flow-reversal process in density oscillators corresponds to the process of switching between the on and off states in a relaxation oscillator (Figures 4.3 and 4.5). Relaxation oscillators can continue their oscillation owing to this process, otherwise the oscillation stops. Therefore, clarifying the flow-reversal mechanism in a density oscillator is crucial for understanding the essential mechanism of relaxation oscillations.

Martin (1970) considered the flow reversal to occur due to the Rayleigh–Taylor instability (Taylor, 1950), in which a perturbation at a static interface between fluids grows when a heavy fluid is located above a light fluid. However, the dynamical process of flow reversal cannot be understood under these terms because the spatiotemporal dynamics during the reversal are extremely complex. For instance, in changing from down- to upflow, the reversal is initiated by an intrusion of light fluid along the inner wall of the pipe. After some time, the intrusion begins to grow rapidly, climbs to the upper end of the pipe, and then the flow reverses completely (Steinbock et al., 1998; Kano and Kinoshita, 2007, 2008; Kano, 2008). Thus, the mechanism is not trivial and clarifying it is a formidable but challenging task.

4.3.1 Hydrodynamic Analysis of Flow Reversal

Steinbock and colleagues (1998) approached this problem from a hydrodynamic viewpoint. They analyzed the stability of downflow inside the pipe and derived the critical height for the flow instability. Since the analysis of 3D flow is difficult, they considered a pipe with a 2D rectangular geometry as shown in Figure 4.17. Here, y and z denote the horizontal and vertical coordinates, respectively, and the pipe length d is assumed to be much larger than the diameter of $2a$. The aim of the analysis is to determine the condition for which the downflow loses stability when an intrusion of light fluid exists steadily at $y=z=0$. Since the analysis is somewhat complicated, readers interested in only the result can skip to Eq. (4-51).

In their analysis, the heavy and light fluids are described by a one-fluid model where the density of the fluid is given by $\rho = \rho_H - \delta\rho\, \Theta$, with Θ being a function of space and time; thus, the conditions $\Theta = 0$ and 1 correspond to the densities of the heavy and light fluid, ρ_H and ρ_L, respectively. Using the steady-state and Boussinesq approximations (the latter is an approximation to neglect the effects of density change except for the buoyancy term), the Navier–Stokes equations are described as

$$w\partial_z w + v\partial_y w = -\frac{1}{\rho_H}\partial_z P + v_H\left(\partial_z^2 w + \partial_y^2 w\right) - g + \frac{\delta\rho g}{\rho_H}\Theta, \tag{4-39}$$

$$w\partial_z v + v\partial_y v = -\frac{1}{\rho_H}\partial_y P + v_H\left(\partial_z^2 v + \partial_y^2 v\right), \tag{4-40}$$

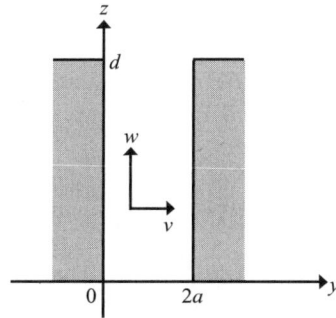

Figure 4.17 Two-dimensional geometry for the rectangular pipe considered in the hydrodynamic analysis of flow reversal. The y and z components of the flow velocity are denoted by v and w, respectively.

where P is the hydrostatic pressure, and v and w are the y and z components of the velocity of flow, respectively, and ∂_z and ∂_y are abbreviations for $\partial/\partial z$ and $\partial/\partial y$, respectively. The left sides of Eqs. (4-39) and (4-40) are the convection terms. The first and second terms on the right sides are the hydrostatic pressure gradient and viscous drag terms, respectively, while the third and fourth terms on the right side in Eq. (4-39) are the gravitational and buoyancy terms, respectively. The continuity condition is given as

$$\partial_z w + \partial_y v = 0, \tag{4-41}$$

and the diffusion equation can be expressed as

$$w\partial_z \Theta + v\partial_y \Theta = D_f \left(\partial_z^2 \Theta + \partial_y^2 \Theta \right), \tag{4-42}$$

where D_f is a diffusion constant. Equations (4-39)–(4-42) are written in the nondimensional form as follows:

$$R_e \varepsilon' \left(\bar{w} \partial_{\bar{z}} \bar{w} + \bar{v} \partial_{\bar{y}} \bar{w} \right) = -\partial_{\bar{z}} \bar{P} + \varepsilon' \partial_{\bar{z}}^2 \bar{w} + \partial_{\bar{y}}^2 \bar{w} - \kappa + \lambda \Theta, \tag{4-43}$$

$$R_e \varepsilon'^3 \left(\bar{w} \partial_{\bar{z}} \bar{v} + \bar{v} \partial_{\bar{y}} \bar{v} \right) = -\partial_{\bar{y}} \bar{P} + \varepsilon'^4 \partial_{\bar{z}}^2 \bar{v} + \partial_{\bar{y}}^2 \bar{v}, \tag{4-44}$$

$$\partial_{\bar{z}} \bar{w} + \partial_{\bar{y}} \bar{v} = 0, \tag{4-45}$$

$$\varepsilon' \left(\bar{w} \partial_{\bar{z}} \Theta + \bar{v} \partial_{\bar{y}} \Theta \right) = -\frac{D_f}{Q} \left(\varepsilon'^2 \partial_{\bar{z}}^2 \Theta + \partial_{\bar{y}}^2 \Theta \right), \tag{4-46}$$

where $\bar{z} = z/d$, $\bar{y} = y/(2a)$, $\bar{w} = 2wa/Q$, $\bar{v} = vd/Q$, $\bar{P} = 8Pa^3/(\rho_H d v_H Q)$, $R_e = Q/v_H$, $\kappa = 8a^3 g/(v_H Q)$, $\lambda = 8a^3 g \delta\rho/(v_H Q \rho_H)$, and $\varepsilon' = 2a/d \ll 1$. The flow rate through the pipe is denoted by Q. Because the flow inside the pipe is regarded to be nearly 2D Poiseuille flow (Schlichting, 1960; see also Section A4.2 in the Appendix), Q is derived as

$$Q = \int_0^{2a} w\,dy = -\frac{2a^3}{3\mu_H} \left[\frac{\rho_H g(x-d) - \rho_L g h}{d} + \rho_H g \right]. \tag{4-47}$$

Here, we have assumed that the hydrostatic pressure gradient is simply derived from the heights of the fluid surfaces when deriving Eq. (4-47). Let a stream function $\Psi(\bar{y}, \bar{z})$ be defined so that both $\bar{w} = \partial_{\bar{y}} \Psi(\bar{y}, \bar{z})$ and $\bar{v} = -\partial_{\bar{z}} \Psi(\bar{y}, \bar{z})$ are satisfied; thus, Eq. (4-45) is automatically fulfilled. The boundary conditions are given by using Ψ as $\partial_{\bar{y}} \Psi = 0$ (no slip) and $-\partial_{\bar{z}} \Psi = 0$ (no penetration of the vertical walls) at $\bar{y} = 0$ and 1. If $\Psi = 0$ at $\bar{y} = 0$ is assumed, $\Psi|_{\bar{y}=1} = \int_0^1 \bar{w}\,d\bar{y} = \int_0^{2a} (w/Q)\,dy = 1$ should be satisfied. In addition, since the heavy fluid flows downward while an intrusion of the light fluid exists steadily at $y=z=0$, the boundary conditions for Θ are given as $\Theta = 1$ at $\bar{z} = \bar{y} = 0$ and $\Theta = 0$ at $\bar{y} = 1$. Using these conditions, Eqs. (4-43), (4-44), and (4-46) are solved for the zeroth order of ε' as follows:

$$\Theta_0(\bar{y},\bar{z}) = \theta_0(\bar{z})(1-\bar{y}), \tag{4-48}$$

$$\Psi_0(\bar{y},\bar{z}) = 3\bar{y}^2 - 2\bar{y}^3 + \frac{\lambda}{24}\theta_0(\bar{z})(1-\bar{y}^2)\bar{y}^2, \tag{4-49}$$

where $\theta_0(0)=1$ should be satisfied.

The downflow loses its stability when $\partial_{\bar{y}}\bar{w}|_{\bar{y}=\bar{z}=0} = \partial_{\bar{y}}^2\Psi|_{\bar{y}=\bar{z}=0} > 0$. This condition is derived from Eq. (4-49) to be $\lambda > 72$, which, by using Eqs. (4-9), (4-15), (4-20), and (4-47), is found to be equivalent to

$$x - x_{de} < \frac{1}{6}\delta x. \tag{4-50}$$

Thus, the critical height for the flow instability $x_{d \to u}$ is derived as

$$x_{d \to u} = x_{de} + \frac{1}{6}\delta x. \tag{4-51}$$

Equation (4-51) suggests that the critical height depends only on the heights of the fluid surfaces at the two hydrostatic equilibria.

This theoretical result agrees with experimental findings in most cases. For example, in the case in Figure 4.10, $(x_{d \to u} - x_{de}^{(i)})/\delta x^{(i)}$ is found to be 0.19 from the average of the second to the fifth cycles, which is close to the theoretical value of 1/6. Steinbock and colleagues (1998) confirmed that the theoretical result was in good agreement with their experiments in which the density of heavy fluid and the pipe length were varied.

4.3.2 Viscosity-Dependent Flow Reversal

Although the theory of Steinbock and colleagues (1998) sufficiently explained their experimental results, several unrealistic assumptions were made in their analysis: they did not take into account the effect of the flow after passing through the pipe, or the temporal evolution of the dynamical behavior during the flow-reversal process (a consequence of the steady-state approximation). Thus, further investigations are necessary to truly understand the mechanism of the flow reversal.

It is quite difficult to solve the flow-reversal process analytically, because the analysis of the flow after it has passed through the pipe is a difficult problem for hydrodynamicists. Thus, it would be ideal to construct a simple model that contains the essential flow-reversal mechanism. To construct such a model, it is important to perform accurate experiments by changing certain key parameters and then evaluating the results quantitatively. From a hydrodynamics point of view, the fluid dynamics of a system consisting of two viscous fluids is an intriguing problem (Saffman and Taylor, 1958; Menikoff et al., 1977). Interestingly, the dynamics are known to be extremely sensitive to the viscosities of the fluids. One well-known example is the Saffman–Taylor instability, where an interface between two fluids tends to become unstable when the less viscous fluid is forced into the more viscous fluid in a porous medium or a Hele-show cell, while it remains stable in the opposite case (Saffman and Taylor, 1958). Thus, it is reasonable to expect that even in a density oscillator, the dynamical behavior inside the connecting pipe will be largely affected by the viscosities of the fluids. Accordingly, we investigated the flow-reversal process by focusing mainly on the viscosities of the fluids (Kano and Kinoshita, 2007; Kano, 2008).

The experimental setup is shown in Figure 4.18. The density oscillator employed here consisted of inner and outer containers with surface areas of 7.7×10^{-4} m^2 and 5.34×10^{-2} m^2 and heights of 0.15 m and 0.32 m, respectively. The glass pipe, with an internal diameter of 0.73 mm and a length of 70 mm, was fixed in the bottom of the inner container. The inner container was filled with the heavy fluid and the outer container with the light fluid. Both fluids were mixtures of water, 1-propanol, and glycerin. By changing their compositions, the viscosities of the fluids can be varied while maintaining their densities, as shown in Figure 4.19. The densities of the heavy and light fluids were set to be constant at $(1.057 \pm 0.003) \times 10^3$ kg·m^{-3} and $(0.996 \pm 0.003) \times 10^3$ kg·m^{-3}, respectively, and therefore the density difference was $(0.061 \pm 0.004) \times 10^3$ kg·m^{-3}. The height of the heavy-fluid surface was monitored by a laser displacement meter, while the lower part of the glass pipe was simultaneously observed by a stereomicroscope equipped with a digital camera. The interface between the two fluids was observed clearly owing to the difference in the refractive indices. The temperature of the experimental system was maintained at 25 ± 0.5 °C.

Figure 4.20(a) shows the temporal evolution of the height of the fluid surface in the case where the viscosities are almost identical, while Figures 4.20(b) and (c) respectively show those in the cases where the viscosity of the heavy fluid is much larger

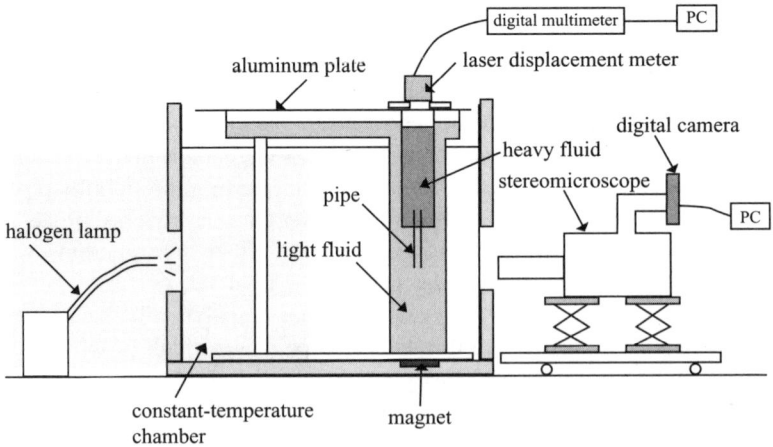

Figure 4.18 Overview of the experimental setup.

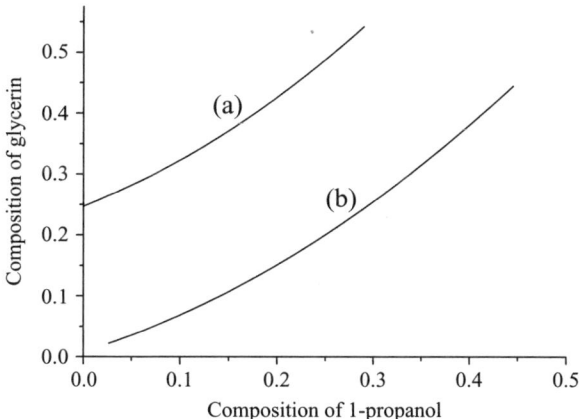

Figure 4.19 The composition curve of 1-propanol, glycerin, and water to give a constant density of (a) 1.057×10^3 kg·m^{-3} and (b) 0.996×10^3 kg·m^{-3}. The horizontal and vertical axes denote the ratio of the weight of 1-propanol and glycerin to that of the summation of the three materials, respectively.

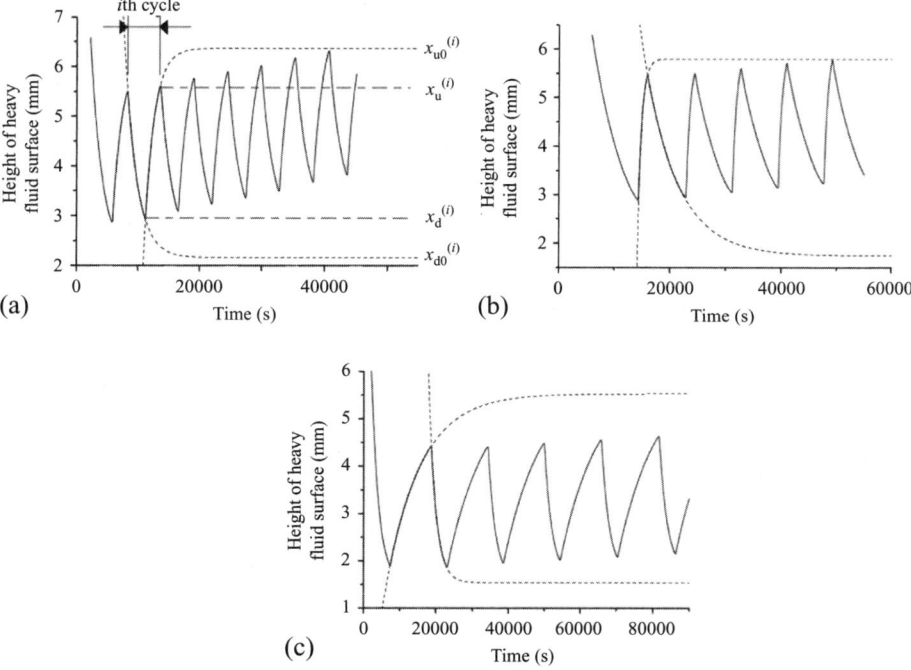

Figure 4.20 Temporal evolution of the height of the heavy-fluid surface in the case of (a) $\mu_H = 2.64 \times 10^{-3}$ Pa·s and $\mu_L = 1.98 \times 10^{-3}$ Pa·s, (b) $\mu_H = 8.59 \times 10^{-3}$ Pa·s and $\mu_L = 0.89 \times 10^{-3}$ Pa·s, and (c) $\mu_H = 2.66 \times 10^{-3}$ Pa·s and $\mu_L = 14.18 \times 10^{-3}$ Pa·s. Each branch of the up- and downflow is well fitted by an exponential curve (dashed line). In (a), definitions of the parameters for the ith cycle are also shown. In the present graph, definitions are shown for $i = 2$. The experimental conditions in (a) are the same as those in Figure 4.10.

than that of the light fluid and vice versa. As expected, the time constant of the exponential curve fitted to each up- and downflow branch varies depending on the viscosities of the fluids. Figure 4.21 shows the viscosity dependence of the time constants. The experimental results are in fairly good agreement with the theoretical values obtained from Eqs. (4-25) and (4-26), although there is a small systematic deviation at high viscosity, which is thought to be caused by the reduction in the viscosity due to the inflow of the low-viscosity fluid. Figure 4.22 shows the viscosity dependence of the ratio of the difference between the asymptotic values for the adjacent exponential curves Δ to its theoretical value δx. We find that $\Delta/\delta x$ remains close to unity regardless of the fluid viscosity. Thus, even when the viscosity of either fluid is varied, each up- and downflow branch is well described by Eqs. (4-23)–(4-26).

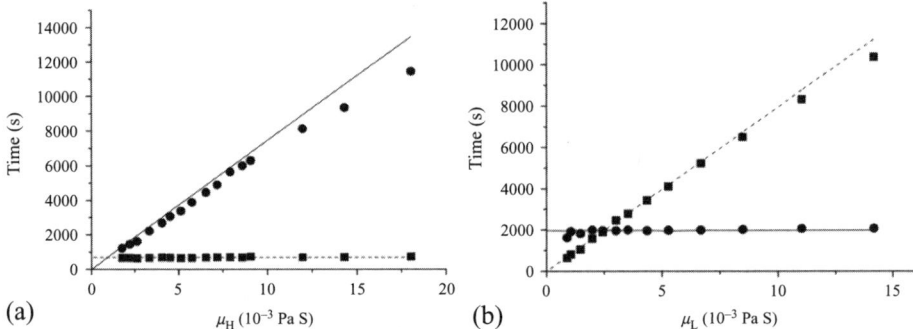

Figure 4.21 Time constants of the exponential curves fitted for each down- and upflow cycle (filled circles and squares, respectively). Each data point is averaged over four cycles from the second to the fifth. In (a), μ_H is varied while μ_L is fixed at 0.89×10^{-3} Pa·s. In (b), μ_L is varied while μ_H is fixed at $(2.63 \pm 0.03) \times 10^{-3}$ Pa·s. The solid and dashed lines indicate the theoretical values for τ_d and τ_u, respectively.

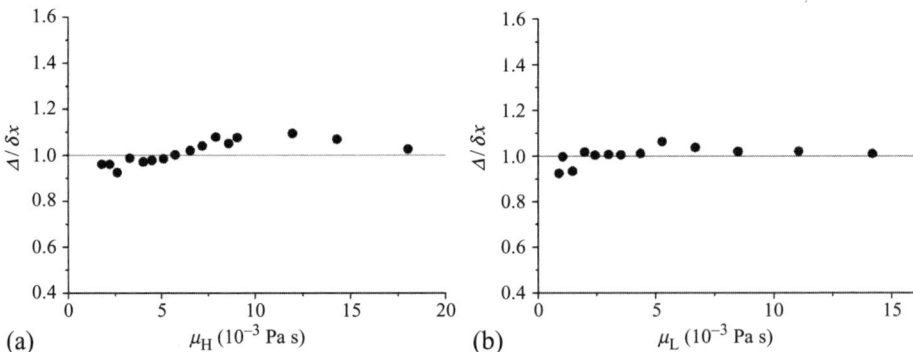

Figure 4.22 Ratio of the difference between the asymptotic values for the exponential curves Δ to the theoretical value δx. Each data point is averaged over four cycles from the second to the fifth. In (a), μ_H is varied while μ_L is fixed at 0.89×10^{-3} Pa·s. In (b), μ_L is varied while μ_H is fixed at $(2.63 \pm 0.03) \times 10^{-3}$ Pa·s.

However, when we compare Figures 4.20(b) and (c), we notice that there is a drastic change in the timing of the flow reversal. In Figure 4.20(b), the flow reverses from down- to upflow when the system is still far from the hydrostatic equilibrium in which the pipe is filled with heavy fluid, while the flow reversal from up- to downflow does not occur until the system is extremely close to the hydrostatic equilibrium. The opposite tendency is observed in Figure 4.20(c).

To evaluate this result quantitatively, we define the following two parameters, s_d and s_u, as

$$s_d = \langle \frac{x_d^{(i)} - x_{d0}^{(i)}}{x_{u0}^{(i-1)} - x_{d0}^{(i)}} \rangle, \tag{4-52}$$

$$s_u = \langle \frac{x_u^{(i-1)} - x_{d0}^{(i)}}{x_{u0}^{(i-1)} - x_{d0}^{(i)}} \rangle, \tag{4-53}$$

where $x_d^{(i)}$ and $x_u^{(i)}$ are the heights of the heavy-fluid surface at the moments when the flow reverses from the i th downflow to the ith upflow and from the i th upflow to the $(i+1)$th downflow, respectively; and $x_{d0}^{(i)}$ and $x_{u0}^{(i)}$ are the asymptotic values of the exponential curves fitted to the i th down- and upflow, respectively (Figure 4.20a). In the present analysis, we averaged the data over four cycles, from the second to the fifth, in a series of experiments. Both s_d and s_u are in fact suitable parameters for quantifying the timing of the flow reversal over the entire process leading to asymptotic equilibrium because $x_{d0}^{(i)}$ and $x_{u0}^{(i-1)}$ are thought to be almost consistent with the hydrostatic equilibria in which the pipe is filled with heavy or light fluid, $x_{de}^{(i)}$ and $x_{ue}^{(i-1)}$, respectively. Figure 4.23 shows the viscosity dependence of s_d and s_u

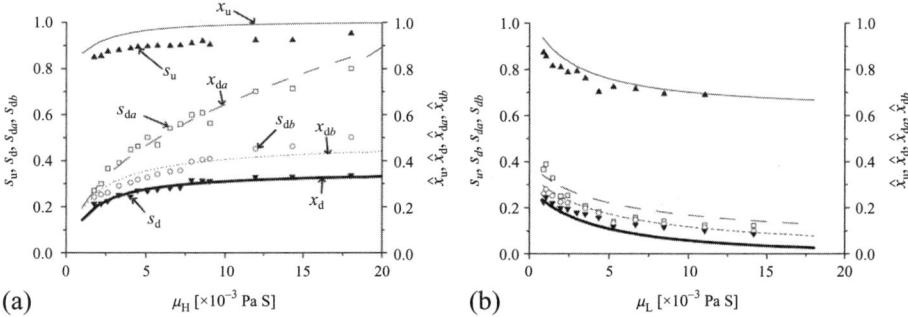

Figure 4.23 Viscosity dependence of s_u (filled triangles), s_d (filled inverted triangles), s_{da} (open squares), and s_{db} (open circles). The data are averaged over four cycles from the second to the fifth. Simulated results for $\hat{x}_u$ (solid line), $\hat{x}_d$ (bold line), $\hat{x}_{da}$ (dashed line), and $\hat{x}_{db}$ (dotted line) are also shown. In (a), μ_H is varied while μ_L is fixed at 0.89×10^{-3} Pa·s. In (b), μ_L is varied while μ_H is fixed at $(2.63 \pm 0.03) \times 10^{-3}$ Pa·s. The parameter values used in the simulation are as follows: $a = 0.365$ mm, $d = 70$ mm, $\rho_H = 1.057 \times 10^3$ kg·m^{-3}, $\rho_L = 0.996 \times 10^3$ kg·m^{-3}, $S = 7.7 \times 10^{-4}$ m^2, $R = 1.44 \times 10^{-4}$, $C_1 = 7.11$, $C_2 = 5.31$, $C_3 = 0.32$, $\alpha = 1.82$, and $\gamma = 0.3$.

when either viscosity is varied. It is clear in the figure that both s_d and s_u increase when the viscosity of the heavy fluid increases (Figure 4.23a), but decrease when that of the light fluid increases (Figure 4.23b). These results differ from the theoretical results of Steinbock and colleagues (1998), in which the critical heights do not depend on the viscosity (Eq. 4-51).

Thus, the flow-reversal process is significantly affected by the fluid viscosity. To investigate this phenomenon in greater detail, we observed the flow reversal from down- to upflow using the stereomicroscope. Snapshots of the process are shown in Figure 4.24. Initially, the light fluid is observed to intrude into the pipe long before the actual moment of flow reversal, and it grows upward rather slowly (Figure 4.24b). When the intrusion length measured from the bottom of the pipe is almost 1 mm, the intrusion suddenly starts to grow rapidly (Figures 4.24c and d), and finally the flow reverses. It is surprising that the temporal evolution of the effective flow volume through the pipe and therefore that of the height of the heavy-fluid surface still obeys an exponential response during the gradual growth of the intrusion. Even after the intrusion starts to grow rapidly, the deviation from an exponential curve is negligibly small compared to the oscillation amplitude (Figure 4.10b).

Figure 4.25 shows the temporal evolution of the intrusion length when the viscosity of either fluid is varied. Here, we set the origin of the time to be at the very moment the flow reverses and plotted the intrusion length against the time leading to the flow reversal. For a large μ_H and small μ_L, the intrusion of the light fluid persists for an extremely long time before the flow reverses, and the growth rate during the rapid-growing process is also relatively low (Figure 4.25a). In contrast, the growth rate of the intrusion does not depend on μ_L to any significant extent (Figure 4.25b).

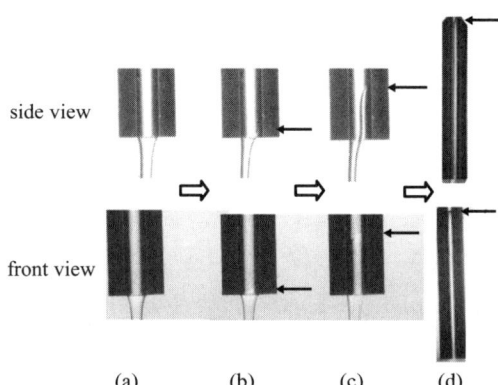

Figure 4.24 Photographs of the lower part of the pipe during the time progress of downflow, (a) → (b) → (c) → (d), taken by the digital camera connected to the stereomicroscope. Upper and lower photographs show the side and front views, respectively. An intrusion of light fluid is seen clearly (indicated by black arrows). In this experiment, glucose solution and water were used as the heavy and light fluids, respectively, with $\rho_H = 1.064 \times 10^3$ kg·m^{-3}, $\rho_L = 0.997 \times 10^3$ kg·m^{-3}, $\mu_H = 1.42 \times 10^{-3}$ Pa·s, and $\mu_L = 0.89 \times 10^{-3}$ Pa·s. Although the conditions are slightly different from those in the other experiments, this difference is not crucial, as the observed behavior is qualitatively identical. For interpretation of the references to color in this figure legend, the reader is referred to the online version of this chapter.

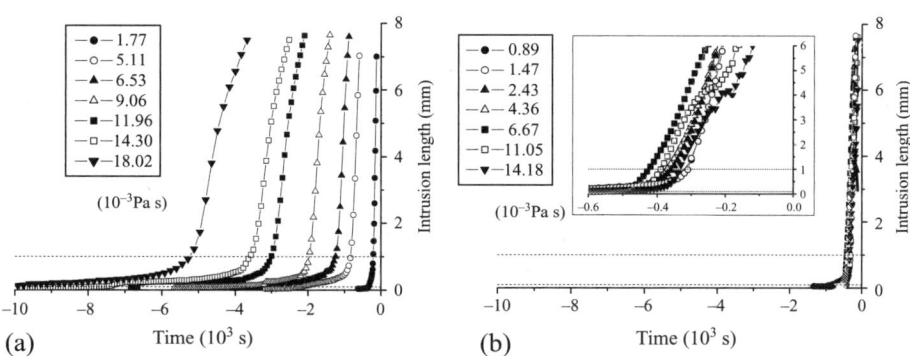

Figure 4.25 Viscosity-dependent temporal evolution of the intrusion of light fluid. In (a), μ_H is varied while μ_L is fixed at 0.89×10^{-3} Pa·s. In (b), μ_L is varied while μ_H is fixed at $(2.63 \pm 0.03) \times 10^{-3}$ Pa·s. Because the visual field under the stereomicroscope is limited, an intrusion length of more than 8 mm is not observed. The dotted lines indicate the intrusion lengths of 0.1 mm and 1 mm. The inset in (b) shows an enlarged view.

We characterize the viscosity-dependent behavior of the intrusion by defining the following two quantities:

$$s_{da} = \langle \frac{x_{da}^{(i)} - x_{d0}^{(i)}}{x_{u0}^{(i-1)} - x_{d0}^{(i)}} \rangle, \tag{4-54}$$

$$s_{db} = \langle \frac{x_{db}^{(i)} - x_{d0}^{(i)}}{x_{u0}^{(i-1)} - x_{d0}^{(i)}} \rangle, \tag{4-55}$$

where $x_{da}^{(i)}$ and $x_{db}^{(i)}$ are the heights of the heavy-fluid surface at the time when the intrusion length exceeds 0.1 mm and 1 mm, respectively. Because the intrusion begins to grow rapidly when its length is almost 1 mm, as shown in Figure 4.25, the two quantities s_{da} and s_{db} roughly characterize the timings for the beginning of the intrusion and the onset of its rapid growth, respectively. With increasing μ_H, significant increases in both s_{da} and s_{db} are observed, which are more remarkable than those of s_d (Figure 4.23a). Conversely, with increasing μ_L, both s_{da} and s_{db} decrease in a similar manner to s_d (Figure 4.23b). Thus, these newly introduced quantities characterize the dynamics of the flow reversal fairly accurately and clarify definitively the viscosity dependence of the density oscillations.

4.3.3 Model Including Flow-Reversal Process

To explain the viscosity-dependent flow-reversal mechanism, we proposed a model for the flow reversal from down- to upflow (Kano and Kinoshita, 2007, 2008, 2009; Kano, 2008). Figure 4.26 shows a schematic illustration of the model. We modeled the flow around the pipe by assuming that the interface between the fluid flowing

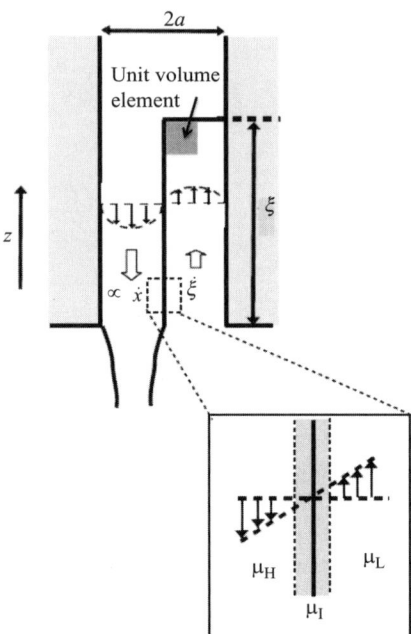

Figure 4.26 Schematic illustration of the flow-reversal model. Forces acting on the unit volume element at the tip of the intrusion (filled gray square) are considered. Velocity profiles of the downflow and the intrusion are also shown. The lower illustration shows a magnified view at the interface between the two fluids. Although an actual interface between the two fluids does not exist, an extremely thin volume element at the boundary between the two fluids is regarded as the interface (gray region), of which the viscosity is denoted by μ_I.

downward and the intrusion fluid is completely parallel to the pipe axis except at the tip of the intrusion, which is parallel to the cross-section of the pipe. We focused on a unit volume element at the tip of the intrusion and described the dynamics by considering the forces acting on this volume. We considered the case of $\beta\sigma^{1/2} \ll 1$— that is, the case where the pipe is sufficiently thin and long—and we assumed that the forces acting on the volume element are parallel to the pipe axis. Furthermore, for simplicity, we assumed that the density of the heavy fluid $\rho_H(t)$ is reset to a fixed value at the beginning of each downflow branch.

We consider the forces concerned with this phenomenon separately. First, there should be a viscous drag force F_1 that acts on the interface between the heavy and light fluids. Although a definite interface does not exist, since the viscosity of the fluid will vary continuously in space because of the miscible nature of fluids, we consider, for convenience, that the mixed region consists of an extremely thin volume element at an interface with a fluid having an appropriate viscosity (inset in Figure 4.26). Since the viscosity of the fluid at the interface μ_I may depend on the viscosities of the two fluids, we write $\mu_I = (\mu_H + \mu_L)/2$ for simplicity. The drag force F_1 should depend both on the viscosity of the fluid and on the velocity gradient at the interface; the latter is proportional to the difference between the downflow velocity and the growth rate of the intrusion. Thus, F_1 is described as

$$F_1 = \mu_1 \left(b_1 \dot{\hat{x}} - b_2 \dot{\hat{\xi}} \right), \tag{4-56}$$

where b_1 and b_2 are positive constants, and $\hat{\xi} \equiv \xi/a$ denotes the nondimensionalized intrusion length with ξ being the intrusion length measured from the bottom of the pipe. We have defined $\hat{x}$ using x_{ue} and x_{de}, the heights of the heavy fluid surface at the two hydrostatic equilibria, as

$$\hat{x} = \frac{x - x_{de}}{x_{ue} - x_{de}}, \tag{4-57}$$

(Figure 4.27), and thus, its derivative characterizes the downflow velocity. Note that the intrusion must be resisted not only by the viscous drag at the interface between the fluids but also by that on the opposite side of the unit volume element. Since we have taken the latter effect into account in the second term on the right side of Eq. (4-56), b_1 and b_2 are considered to be independent parameters. Here, we find from Eqs. (4-24) and (4-26) that $\dot{\hat{x}}$ is proportional to $-\hat{x}/\mu_H$ under the assumption that Eq. (4-24) holds until the very moment the flow reverses from down- to upflow. This assumption is valid because the deviation from an exponential curve of the height of the heavy-fluid

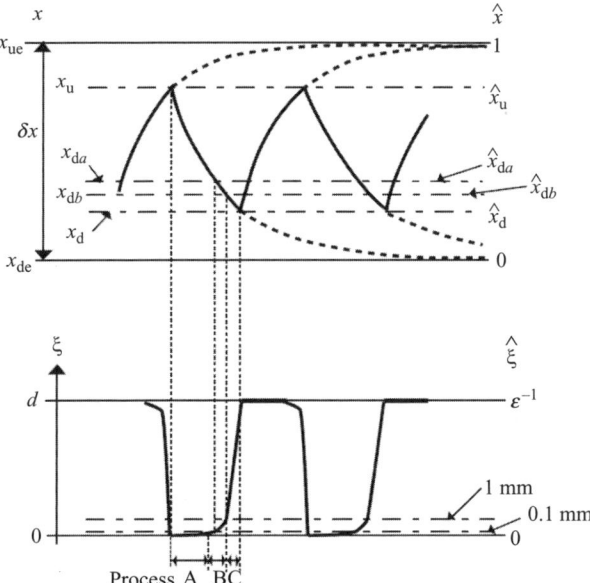

Figure 4.27 Definitions of the parameters used in the simulation. The parameters with dimensions are shown on the left, while nondimensionalized parameters are shown on the right. The upper and lower figures represent the temporal evolutions of the heavy-fluid surface and the tip of the intrusion, respectively.

surface is sufficiently small compared to the amplitude of the oscillation (Figure 4.10b). Thus, Eq. (4-56) is rewritten as

$$F_1 = -\frac{c\mu_I \hat{x}}{\mu_H} - b_2 \mu_1 \dot{\hat{\xi}}, \qquad (4\text{-}58)$$

where c is a positive constant.

Second, there must be a gravitational force and a force due to the gradient of the hydrostatic pressure. We denote these hydrostatic forces as F_2. When the intrusion exists to a certain extent within the interior of the pipe, F_2 will satisfy the relation

$$F_2 = -\frac{P_u - P_d}{d} - \rho_L g, \qquad (4\text{-}59)$$

where P_u and P_d respectively denote the hydrostatic pressures at the upper and lower ends of the pipe, and the gradient of the hydrostatic pressure inside the pipe is assumed to be homogeneous. If we then assume that P_u and P_d are simply derived from the heights of the fluid surfaces as $(\rho_L + \delta\rho)g(x - d)$ and $\rho_L g h$, respectively, we find from Eqs. (4-9), (4-15), (4-20), and (4-57) that $F_2 = \delta\rho g(1 - \hat{x})$. If the volume element of the light fluid exists outside the pipe, then the hydrostatic pressure gradient balances the gravitational force. Therefore, when there is no intrusion ($\hat{\xi} = 0$), the relation $F_2 = 0$ should be automatically satisfied. Since F_2 is thought to be a continuous function of ξ, it seems appropriate to describe F_2 in the following way:

$$F_2 = \delta\rho g(1 - \hat{x})\left(1 - e^{-\hat{\xi}/\alpha}\right). \qquad (4\text{-}60)$$

The term $(1 - \exp[-\hat{\xi}/\alpha])$ is introduced to connect $F_2 = \delta\rho g(1 - \hat{x})$ for $\hat{\xi} \gg \alpha$ to $F_2 = 0$ for $\hat{\xi} = 0$. Here, α characterizes the spatial range for which F_2 takes a value between 0 and $\delta\rho g(1 - \hat{x})$.

Third, we consider the effect of acceleration of the fluid outside the pipe. Let us consider arbitrarily two planes inside and outside the pipe as PL$_1$ and PL$_2$, as shown in Figure 4.28. We define the absolute values of the mean flow velocity at PL$_1$ and PL$_2$ as V_1 and V_2, respectively; thus, $V_1 \propto -\dot{\hat{x}} \propto \hat{x}/\mu_H$ is obtained from Eqs. (4-24) and (4-26). Since the heavy fluid that has passed through the pipe is accelerated due to the hydrostatic pressure gradient, the following relation is expected to hold:

$$V_2 = V_1 + \zeta, \qquad (4\text{-}61)$$

where ζ expresses the effect of the acceleration. The continuity condition naturally holds, which results in

$$V_1 \pi a^2 = V_2 S', \qquad (4\text{-}62)$$

where S' denotes the cross-section of the downflow at PL$_2$. When S' is sufficiently small, it is expected that the intrusion of the light fluid is enforced because the contraction of the flow causes a detachment of the downflow from the pipe wall. We define F_3 as the force due to this detachment. By substituting $V_1 \propto \hat{x}/\mu_H$ into Eqs. (4-61) and (4-62), we find that $(\pi a^2/S' - 1) \propto \mu_H/\hat{x}$. Since F_3 increases as S'

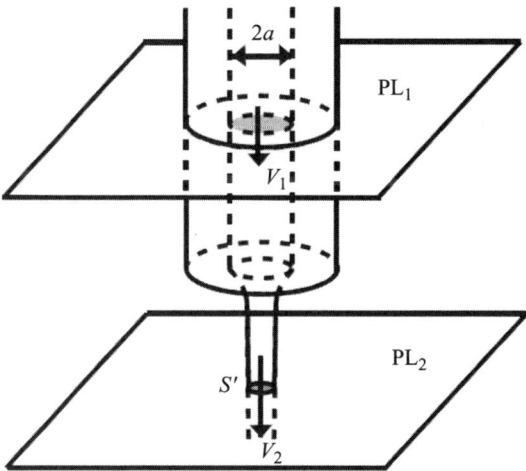

Figure 4.28 Schematic illustration of two planes, PL_1 and PL_2. Plane PL_1 intersects the pipe, while PL_2 is below but not too far from the lower end of the pipe. The area S' is the surface area of downflow at PL_2. The absolute values of the mean flow velocity at PL_1 and PL_2 are denoted by V_1 and V_2, respectively.

decreases, F_3 should increase as $\mu_H/\hat{x}$ increases. In addition, F_3 is also expected to be large when $\hat{\xi}$ is small. Therefore, F_3 is phenomenologically described as

$$F_3 = k e^{-\hat{\xi}/\gamma} \frac{\mu_H}{\hat{x}}, \tag{4-63}$$

where k is a positive constant, and γ expresses the spatial range over which the force works effectively.

From Eqs. (4-58), (4-60), and (4-63), we can describe the total force F acting on the unit volume element located at the tip of the intrusion as

$$\begin{aligned} F &= F_1 + F_2 + F_3 \\ &= -\frac{c\mu_I}{\mu_H}\hat{x} - b_2\mu_I\dot{\hat{\xi}} + \delta\rho g(1-\hat{x})\left(1 - e^{-\hat{\xi}/\alpha}\right) + k e^{-\hat{\xi}/\gamma}\frac{\mu_H}{\hat{x}}. \end{aligned} \tag{4-64}$$

Thus, the equation of motion for the unit volume element, $\rho_L \ddot{\xi} = F$, is rewritten by using Eq. (4-64) as

$$\rho_L a \ddot{\hat{\xi}} = -\frac{c\mu_I}{\mu_H}\hat{x} - b_2\mu_I\dot{\hat{\xi}} + \delta\rho g(1-\hat{x})\left(1 - e^{-\hat{\xi}/\alpha}\right) + k e^{-\hat{\xi}/\gamma}\frac{\mu_H}{\hat{x}}, \tag{4-65}$$

where $\hat{x}$ satisfies the following relation:

$$\dot{\hat{x}} = -\frac{\hat{x}}{\tau_d}. \tag{4-66}$$

Equations (4-65) and (4-66) describe the dynamical behavior of the density oscillator during downflow. Similarly, its dynamical behavior during upflow is described as

$$\rho_H a \ddot{\hat{\xi}} = \frac{c\mu_I}{\mu_L}(1-\psi\hat{x}) - b_2\mu_I\dot{\hat{\xi}} - \delta\rho g\psi\hat{x}\left(1-e^{-(\varepsilon^{-1}-\hat{\xi})/\alpha}\right) \\ - ke^{-(\varepsilon^{-1}-\hat{\xi})/\gamma}\frac{\mu_L}{1-\psi\hat{x}}, \qquad (4\text{-}67)$$

$$\dot{\hat{x}} = -\frac{\psi^{-1} - \hat{x}}{\tau_u}, \qquad (4\text{-}68)$$

where $\psi = D(1+R)/(1+DR)$ and $\varepsilon = a/d$ (Kano, 2008; Kano and Kinoshita, 2009). Here, $\hat{\xi}$ is defined as the nondimensional distance between the bottom of the pipe and the tip of the intrusion fluid.

Equations (4-65)–(4-68) can be expressed simply using the nondimensionalized variables. By neglecting the inertia terms through order estimation (Kano, 2008; Kano and Kinoshita, 2009), the following equations are obtained:

$$\phi^{-1}\frac{d\hat{x}}{d\hat{t}} = -\frac{\hat{x}}{\hat{\mu}_H}, \qquad (4\text{-}69)$$

$$\varepsilon\frac{d\hat{\xi}}{d\hat{t}} = -C_1\frac{\hat{x}}{\hat{\mu}_H} + C_2\frac{1-\hat{x}}{\hat{\mu}_I}\left(1-e^{-\hat{\xi}/\alpha}\right) + C_3\frac{\hat{\mu}_H}{\hat{\mu}_I}\frac{e^{-\hat{\xi}/\gamma}}{\hat{x}} \left(\equiv f_d(\hat{\xi})\right), \qquad (4\text{-}70)$$

for the downflow, and

$$\phi^{-1}\frac{d\hat{x}}{d\hat{t}} = -\frac{1-\psi\hat{x}}{\hat{\mu}_L}, \qquad (4\text{-}71)$$

$$\varepsilon\frac{d\hat{\xi}}{d\hat{t}} = C_1\frac{1-\psi\hat{x}}{\hat{\mu}_L} - C_2\frac{\psi\hat{x}}{\hat{\mu}_I}\left(1-e^{-(\varepsilon^{-1}-\hat{\xi})/\alpha}\right) - C_3\frac{\hat{\mu}_L}{\hat{\mu}_I}\frac{e^{-(\varepsilon^{-1}-\hat{\xi})/\gamma}}{1-\psi\hat{x}} \left(\equiv f_u(\hat{\xi})\right), \qquad (4\text{-}72)$$

for the upflow, where $\hat{\mu}_I = \mu_I/\mu_w$, $\hat{\mu}_H = \mu_H/\mu_w$, $\hat{\mu}_L = \mu_L/\mu_w$, $\phi = 1+DR$, $\hat{t} = \eta\varepsilon t$, $\eta = ga^3\pi\rho_H/(8S\mu_w)$, $C_1 = c/(b_2\mu_w\eta)$, $C_2 = \delta\rho g/(b_2\mu_w\eta)$, and $C_3 = k/(b_2\eta)$, with μ_w being the viscosity of water at 25 °C (0.89×10^{-3} Pa·s). From experimental observations, we found that the interval between the time when the intrusion fluid reaches the end of the pipe and the time when the flow completely reverses is generally less than 2% of the oscillation period. Therefore, we assume in our model that the flow instantaneously switches from down- to upflow—from Eqs. (4-69) and (4-70) to (4-71) and (4-72)—when $\hat{\xi} = \varepsilon^{-1}$, and from up- to downflow—from Eqs. (4-71) and (4-72) to (4-69) and (4-70)—when $\hat{\xi} = 0$. In this way, the whole oscillatory behavior can be described in a unified way.

To confirm the validity of our model, we have performed a series of simulations. Figure 4.29(a) shows the results for the case in which the viscosities of the two fluids are identical. As expected, the temporal evolution of $\hat{x}$ shows an exponential response for each up- and downflow branch, and $\hat{\tilde{\xi}}$ is found to behave in the following way (right side graphs in Figure 4.29a). During downflow, $\hat{\tilde{\xi}}$ remains zero when the value of $\hat{x}$ is large (hereafter, referred to as process A). As $\hat{x}$ decreases, $\hat{\tilde{\xi}}$ begins to increase gradually (process B). When $\hat{\tilde{\xi}}$ exceeds a certain threshold, it suddenly begins to increase rapidly (process C). Then, when the intrusion reaches the upper end of the pipe (i.e., $\hat{\tilde{\xi}}$ is equal to ε^{-1}), the flow reverses. Conversely, during upflow, $\hat{\tilde{\xi}}$ remains equal to ε^{-1} for small $\hat{x}$ values (process A), begins to decrease gradually as $\hat{x}$ increases (process B), begins to decrease rapidly (process C) when the intrusion length exceeds a threshold, and then the flow reverses when the intrusion reaches the lower end of the pipe ($\hat{\tilde{\xi}} = 0$). Thus, the behavior observed in the experiments is well reproduced by the simulation.

The change in the fluid viscosity drastically alters the behavior. Figure 4.29(b) shows the results for the case in which μ_H is much larger than μ_L. Again, as expected, the time constant of the exponential curve of $\hat{x}$ is larger for downflow than for upflow. The temporal evolution of $\hat{\tilde{\xi}}$ is also strongly affected by the viscosity. For the flow reversal from down- to upflow, processes B and C begin even when $\hat{x}$ is still large, and therefore, the flow reverses for a relatively large $\hat{x}$ compared to the case shown in Figure 4.29(a). For the flow reversal from up- to downflow, however, processes B and C do not begin until the value of $\hat{x}$ is close to the asymptotic value of the exponential curve, and as a consequence, the flow reverses for a relatively large $\hat{x}$ compared to the case shown in Figure 4.29(a).

Figure 4.29(c) shows the results for the case in which μ_L is much larger than μ_H. The opposite trend to that seen in Figure 4.29(b) is, of course, apparent here: the time constant of the exponential curve of $\hat{x}$ is larger during upflow than during downflow. For the flow reversal from down- to upflow, processes B and C do not begin until the value of $\hat{x}$ is close to the asymptotic value of the exponential curve, and as a consequence, the flow reverses for a relatively small $\hat{x}$. For the flow reversal from up- to downflow, processes B and C begin when $\hat{x}$ is still small, and therefore, the flow reverses for a relatively small $\hat{x}$. These findings are in good agreement with the experimental results (Figure 4.20).

In the same way as in the experiment, we have defined the parameters $\hat{x}_{da}$ and $\hat{x}_{db}$ as the values of $\hat{x}$ when the intrusion length during downflow is 0.1 mm and 1 mm, which correspond to $\hat{\tilde{\xi}} = 0.27$ and 2.74 and characterize the timings for the beginning of the intrusion and the onset of its rapid growth, respectively. In addition, we have defined $\hat{x}_d$ and $\hat{x}_u$ as the value of $\hat{x}$ at the flow reversal from downflow to upflow and upflow to downflow, respectively (Figure 4.27). The calculated result is shown in Figure 4.23. We find that $\hat{x}_{da}$, $\hat{x}_{db}$, $\hat{x}_d$, and $\hat{x}_u$ are in surprisingly good agreement with the experimental results of s_{da}, s_{db}, s_d, and s_u, respectively.

This behavior can be easily understood by considering the $\hat{x}$-dependent functional form of $f_d(\hat{\tilde{\xi}})$ in Eq. (4-70). Since $\hat{x}$ is slowly varying compared to $\hat{\tilde{\xi}}$ with respect to time (Eqs. 4-69–4-72), $\hat{x}$ can be regarded as a parameter characterizing the functional form of $f_d(\hat{\tilde{\xi}})$. Figure 4.30 shows the $\hat{x}$ dependence of $f_d(\hat{\tilde{\xi}})$

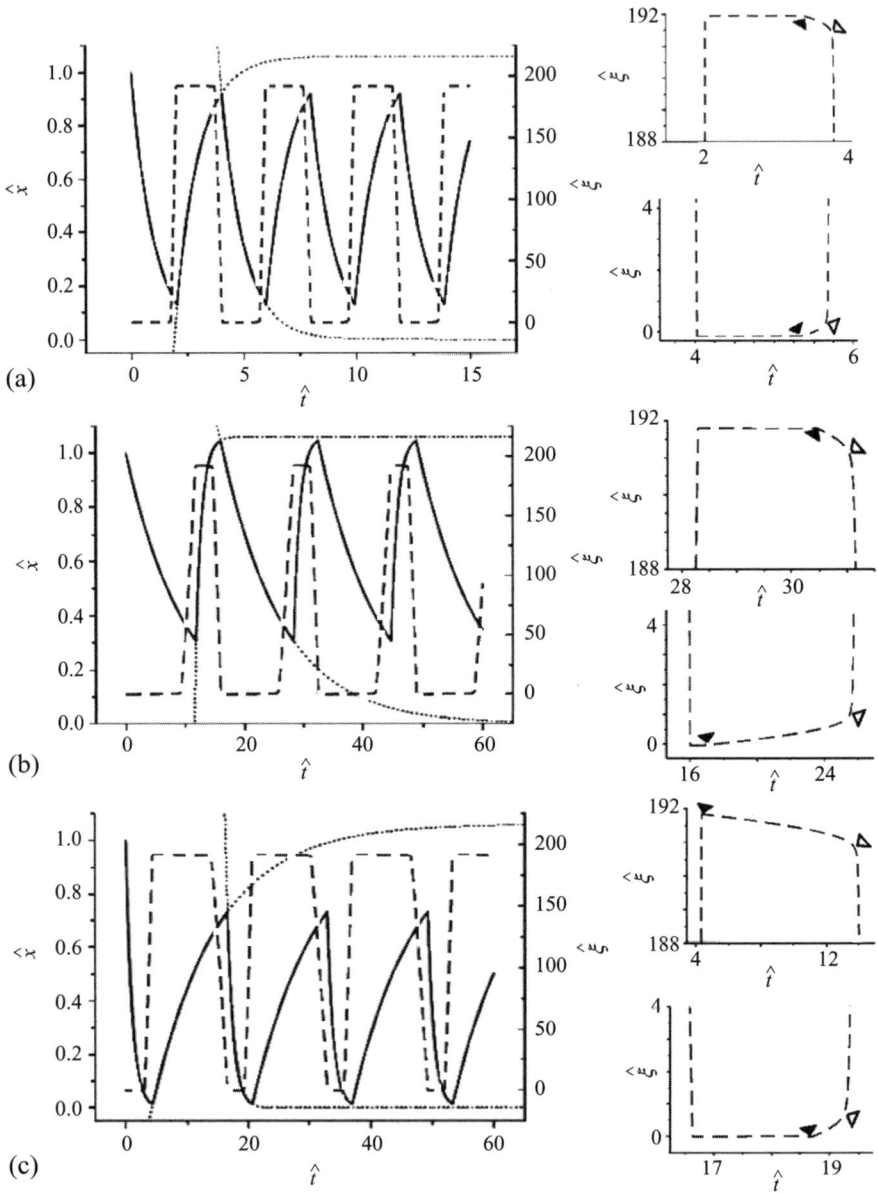

Figure 4.29 Simulation results for the temporal evolutions of the height of the heavy-fluid surface $\hat{x}$ (solid lines) and the intrusion length $\hat{\xi}$ (dashed lines): (a) $\mu_H = 0.89 \times 10^{-3}$ Pa·s and $\mu_L = 0.89 \times 10^{-3}$ Pa·s, (b) $\mu_H = 8.9 \times 10^{-3}$ Pa·s and $\mu_L = 0.89 \times 10^{-3}$ Pa·s, and (c) $\mu_H = 0.89 \times 10^{-3}$ Pa·s and $\mu_L = 8.9 \times 10^{-3}$ Pa·s. The graphs on the right show the magnified views of the temporal evolution of $\hat{\xi}$. Filled and empty arrowheads indicate the onset of gradual and rapid growth, respectively. The other parameter values are as follows: $a = 0.365$ mm, $d = 70$ mm, $\rho_H = 1.057 \times 10^3$ kg·m^{-3}, $\rho_L = 0.996 \times 10^3$ kg·m^{-3}, $S = 7.7 \times 10^{-4}$ m^2, $R = 1.44 \times 10^{-4}$, $C_1 = 7.11$, $C_2 = 5.31$, $C_3 = 0.32$, $\alpha = 1.82$, and $\gamma = 0.3$.

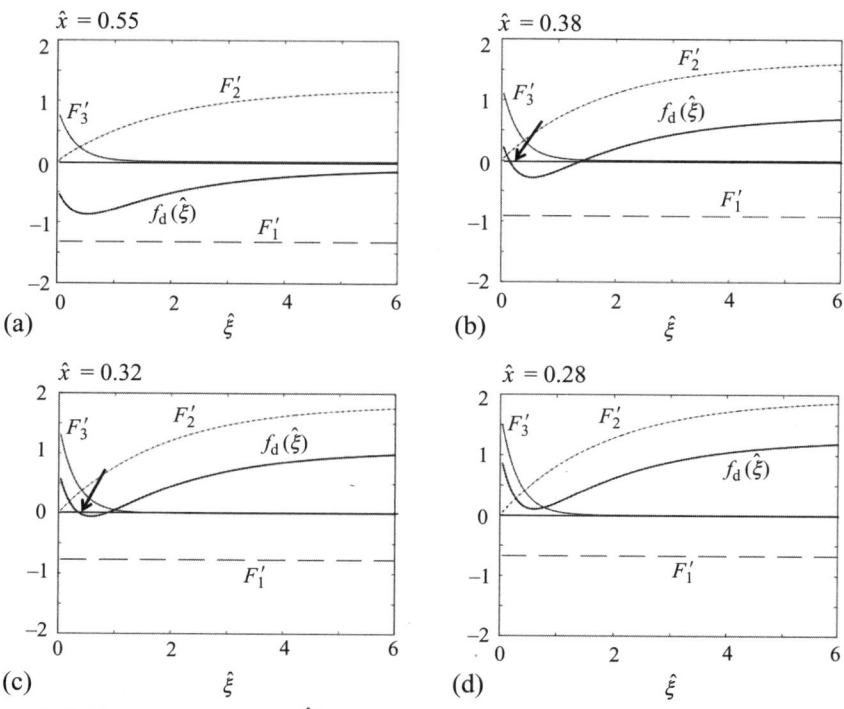

Figure 4.30 Functional form of $f_d(\hat{\xi})$ (thick solid lines), F'_1 (dashed lines), F'_2 (dotted lines), and F'_3 (solid lines): (a) $\hat{x} = 0.55$, (b) $\hat{x} = 0.38$, (c) $\hat{x} = 0.32$, and (d) $\hat{x} = 0.28$. The arrows denote the stable solution of Eq. (4-70)—that is, $\hat{\xi} = \hat{\xi}_0$. The parameter values are as follows: $a = 0.365$ mm, $d = 70$ mm, $\rho_H = 1.057 \times 10^3$ kg·m^{-3}, $\rho_L = 0.996 \times 10^3$ kg·m^{-3}, $S = 7.7 \times 10^{-4}$ m^2, $R = 1.44 \times 10^{-4}$, $\mu_H = 2.63 \times 10^{-3}$ Pa·s, $\mu_L = 0.89 \times 10^{-3}$ Pa·s, $C_1 = 7.11$, $C_2 = 5.31$, $C_3 = 0.32$, $\alpha = 1.82$, and $\gamma = 0.3$.

and its components, $F'_1 \equiv -C_1 \hat{x}/\hat{\mu}_H$, $F'_2 \equiv C_2(1-\hat{x})(1-\exp[-\hat{\xi}/\alpha])/\hat{\mu}_I$, and $F'_3 \equiv C_3 \hat{\mu}_H \exp[-\hat{\xi}/\gamma]/(\hat{\mu}_I \hat{x})$. Remember that F'_1, F'_2, and F'_3 originate from the viscous drag force, hydrostatic pressure gradient, and effect of the acceleration of the fluid that has passed through the pipe, respectively. When the value of $\hat{x}$ is large, the intrusion does not occur and $\hat{\xi}$ remains zero because the relation $f_d(0) \leq 0$ holds owing to the large contribution of F'_1 (Figure 4.3a; process A). However, when $\hat{x}$ decreases, $f_d(0)$ becomes positive because the contribution of F'_3 becomes large and also because the contribution of F'_1 decreases at the same time. Then, the relation $f_d(\hat{\xi}) = 0$ leads to a positive solution $\hat{\xi}_0$ with $f'_d(\hat{\xi}_0) < 0$; therefore, $\hat{\xi} = \hat{\xi}_0$ becomes a stable solution of Eq. (4-70). The light fluid then begins to intrude into the pipe (Figure 4.30b; process B). As $\hat{x}$ continues to decrease, $\hat{\xi}_0$ gradually increases (Figure 4.30c), and when the summation of F'_2 and F'_3 is larger than F'_1, the solution of $f_d(\hat{\xi}) = 0$ no longer exists, which leads to $f_d(\hat{\xi}) > 0$ for all $\hat{\xi}$ (Figure 4.30d). Thus, the intrusion suddenly accelerates (process C), and the flow reverses completely when the tip of the intrusion reaches the upper end of the pipe ($\hat{\xi} = \varepsilon^{-1}$).

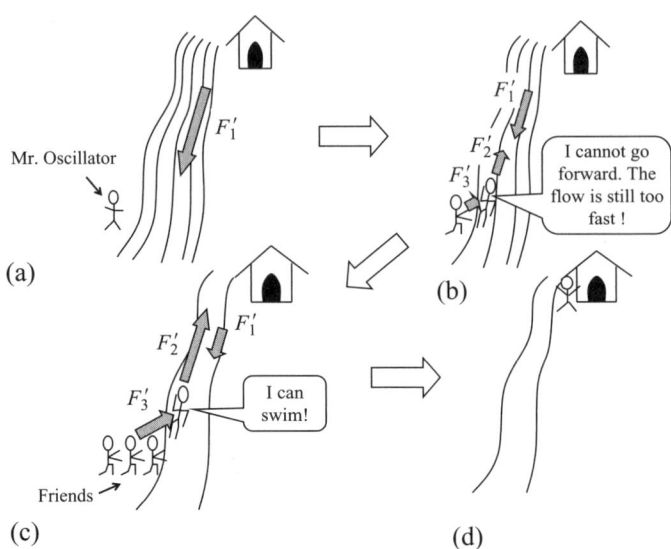

Figure 4.31 Intuitive interpretation of the flow-reversal process.

Intuitively, this flow-reversal process can be understood in the following way. Let us consider a case in which "Mr. Oscillator" wants to swim upstream in a river, as shown in Figure 4.31. At first, he cannot enter the river because the flow is too fast (Figure 4.31a). After a while, his friend pushes him into the river, but he cannot move forward because the flow is still too fast (Figure 4.31b). As time passes, the flow rate decreases, he gets used to swimming, and he is then able to swim forward (Figures 4.31c and d). In this example, the resistance force of the flow, his swimming ability, and his friends' "push" effect represent F'_1, F'_2, and F'_3, respectively. Thus, as is expected from this example, F'_1, F'_2, and F'_3 function as an *inhibitor*, *promoter*, and *trigger* for the flow reversal of the density oscillator.

The viscosity dependence of $\hat{x}_{da}$, $\hat{x}_{db}$, and $\hat{x}_d$ is also easily understood from the preceding consideration. For a large μ_H, $f_d(\hat{\xi})$ is relatively large, especially for small $\hat{\xi}$, owing to the large contribution of F'_3 (Eq. 4-70). Thus, the intrusion and thus the onset of its rapid growth occur even when $\hat{x}$ is still large. In contrast, for a large μ_L, the value of $f_d(\hat{\xi})$ is generally small because the contribution of F'_1 is relatively large compared to those of F'_2 and F'_3 (Eq. 4-70). Therefore, the intrusion and onset of rapid growth do not occur until $\hat{x}$ becomes sufficiently small. Thus, the present model captures the essence of the viscosity-dependent flow-reversal process fairly well.

4.4 Concluding Remarks

In this chapter, we introduced studies of the density oscillator by focusing particularly on its fundamental oscillation mechanism. Although the phenomenological model (Eq. 4-28) described the oscillatory behavior qualitatively, it was not sufficient to fully understand the oscillation mechanism since it did not properly describe the flow-reversal process, the most important process of the density oscillation. Through detailed experiments, we found that the flow-reversal process depends strongly on the viscosities of the fluids. This viscosity-dependent behavior was well explained by a simple model that takes into account the forces acting on a unit volume element located at the tip of the intrusion. We concluded from the model that the flow-reversal process occurs in the following way:

1. The acceleration of the flow that has passed through the pipe triggers the fluid to intrude into the pipe (F'_3).
2. After the intrusion occurs, it does not grow immediately but grows slowly because the viscous drag force (F'_1) hinders the growth.
3. As the height of the heavy-fluid surface approaches the hydrostatic equilibrium, the viscous drag force (F'_1) decreases while the hydrostatic pressure gradient (F'_2) that accelerates the intrusion increases. Then, at a threshold, the intrusion is accelerated suddenly and begins to grow rapidly.
4. The flow reverses completely when the intrusion reaches the upper (or lower) end of the pipe.

These findings from density oscillators could also be applicable to the relaxation oscillators found in nature. As described in Section 4.1.2, relaxation oscillators exhibit a repetitive change of slow and fast processes; the latter alters the state that the system approaches (Figure 4.3). Our findings suggest that the three factors corresponding to F'_1, F'_2, and F'_3 considered in the flow-reversal process are also essential factors for the onset of the fast process in general relaxation oscillators: F'_3 triggers the change from the slow to fast process and F'_2 promotes the onset of the fast process when it overwhelms F'_1.

Let us take the respiratory rhythms in mammals as an example. A respiratory system exhibits typical relaxation oscillations consisting of inspiration and expiration processes. In switching from inspiration to expiration, lung inflation activates the pulmonary stretch receptors (PSRs), and the afferent activity from the PSRs is carried to the brainstem by the vagus nerves (Botros and Bruce, 1990; Hayashi et al., 1996). This sensory feedback mechanism triggers the switching, and thus, it may correspond to the F'_3 factor in the density oscillator. Therefore, we expect that the findings from density oscillators will impart a deep insight into the mechanisms underlying systems that exhibit relaxation oscillations.

Acknowledgments

The author thanks Professor Shuichi Kinoshita of Osaka University for his valuable suggestions.

References

Adamčiková, L., 1992. Chemical oscillatory reaction in a hydrodynamic oscillator. React. Kin. Cat. Lett. 48, 649–654.
Alfredsson, P.H., Lagerstedt, T., 1981. The behavior of the density oscillator. Phys. Fluids 24, 10–14.

Andronov, A.A., Chaikin, C.C., 1949. Theory of Oscillations. Princeton University Press, Princeton, NJ.
Aoki, K., 2000. Mathematical model of a saline oscillator. Physica D 147, 187–203.
Bennett, M.V.L., Zukin, R.S., 2004. Electrical coupling and neuronal synchronization in the mammalian brain. Neuron 41, 495–511.
Blasius, B., Beck, F., Lüttge, U., 1997. A model for photosynthetic oscillations in crassulacean acid metabolism (CAM). J. Theor. Biol. 184, 345–351.
Botros, S.M., Bruce, E.N., 1990. Neural network implementation of a three-phase model of respiratory rhythm generation. Biol. Cybern. 63, 143–153.
Cervellati, R., Solda, R., 2001. Alternating voltage battery with two salt-water oscillators. Am. J. Phys. 69, 543–545.
Chua, L.O., Desoer, C.A., Kuh, E.S., 1987. Linear and Nonlinear Circuits. McGraw-Hill, New York.
Das, A.K., Srivastava, R.C., 1993. Oscillations of electrical potential differences in the non-electrolyte analogue of the salt-water oscillator. J. Chem. Soc., Faraday Trans. 89, 905–908.
DiFrancesco, D., 1993. Pacemaker mechanisms in cardiac tissue. Ann. Rev. Physiol. 55, 455–472.
Field, R.J., Noyes, R.M., 1974. Oscillations in chemical systems. IV. Limit cycle behavior in a model of a real chemical reaction. J. Chem. Phys. 60, 1877–1884.
González, H., Arce, H., Guevara, M.R., 2008. Phase resetting, phase locking, and bistability in the periodically driven saline oscillator: Experiment and model. Phys. Rev. E 78, 036217.
Hayashi, F., Coles, S.K., McCrimmon, D.R., 1996. Respiratory neurons mediating the Breuer–Hering reflex prolongation of expiration in rat. J. Neurosci. 16, 6526–6536.
Hong, Y.W., Scaglione, A., 2005. A scalable synchronization protocol for large scale sensor networks and its applications. IEEE J. Sel. Areas Commun. 23, 1085–1099.
Ijspeert, A.J., 2008. Central pattern generators for locomotion control in animals and robots: A review. Neural Netw. 21, 642–653.
Kaempfer, E., 1727. The History of Japan (with a description of the Kingdom of Siam). Sloane, London.
Kano, T., Kinoshita, S., 2007. Viscosity-dependent flow reversal in a density oscillator. Phys. Rev. E 76, 046208.
Kano, T., Kinoshita, S., 2008. Modeling of the flow-reversal process in a density oscillator. J. Korean Phys. Soc. 53, 1273–1279.
Kano, T., Kinoshita, S., 2009. Modeling of a density oscillator. Phys. Rev. E 80, 046217.
Kano, T., 2008. Experimental and theoretical study on density oscillator, Ph.D. Osaka University, Japan.
Kiss, I.Z., Rusin, C.G., Kori, H., Hudson, J.L., 2007. Engineering complex dynamical structures: Sequential patterns and desynchronization. Science 316, 1886–1889.
Kuramoto, Y., 1984. Chemical Oscillations, Waves, and Turbulence. Springer-Verlag, Berlin.
Lagomarsino, M.C., Jona, P., Bassetti, B., 2003. Metachronal waves for deterministic switching two-state oscillators with hydrodynamic interaction. Phys. Rev. E 68, 021908.
Landa, P.S., 1996. Nonlinear Oscillations and Waves in Dynamical Systems. Kluwer Academic Publishers, Dordrecht.
Manrubia, S.C., Mikhailov, A.S., Zanette, D.H., 2004. Emergence of Dynamical Order: Synchronization Phenomena in Complex Systems. World Scientific, Singapore.
Martin, S., 1970. A hydrodynamic curiosity: The salt oscillator. Geophys. Fluid Dyn. 1, 143–160.
Menikoff, R., Mjolsness, R.C., Sharp, D.H., Zemach, C., 1977. Unstable normal mode for Rayleigh–Taylor instability in viscous fluids. Phys. Fluids 20, 2000–2004.
Minorsky, N., 1974. Nonlinear Oscillations. Robert E. Krieger Publishing Company, Huntington, NY.

Miyakawa, K., Yamada, K., 1999. Entrainment in coupled salt-water oscillators. Physica D 127, 177–186.
Miyakawa, K., Yamada, K., 2001. Synchronization and clustering in globally coupled salt-water oscillators. Physica D 151, 217–227.
Miyazaki, J., Kinoshita, S., 2006. Method for determining a coupling function in coupled oscillators with application to Belousov–Zhabotinsky oscillators. Phys. Rev. E 74, 056209.
Nakata, S., Miyata, T., Ojima, N., Yoshikawa, K., 1998. Self-synchronization in coupled salt-water oscillators. Physica D 115, 313–320.
Noyes, R.M., 1989. A simple explanation of the salt-water oscillator. J. Chem. Edu. 66, 207–209.
Okamura, M., Yoshikawa, K., 2000. Rhythm in a saline oscillator. Phys. Rev. E 61, 2445–2552.
Pikovsky, A., Rosenblum, M., Kurths, J., 2001. Synchronization: A Universal Concept in Nonlinear Sciences. Cambridge University Press, Cambridge.
Poincaré, H., 1892. Lés méthodes nonvelles de la mécaniqae celeste. Kris, Gauthier-villars.
Rastogi, R.P., Srivastava, R.C., Kumar, S., 2005. Oscillatory phenomena at liquid–liquid interfaces. J. Colloid Interface Sci. 283, 139–143.
Rayleigh, J.W.S., 1877. The Theory of Sound. Macmillan, London.
Saffman, P.G., Taylor, G.I., 1958. The penetration of a fluid into a porous medium or Hele-show cell containing a more viscous fluid. Proc. R. Soc. Lond. A 245, 312–329.
Sato, T., Kano, T., Ishiguro, A., 2011. On the applicability of the decentralized control mechanism extracted from the true slime mold: A robotic case study with a serpentine robot. Bioinspir. Biomim. 6, 026006.
Schlichting, H., 1960. Boundary Layer Theory. McGraw-Hill, New York.
Srivastava, R.C., 1994. Salt-water oscillator and its non-electrolyte analogues. Pure Appl. Chem. 66, 455–460.
Srivastava, R.C., Agarwala, V., Das, A.K., Upadhyaya, S., 1994. Mimicking sensing mechanism of taste salt-water oscillator and its non-electrolyte analogues: Experiments with compounds belonging to different taste categories. Ind. J. Chem. A 33, 978–984.
Steinbock, O., Lange, A., Rehberg, I., 1998. Density oscillator: Analysis of flow dynamics and stability. Phys. Rev. Lett. 81, 798–801.
Stern, K., McClintock, M.K., 1998. Regulation of ovulation by human pheromones. Nature 392, 177–179.
Takamatsu, A., Tanaka, R., Yamada, H., Nakagaki, T., Fujii, T., Endo, I., 2001. Spatiotemporal symmetry in rings of coupled biological oscillators of *Physarum* plasmodial slime mold. Phys. Rev. Lett. 87, 078102.
Tass, P.A., 1999. Phase Resetting in Medicine and Biology: Stochastic Modeling and Data Analysis. Springer-Verlag, Berlin.
Taylor, G.I., 1950. The instability of liquid surfaces when accelerated in a direction perpendicular to their planes. I. Proc. R. Soc. Lond. A 201, 192–196.
Ueno, M., Uehara, F., Narahara, Y., Watanabe, Y., 2006. Hydrodynamic analysis of density oscillator. Jap. J. Appl. Phys. 45, 8928–8938.
Umedachi, T., Takeda, K., Nakagaki, T., Kobayashi, R., Ishiguro, A., 2010. Fully decentralized control of a soft-bodied robot inspired by true slime mold. Biol. Cybern. 102, 261–269.
Upadhyay, S., Das, A.K., Agarwala, V., Srivastava, R.C., 1992. Oscillations of electrical potential differences in the salt-water oscillator. Langmuir 8, 2567–2571.
van der Pol, B., 1920. A theory of the amplitude of free and forced triode vibrations. Radio Rev 1, 701–710.
van der Pol, B., 1926. On "relaxation-oscillations" The London, Edinburgh, and Dublin Phil. Mag. J. Sci. Ser.7, 2, 978–992.

van der Pol, B., 1927. Forced oscillations in a circuit with non-linear resistance (reception with reactive triode). The London, Edinburgh, and Dublin Phil. Mag. J. Sci. Ser.7, 3, 65–80.

van der Pol, B., van der Mark, J., 1927. Frequency demultiplication. Nature 120, 363–364.

van der Pol, B., van der Mark, J., 1928. The heartbeat considered as a relaxation oscillation, and an electrical model of the heart. The London, Edinburgh, and Dublin Phil. Mag. J. Sci. Ser.7, 6, 763–775.

Villarreyes, J.A.M., da Costa, H.J.B., Kokubun, F., Schmitz, L.C., Castro, J.A., 2000. Some applications of salt-water oscillator in chemical engineering teaching and process equipment design. Comp. Chem. Eng. 24, 1753–1757.

Wiesenfeld, K., Colet, P., Strogatz, S.H., 1998. Frequency locking in Josephson arrays: Connection with the Kuramoto model. Phys. Rev. E 57, 1563–1569.

Yoshikawa, K., Fukunaga, K., Kawakami, H., 1990. A tri-phasic mode is stable when three nonlinear oscillators interact with each other. Chem. Phys. Lett. 174, 203–207.

Yoshikawa, K., Maeda, S., Kawakami, H., 1988. Various oscillatory regimes and bifurcations in a dynamic chemical system at an interface. Ferroelectrics 86, 281–298.

Yoshikawa, K., Nakata, S., 1990. Rhythmic phenomena at interface and membrane self-organization of spatio-temporal structure. React. Kin. Cat. Lett. 42, 333–338.

Yoshikawa, K., Nakata, S., Yamakata, M., Waki, T., 1989. Amusement with a salt-water oscillator. J. Chem. Edu. 66, 205–207.

Yoshikawa, K., Oyama, N., Shoji, M., Nakata, S., 1991. Use of a saline oscillator as a simple nonlinear dynamical system: Rhythms, bifurcation, and entrainment. Am. J. Phys. 59, 137–141.

Appendix

A4.1 Three-dimensional Hagen–Poiseuille Flow

Three-dimensional Hagen–Poiseuille flow is the steady flow of an incompressible fluid through a straight pipe of circular cross-section with rotational symmetry (Schlichting, 1960). Consider a long pipe of radius a. Let the z axis define the axis of the pipe and r denote the radial coordinate measured from the axis outwards. The flow is assumed to be parallel to the pipe, and thus, the velocity components in the tangential and radial directions are assumed to be zero. The z component of the velocity is denoted by u.

Then, from the continuity condition, we obtain $\partial u/\partial z = 0$. Thus, u does not depend on z and is described as $u = u(r)$. We find that the hydrostatic pressure P does not depend on r and is described as $P = P(z)$, which is obtained from the radial and tangential components of the Navier–Stokes equation. Therefore, the z component of the Navier–Stokes equations in cylindrical coordinates becomes

$$\mu \left(\frac{\partial^2 u}{\partial r^2} + \frac{1}{r} \frac{\partial u}{\partial r} \right) = \frac{\partial P}{\partial z}, \qquad (A4-1)$$

where μ is the viscosity of the fluid. The boundary condition is given as $u = 0$ for $r = a$ (no-slip condition). By solving Eq. (A4-1) with this boundary condition, we obtain

$$u(r) = -\frac{1}{4\mu}\frac{dP(z)}{dz}\left(a^2 - r^2\right). \tag{A4-2}$$

Here, $dP(z)/dz$ should be a constant, because $u(r)$ does not depend on z. The mean velocity $\bar{u}$ is calculated from Eq. (A4-2) as

$$\bar{u} = \frac{1}{\pi a^2}\int_0^a 2\pi r\,dr\,u(r) = \frac{2}{a^2}\int_0^a ru(r)dr = -\frac{a^2}{8\mu}\frac{dP(z)}{dz}. \tag{A4-3}$$

From Eqs. (A4-2) and (A4-3), the following relation is obtained:

$$\left.\frac{\partial u}{\partial r}\right|_{r=a} = -\frac{1}{4\mu}\frac{dP(z)}{dz}\cdot(-2a) = -\frac{4\bar{u}}{a}. \tag{A4-4}$$

A4.2 Two-dimensional Hagen–Poiseuille Flow

The steady flow between two parallel flat walls, known as two-dimensional Poiseuille flow, is considered in a similar manner to the three-dimensional Poiseuille flow (Schlichting, 1960). Let the y and z axes be perpendicular and parallel to the flat walls, respectively. The flat walls are assumed to be located at $y=0$ and $2a$. When we assume that the flow is parallel to the flat walls, the y component of the velocity becomes zero. Here, w defines the z component of the velocity.

From the continuity condition, we obtain $\partial w/\partial z = 0$. Thus, w is described as $w = w(y)$. From the y component of the Navier–Stokes equation, we find that the hydrostatic pressure gradient P does not depend on y, and thus, P is written as $P = P(z)$. Therefore, the z component of the Navier–Stokes equation becomes

$$\mu\frac{\partial^2 w}{\partial y^2} = \frac{dP(z)}{dz}, \tag{A4-5}$$

The boundary conditions are given by $w=0$ for $y=0$ and $2a$ (no-slip condition). By solving Eq. (A4-5) with these boundary conditions, $w(y)$ is obtained as

$$w(y) = -\frac{1}{2\mu}\frac{dP(z)}{dz}\left[a^2 - (y-a)^2\right], \tag{A4-6}$$

where $dP(z)/dz$ should be a constant, because $w(y)$ does not depend on z. Thus, the flow rate Q is derived as

$$Q = \int_0^{2a} dy\,w(y) = -\frac{2a^3}{3\mu}\frac{dP(z)}{dz}. \tag{A4-7}$$

5 Colloidal Crystals

Junpei Yamanaka, Tohru Okuzono, Akiko Toyotama

Chapter Contents
5.1 Introduction   165
   5.1.1 Order Induced by Entropy   165
   5.1.2 Structures of Colloidal Dispersions   166
   5.1.3 Interactions between Colloidal Particles   170
5.2 Samples and Methodology   175
   5.2.1 Colloidal Samples   176
   5.2.2 Characterization of Crystal Structures   177
5.3 Crystallization of Charged Colloids   181
   5.3.1 Charge-Induced Crystallization   181
   5.3.2 Unidirectional Crystallization   183
   5.3.3 Gel Immobilization   189
   5.3.4 Exclusion of Impurity Particles   189
5.4 Current Topics   193
   5.4.1 Opal-Type Crystals   193
   5.4.2 Complex Structures   194

5.1 Introduction

5.1.1 Order Induced by Entropy

Entropy is usually considered to be a measure of the disorder of thermodynamic states, but it can lead to the emergence of ordered states. Indeed, it occurs in a very simple system as follows. Let us imagine many beads (hard spheres) confined in a box in a space where gravity and air do not exist. There are no interactions between beads except for the excluded volume interaction; that is, the distance between the centers of any pair of beads is not less than $2a$, where a is the radius of a bead, because the beads cannot overlap. How are the beads arranged in the box? It depends on their packing fraction or volume fraction ϕ, defined as $\phi = (4/3)\pi a^3 N/V$, where N is the number of beads, and V is the volume of the box. The arrangements of the beads for various volume fractions are illustrated in Figure 5.1. When the packing fraction is small, say 1% ($\phi = 0.01$), the beads are randomly distributed and move through the entire region inside the box, colliding with other beads. In this case, the system is in a fluid phase—that is, a gas/liquid phase (gas or liquid are indistinguishable in this system). As the

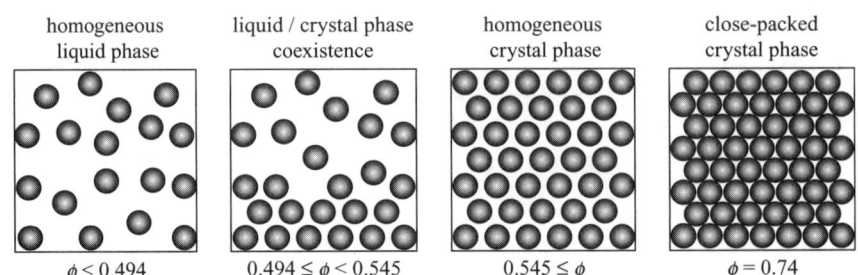

Figure 5.1 Illustrations of crystallization of hard-sphere colloid.

packing fraction increases, the beads collide more frequently, and eventually their positions become localized; that is, the long-term averages of all the bead positions are fixed and form a periodic structure. In other words, a fluid-to-crystal transition occurs. This phase transition, called the Alder transition, was discovered in computer experiments by Alder and Wainwright (1957). The results of numerical simulations show that the volume fraction at the freezing point (from fluid to crystal) is $\phi_F = 0.494$, and that at the melting point (from crystal to fluid) it is $\phi_M = 0.545$. For $\phi_F < \phi < \phi_M$, the fluid and crystal phases coexist, suggesting a first-order phase transition. Because there is no interaction energy, the free energy F of the preceding system is given by $F = -TS$, where T is the absolute temperature, and S is the entropy of the system. Hence, the equilibrium state that minimizes F is governed by S. Therefore, a system in the fluid state becomes a crystal with increasing entropy.

The Alder transition is a remarkable feature of hard-sphere systems and gives us a deep insight into the fluid–crystal transition in a simple system. However, the hard-sphere system is an ideal system that is easy to generate on computers. A colloidal dispersion is an experimental system that closely resembles a hard-sphere system. The phase behavior of such systems was studied in detail by Pusey and van Megen (1986). An important point here is that to generate crystalline order, only repulsive interactions, including the excluded volume interaction, are needed, and attractive interactions are not always required.

5.1.2 Structures of Colloidal Dispersions

A colloidal dispersion system consists of colloidal particles (dispersed phase, e.g., silica, polystyrene [PS]) and a dispersion medium (continuous phase, e.g., water, organic solvent) in which they are immersed (Russel et al., 1989). A colloidal particle is typically 1–1000 nm in diameter, whereas the dispersion medium is composed of molecules smaller than the colloidal particles and can be regarded as a continuum. Although the physical or material properties of colloidal systems as well as the size of their building blocks are quite different from those of atomic or molecular systems, their equilibrium or nonequilibrium structures (e.g., liquids, crystals, glasses) are similar to those of atomic systems. Therefore, colloidal dispersions are often studied as model systems of atomic systems.

Because the hard-sphere systems just presented have no interaction energy between particles, the volume fraction ϕ of particles is the only parameter that determines their phase behaviors. The volume fraction is also a primary parameter for characterizing the structures of colloidal dispersions. For relatively small volume fractions ($\phi \lesssim 0.5$ for the hard-sphere systems), a colloidal dispersion takes the state of a fluid in which the positions of particles are disordered, and each particle moves randomly owing to the thermal motion of small molecules surrounding it. Such random motion is called Brownian motion. Because the particle density is uniform in the fluid phase after taking the statistical average, the fluid state has continuous translation and rotation symmetries. For larger volume fractions ($\phi \gtrsim 0.5$), these continuous symmetries are broken, and some discrete symmetries corresponding to a periodic arrangement of particles hold—that is, a crystal structure appears. Colloidal crystals that take the close-packed structures ($\phi = 0.74$ for face-centered cubic [fcc] or hexagonal close-packed [hcp] structures) are known as opals (Figure 5.2).

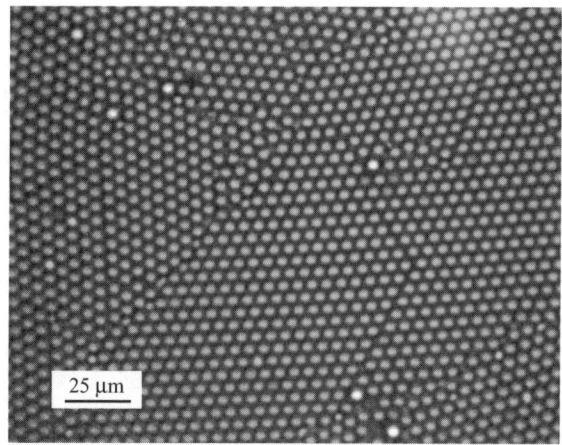

Figure 5.2 Optical micrograph of an opal-type colloidal crystal prepared by evaporating an aqueous medium from a charge-stabilized dispersion of polystyrene particles; diameter = 3.1 μm.

Box 5.1 How to Make Colloidal Crystals

Here we introduce simple methods of making colloidal crystals. We use an aqueous dispersion of colloidal particles having a narrow size distribution (standard deviation of ±10%). Polystyrene or silica particles having narrow size distributions are widely used as size-standard materials for electron microscopy and are also used for various biological applications. They are commercially available in the form of aqueous dispersions.

Continued

> **Box 5.1 How to Make Colloidal Crystals—cont'd**
>
> - **Opal crystals:** To prepare opal-type crystals (see Figure 5.2) showing Bragg diffraction of visible light, a suitable size for the colloidal particles is about 200–300 nm. Further purification of the sample is usually not necessary. Prepare an aqueous dispersion having $\phi \approx 0.05$; drop about 0.1 ml of the sample onto a flat surface (e.g., a glass plate), and then dry it gradually at room temperature (taking one day). The key to preparing good opals is to evaporate the water sufficiently slowly (so that the particles can reach stable positions—crystal lattice points—during evaporation). The evaporation rate can be controlled by adjusting the humidity around the sample. The appearance of brilliant diffraction colors is indicative of opals. We can tune the diffraction colors by changing the particle size and refractive index of the particles used.
> - **Charged colloidal crystals:** Commercially available colloidal particles are more or less charged. We recommend the use of polystyrene particles having strong acid groups on the surfaces. To make charged crystal structures, we need to reduce the ionic impurity level of the medium to 1–10 μM. A simple way to deionize the colloid is to apply a mixed bed of cation and anion exchange resin beads (about 1 mm in diameter) to the sample. The Bragg wavelengths of charged colloidal crystals are continuously tunable by changing ϕ. For particles 100 nm in diameter, the colloidal crystals exhibit red, green, and blue diffraction colors at $\phi \approx 0.02$, 0.03, and 0.04, respectively. Use 0.1–0.5 ml of the ion exchange resin beads to 1 ml of the colloid and shake it well. Over time, the sample should become iridescent owing to the formation of colloidal crystals.

A spatial distribution of particles in uniform systems is characterized by a radial distribution function (also called a pair correlation function), $g(r)$, defined as

$$\rho(r) = \bar{\rho} g(r), \tag{5-1}$$

where $\rho(r)$ is the average number density of particles of which the distance l from a (reference) particle is within the range $r \leq l < r + dr$, and $\bar{\rho} = N/V$ is the particle number density averaged over the entire system (Chandler, 1987; Barrat and Hansen, 2003; Hansen and McDonald, 2006). This quantity, $g(r)$, is proportional to the probability of finding another particle at a distance r from the position of a reference particle. In other words, $g(r) - 1$ represents the correlation of two particle positions of which the distance is r. Typical behaviors of $g(r)$ for the liquid and crystal phases are shown in Figure 5.3. In the liquid phase, $g(r)$ has the first peak at $r \sim 2a$, corresponding to the positions of its nearest neighbors. Note that $g(r)$ vanishes for r less than the particle diameter ($2a$) because of the excluded volume effect. There are a few peaks at $r > 2a$ that decay rapidly and $g(r) \to 1$ as $r \to \infty$, because $\rho(r)$ should approach $\bar{\rho}$ for large r (Figure 5.3a). These behaviors of $g(r)$ imply that the order of the liquid is short-range and isotropic at equilibrium. In the crystal phase, $g(r)$ has many sharp peaks located over a wide range corresponding to the interparticle distances that take discrete values in the crystal (Figure 5.3b). Therefore, the crystal order is long-range and anisotropic.

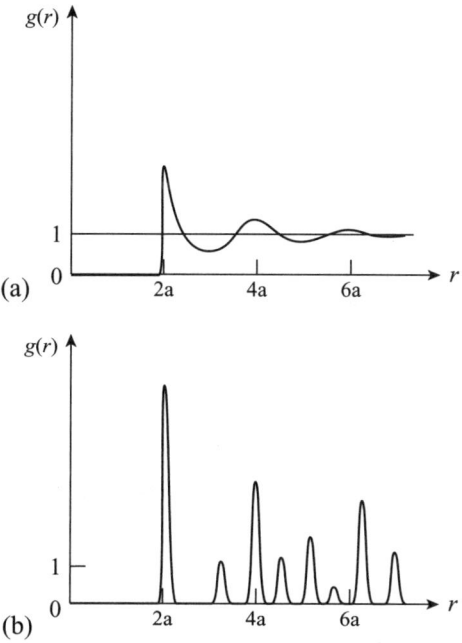

Figure 5.3 Radial distribution function $g(r)$ versus r for (a) liquid and (b) crystal states; r is the center-to-center distance between two particles.

The structures of particle positions in colloidal dispersion systems can be identified by diffraction experiments using X-rays or visible light. In these experiments, we can observe the scattering intensity, which is proportional to the static structure factor (Chandler, 1987; Kittel, 2005; Hansen and McDonald, 2006), defined as

$$S(\boldsymbol{k}) = \frac{1}{N}\langle \rho_k \, \rho_{-k}\rangle, \tag{5-2}$$

where $\boldsymbol{k} = \boldsymbol{k}_s - \boldsymbol{k}_i$ is the difference between the wavevectors $\boldsymbol{k}_i$ of incident light and $\boldsymbol{k}_s$ of scattered light (Figure 5.4), N is the number of particles, and the angle brackets represent the statistical average. ρ_k is a Fourier component of the particle number density $\rho(\boldsymbol{r}) = \Sigma_{i=1}^{N}\delta(\boldsymbol{r} - \boldsymbol{r}_i)$ at position $\boldsymbol{r}$ with $\boldsymbol{r}_i$ being the position of particle i, where $\delta(\boldsymbol{r})$ is the Dirac delta function. Note that in elastic scattering, the wave number of

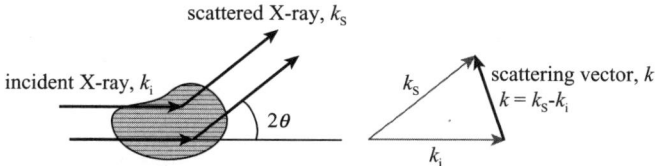

Figure 5.4 Schematic drawing of scattering of an incident X-ray beam. The scattering vector is defined on the right. 2θ is the scattering angle.

incident light is equal to that of scattered light; that is, $|\boldsymbol{k}_i|=|\boldsymbol{k}_s|=2\pi/\lambda$, where λ is the wavelength of the light. Because in a liquid state the particle density is homogeneous and isotropic, $S(\boldsymbol{k})$ depends only on $k\equiv|\boldsymbol{k}|$ and is related to $g(r)$ as

$$S(k) = 1 + \bar{\rho}\int g(r)e^{i\boldsymbol{k}\cdot\boldsymbol{r}}d\boldsymbol{r}. \tag{5-3}$$

Therefore, in a diffraction experiment we observe a halo pattern indicating the structure factor $S(\boldsymbol{k})$ of a liquid as well as a polycrystal (see Figure 5.8 later). In the single-crystal state, $S(\boldsymbol{k})$ has sharp peaks, called Bragg spots, at $\boldsymbol{k}=\boldsymbol{G}$, where $\boldsymbol{G}$ is the reciprocal lattice vector of the crystal structure (Kittel, 2005). We can identify the crystal structure by detecting the peak positions in the reciprocal space. This argument is a simple extension of the Bragg law. Because the colloidal particles are regularly arrayed in a crystal structure, the light waves scattered by the particles interfere constructively when the optical path difference of two scattered light waves is equal to multiples of the wavelength of the light. We can often see iridescent colors in a colloidal crystal, because the distances between neighboring particles in the crystal are usually similar to the wavelengths of visible light.

5.1.3 Interactions between Colloidal Particles

Several types of interactions occur between colloidal particles: the van der Waals interaction, steric interaction, electrostatic interaction, depletion interaction, and so on (Israelachivili, 1992). The van der Waals interaction in particular always exists not only between molecules but also between colloidal particles or macroscopic bodies, which are assemblies of molecules. The magnitude of the interparticle van der Waals force can be estimated by integrating all pairs of volume units in the two particles (the Derjaguin approximation). Because the van der Waals interaction yields attractive forces between colloidal particles, the particles tend to come into contact with each other. Therefore, a colloidal dispersion system in which the particles interact with each other via attractive forces alone cannot maintain a homogeneously dispersed state and becomes unstable—that is, coagulation occurs (Everett, 1988).

To avoid coagulation due to van der Waals attraction, the following methods are used:

1. The refractive index of the dispersion medium is matched to that of the dispersion phase (particles) to vanish the van der Waals attractive forces (Figure 5.5a). The magnitude of van der Waals attraction is nearly proportional to a difference in the refractive indices of the dispersion medium and dispersion phase (Israelachivili, 1992). Using this method, we can realize nearly hard-sphere systems.
2. The surfaces of the particles are modified with appropriate polymers or surfactants (Figure 5.5b). Using this technique, we can stabilize the dispersion systems because of the short-range repulsive interaction between particles (steric interaction) induced by the polymers.

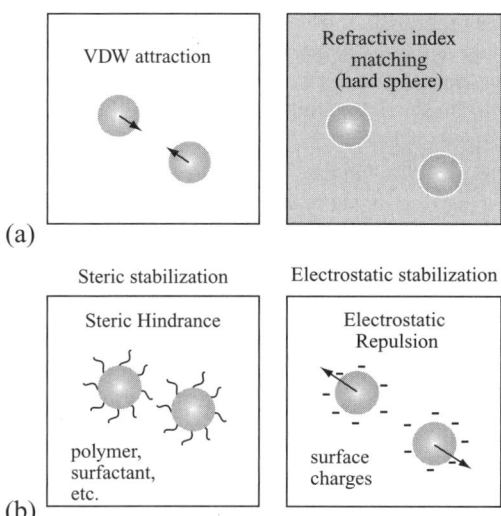

Figure 5.5 Illustrations of (a) van der Waals (VDW) attraction between the particles and reduction of the VDW force, and (b) steric and electrostatic stabilization of colloids.

3. Surface charges are introduced by a surface modification technique (Figure 5.5b). The long-range electrostatic repulsive interactions between likely charged particles, which are much larger than the van der Waals interactions, stabilize the dispersion systems.

Charged colloidal systems obtained by the last method are widely used to study the structures and phases of colloidal systems, because various experimental parameters (e.g., the particle volume fraction, salt concentration, and charge number) can be used to control the systems, and the systems respond to external fields (e.g., the electric field, flow field, and gravity). We will see below how the charged particles interact with each other in an aqueous solution in a simple case.

In the classical Derjaguin–Landau–Verwey–Overbeek theory (Verwey and Overbeek, 1948), the interaction between particles is given as a sum of the attractive van der Waals interaction and the repulsive electrostatic interaction. However, in electrostatically stabilized systems, the electrostatic interaction is much larger than the van der Waals interaction. Next, we consider electrostatic interactions only (for more details, see van Roij et al., 1999; Belloni, 2000; Hansen and Löwen, 2000; Ise and Sogami, 2005).

The electrostatic (Coulomb) interaction is generally long-ranged, and the interaction energy between two charged particles in a vacuum is proportional to $1/r$, where r is the distance between the particles. When immersed in an electrolyte solution, however, the charged particles are surrounded by small ions of the opposite charge (counterions), as shown in Figure 5.6. This results in screening of the electrostatic interaction in proportion to $\exp(-\kappa r)/r$, where $1/\kappa$ is the screening length (also called the Debye length). The interaction is significantly reduced for $r > 1/\kappa$, so the

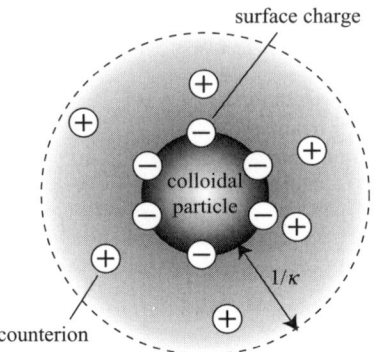

Figure 5.6 Illustration of charged colloidal particle.

parameter κ determines the interaction range. The Coulomb interaction is generally screened near a charged surface immersed in an electrolyte solution. The narrow region near the surface where the ion density is high is called an electric double layer. This is a key concept for understanding charged colloidal phenomena.

Now, we derive a detailed expression for the interaction between charged colloidal particles based on the Poisson–Boltzmann theory (see later). Consider a charged colloidal system of volume V that consists of N identical spherical particles of radius a and an ionic solution regarded as a continuum with a permittivity of $\varepsilon = \varepsilon_0 \varepsilon_r$, where ε_0 is the permittivity of a vacuum, and ε_r is the dielectric constant of the solution ($\varepsilon_r \approx 80$ for aqueous solution at 20°C). A particle carries the negative charge $-Ze = 4\pi a^2 \sigma$, where Z is the charge number, e is the elementary positive charge, and σ is the surface charge density distributed uniformly on the surface of the particle. The electrolyte solution contains N_+ cations (counterions) and N_- anions (coions). In a charge-neutral system, the charge-neutrality condition $N_+ = ZN + N_-$ holds if the added salt (electrolyte) completely dissociates in the solution, and all the ions are monovalent. Because the added salt (electrolyte) density C_s is given by $C_s = \bar{n}_- \equiv N_-/V$, the average number density of counterions $\bar{n}_+ \equiv N_+/V$ is given as

$$\bar{n}_+ = Z\bar{\rho} + C_s, \tag{5-4}$$

using the average number density of particles $\bar{\rho} \equiv N/V$ and the salt density C_s.

Suppose that the colloidal dispersion system is in contact with a large salt reservoir via a membrane that is permeable only to small ions. The reservoir contains cations and anions with densities n_+^0 and n_-^0, respectively. We assume here that $n_+^0 = n_-^0 = n_r$, where n_r is the salt density in the reservoir. The system is in osmotic equilibrium with the reservoir (called Donnan equilibrium), so the chemical potentials of ions in the dispersion system are equal to those in the reservoir. This equilibrium condition produces the Boltzmann distribution of the ions under an electric field expressed by using the electrostatic potential $\psi(r)$ at position r for fixed colloidal particle positions. Therefore, the densities of cations and anions denoted by $n_+(r)$ and $n_-(r)$, respectively, are expressed as

$$n_+(\mathbf{r}) = n_r \exp\left[-\frac{e\psi(\mathbf{r})}{k_B T}\right], \quad n_-(\mathbf{r}) = n_r \exp\left[\frac{e\psi(\mathbf{r})}{k_B T}\right], \tag{5-5}$$

where k_B is the Boltzmann constant, and T is the absolute temperature. The electrostatic potential satisfies Poisson's equation,

$$\nabla^2 \psi(\mathbf{r}) = -\frac{\rho_c(\mathbf{r})}{\varepsilon}, \tag{5-6}$$

for the charge density

$$\rho_c(\mathbf{r}) = e[n_+(\mathbf{r}) - n_-(\mathbf{r}) - Z\rho(\mathbf{r})], \tag{5-7}$$

where $\rho(\mathbf{r}) = \sum_{i=1}^{N} \delta(\mathbf{r} - \mathbf{r}_i)$ is the particle density in the point particle limit for the fixed particle positions $\{\mathbf{r}_i\}$ ($i = 1, 2, \ldots, N$). Equations (5-5)–(5-7) yield the so-called (nonlinear) Poisson–Boltzmann equation,

$$\nabla^2 \varphi(\mathbf{r}) - \kappa^2 \sinh \varphi(\mathbf{r}) = 4\pi l_B Z \rho(\mathbf{r}), \tag{5-8}$$

where $\varphi(\mathbf{r}) \equiv e\psi(\mathbf{r})/k_B T$ is the dimensionless electrostatic potential, $\kappa^2 \equiv 8\pi l_B n_r$ is the square of the inverse Debye screening length, and $l_B \equiv e^2/4\pi\varepsilon k_B T$ is the Bjerrum length (about 0.7 nm in water at 20°C). Equations (5-5) and (5-8) determine $n_+(\mathbf{r})$, $n_-(\mathbf{r})$, and $\psi(\mathbf{r})$ for a given $\rho(\mathbf{r})$.

Equation (5-8) is a nonlinear partial differential equation of φ and can be solved numerically. However, when the condition $|\varphi| \ll 1$ is satisfied, we can use the approximations

$$n_+(\mathbf{r}) \approx n_r[1 - \varphi(\mathbf{r})], \quad n_-(\mathbf{r}) \approx n_r[1 + \varphi(\mathbf{r})]. \tag{5-9}$$

Substituting these expressions into Eqs. (5-6) and (5-7), or using the approximation $\sinh \varphi \approx \varphi$, we obtain the linearized Poisson–Boltzmann equation,

$$\nabla^2 \varphi(\mathbf{r}) - \kappa^2 \varphi(\mathbf{r}) = 4\pi l_B Z \rho(\mathbf{r}). \tag{5-10}$$

Because $\bar{n}_+ + \bar{n}_- = 2n_r$ from Eq. (5-9), we have another expression for κ:

$$\kappa^2 = 4\pi l_B (\bar{n}_+ + \bar{n}_-) = 4\pi l_B (Z\bar{\rho} + 2C_s). \tag{5-11}$$

Here we have used the charge-neutrality condition, Eq. (5-4). Fourier transformation of Eq. (5-10) yields

$$\varphi_k = -4\pi l_B Z \frac{\rho_k}{k^2 + \kappa^2}, \tag{5-12}$$

where φ_k and ρ_k are the Fourier components of $\psi(\mathbf{r})$ and $\rho(\mathbf{r})$, respectively. Using the convolution theorem and the formula

$$\int \frac{e^{-i\mathbf{k}\cdot\mathbf{r}}}{k^2 + \kappa^2} \frac{d\mathbf{k}}{(2\pi)^3} = \frac{e^{-\kappa|\mathbf{r}|}}{4\pi|\mathbf{r}|}, \tag{5-13}$$

we obtain the following after inverse Fourier transformation of Eq. (5-12):

$$\varphi(\mathbf{r}) = -l_\mathrm{B} Z \sum_i \frac{\exp(-\kappa|\mathbf{r} - \mathbf{r}_i|)}{|\mathbf{r} - \mathbf{r}_i|}. \tag{5-14}$$

Equation (5-14) indicates that the electrostatic potential is given by the superposition of the screened Coulomb potentials due to the charged particles. Substituting Eq. (5-14) into Eq. (5-7) with Eq. (5-9), we can express the charge density as

$$\rho_c(\mathbf{r}) = \sum_i \rho_c^{(i)}(\mathbf{r}), \tag{5-15}$$

with

$$\rho_c^{(i)}(\mathbf{r}) \equiv Ze\left[\kappa^2 \frac{\exp(-\kappa|\mathbf{r} - \mathbf{r}_i|)}{4\pi|\mathbf{r} - \mathbf{r}_i|} - \delta(\mathbf{r} - \mathbf{r}_i)\right]. \tag{5-16}$$

This expression implies that the charge density due to small ions is distributed around each particle in the functional form $\exp(-\kappa r)/r$.

So far, we have treated a colloidal particle as a point charge. However, small ions cannot penetrate a particle of finite size and should be distributed outside the particle. This defect can be removed by changing the charge number Z to Z' which satisfies the condition

$$\int_{|\mathbf{r} - \mathbf{r}_i| > a} \rho_c^{(i)}(\mathbf{r}) d\mathbf{r} = Z, \tag{5-17}$$

where a is the particle radius. We find

$$Z' = \frac{\exp(\kappa a)}{1 + \kappa a} Z. \tag{5-18}$$

From Eqs. (5-14)–(5-16), if we replace Z with Z', we obtain the interaction energy $U(\{\mathbf{r}_i\})$ between particles up to linear order in φ as a function of the particle positions $\{\mathbf{r}_i\}$,

$$U(\{\mathbf{r}_i\}) = \frac{1}{2}\int \rho_c(\mathbf{r})\psi(r)d\mathbf{r} \approx U_0 + \sum_{\langle ij \rangle} U_p(\mathbf{r}_i - \mathbf{r}_j), \tag{5-19}$$

with

$$U_p(r) = \frac{(Z'e)^2}{4\pi\varepsilon} \frac{\exp(-\kappa|r|)}{|r|}, \qquad (5\text{-}20)$$

where the sum is taken over all pairs of i and j ($i \neq j$). The constant U_0 in Eq. (5-19) arises from the self-energy term ($i=j$). For a detailed discussion of this term, see van Roij et al. (1999). We see from Eqs. (5-19) and (5-20) that the effective pairwise interaction $U_p(r)$ between particles is described by the screened Coulomb, or Yukawa-type, potential, which includes Z, a, and κ as parameters. This implies that we can control the phase behaviors of the system by changing the charge number Z, the volume fraction ϕ, and the salt density C_s, taking Eq. (5-11) into account.

Note that the linearized Poisson–Boltzmann equation (5-10) is no longer valid for a highly charged colloidal system where $|\varphi| > 1$. In that case, the small ions are adsorbed on the surfaces of the particles because of the strong electric field, and thin ionic condensate layers having an opposite charge to the surface charge of the particles are formed. Those layers reduce the total charge of the particles to the effective or renormalized charge, which is much smaller than the bare charge. Therefore, even for such a highly charged system, Eq. (5-19) is still valid asymptotically as in the limit $\kappa|r_i - r_j| \to \infty$, if we use the effective or renormalized charge instead of the bare charge Z. Alexander et al. (1984) discussed charge renormalization theoretically for the cell model of charged colloidal crystals.

5.2 Samples and Methodology

Disordered charged colloids undergo a phase transition to the ordered crystal phase with increasing magnitude of the electrostatic interaction (Pieranski, 1983; Sood, 1991). Crystallization occurs at much lower particle concentrations than that for the hard-sphere colloids described in Section 5.1.1. Figures 5.7(a) and (b) show optical micrographs of aqueous dispersions of PS particles ($2a = 300$ nm) in the liquid and crystal phases, respectively. Here, crystallization is obtained by increasing ϕ from 0.006 (a) to 0.007 (b). In the liquid state, the spatial distribution of the colloidal particles is almost disordered and homogeneous, whereas they are regularly arranged in

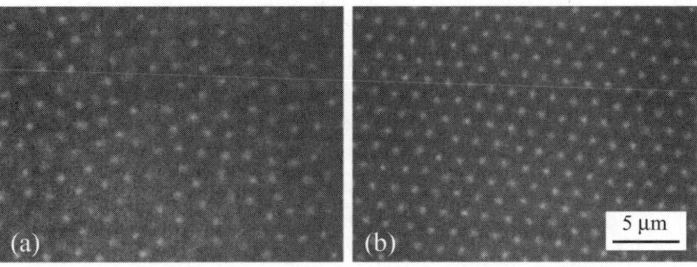

Figure 5.7 Optical micrograph of charged colloidal crystals in the (a) liquid and (b) crystal phases. Polystyrene particles (diameter = 300 nm), no added salt, $\phi =$ (a) 0.006 and (b) 0.007.

the crystal state. In this section, we describe typical experimental methods of studying colloidal crystallization, including sample preparation and purification, and crystal structure characterization.

5.2.1 Colloidal Samples

To obtain both hard-sphere and charged colloidal crystals, it is necessary to use particles having a sufficiently narrow size distribution. A distribution having a standard deviation of less than 10% is usually required. A technique for synthesizing such highly size-monodisperse polymer particles was established in the 1950s. Luck and colleagues (1963) observed that dispersions of such monodisperse particles show brilliant iridescent color and found that it was caused by Bragg diffraction of visible light from the ordered crystal structures. Syntheses of uniform particles have been developed for polymer particles (e.g., PS) (Chonde and Krieger, 1981) and poly(methyl methacrylate) (Bosma et al., 2002) particles, and colloidal silica (SiO_2) particles (Iler, 1979), which are frequently used in crystallization experiments.

In many cases, the colloidal particles bear electric charges on their surfaces. The polymer particles sometimes have strong acid groups, such as sulfonic acid (SO_3H) and sulfuric acid (SO_4H) groups, which are introduced by copolymerizing a charged monomer (e.g., styrenesulfonate) and by a polymerization initiator (e.g., potassium persulfate). In aqueous dispersions, these groups dissociate (e.g., $SO_3H \rightarrow SO_3^- + H^+$) to produce negative surface charges. Positively charged particles have been synthesized by copolymerizing cationic monomers (e.g., vinylpyridine) and cationic polymerization initiators (Bazin and Zhu, 2010).

The surfaces of metal oxides, including silica, are generally covered with dissociable groups, M-OH (M: metal atom), when they are in contact with water (Iler, 1979). The dissociation state and degree of dissociation of the M-OH groups vary significantly with changes in the pH. At low pH, they dissociate and provide positive charges ($M^+ + OH^-$), whereas negative surface charges are produced at high pH ($MO^- + H^+$). At an intermediate pH, a condition exists in which the net surface charge is zero. This pH value is referred to as the isoelectric point (iep). Silica particles have Si-OH (silanol) groups on their surfaces. The iep of the silica is at pH$\approx$2. The tuning of surface charges on silica particles will be described in Section 5.3.1.

The electrostatic interaction between charged colloidal particles is significantly screened by the small ions present in the dispersion medium, as described in Section 5.1.3. Thus, the crystal state is favored under low-salt conditions. Therefore, purification (in particular, deionization) of the colloid samples is a crucial process in crystallization experiments (Murai et al., 2007). Samples are typically purified by dialysis against pure water or ultrafiltration. The completeness of the purification is judged by measuring the electrical conductivity of the water. The samples are further deionized by using a mixed bed of cation and anion exchange resin beads. These resin beads exchange cations and anions in the aqueous media with H^+ and OH^- ions, respectively. The resulting H^+ and OH^- ions yield water molecules. We note that glassware should not be used for experiments on charged colloids because impurity ions (e.g., Na^+ and K^+ ions) are eluted from the glass walls. For silica particles, the elution of these alkali ions sometimes results in slight crystallization, because the Z values increase with the addition of a base (see Section 5.1.3). Quartz and plastic apparatus should be used instead of glassware; Teflon® is one of the most desirable materials. Airborne carbon dioxide generates carbonic acid when dissolved in water, which also acts as an ionic impurity. Thus, it is advisable to perform the experiments under a nitrogen or argon atmosphere or in a closed-circuit system.

Colloid characterization includes determination of the particle size, charge number, and particle concentration. The particle size and its distribution can be determined from the diffusion coefficients of the particles by using the dynamic light-scattering method (Johnson and Gabriel, 1981) or measured directly using transmission electron microscopy. Ultra-small-angle X-ray scattering measurements (USAXS, see later) are also applicable. The charge number can be obtained by measuring the electrical conductivity (Sood, 1991; Palberg et al., 1999) or zeta potential (Hunter, 1981; Russel et al., 1989). The particle concentration can be determined by a drying method.

5.2.2 Characterization of Crystal Structures

As in ordinary crystalline materials, the structures of colloidal crystals can be characterized by scattering (X-ray, neutron, and laser light) experiments and by microscopy. Bragg diffraction from colloidal crystals is detectable by spectroscopy. Scattering methods and spectroscopy yield statistically averaged information from a macroscopic region, whereas microscopy provides insight into the local structural characteristics.

X-ray Scattering

Light scattering is a commonly applied scattering method for studying systems having micron-length to submicron-length scales. However, because colloids are often highly turbid, light scattering is not always applicable. Therefore, scattering of X-rays, which arises from spatial fluctuations of the electron density in the sample, is applied to colloidal systems. Because X-rays scatter from micron-sized fluctuations at very small scattering angles (2θ is on the order of 0.01 degree of arc), USAXS measurements are required. Specially designed optics referred to as a Bonse–Hart camera (Bonse and Hart, 1966; Koga et al., 1996; Ise and Sogami, 2005) has been developed; it consists of two triple-bounce channel-cut crystals acting as a monochromator and an analyzer. This setup enables measurements at very low angles. USAXS measurements can also be made by using synchrotron radiation having very high brilliance because the sample-to-detector distance can be long enough to obtain accurate scattering patterns at low angles. Figure 5.8 shows the scattering pattern for colloidal crystals (sample: silica colloid, particle size = 105 nm, $\phi = 0.03$) obtained by the USAXS apparatus set up at the SPring-8 synchrotron radiation facility, JASRI, Japan.

The general principle of the scattering method was outlined in Section 5.1.2. The crystals contain a number of lattice planes characterized by the Miller indices (ν_1, ν_2, ν_3) (see Box 5.2). Calculations of $S(k)$ for various lattice planes based on the G values show that not all the lattice planes cause diffractions, but for some

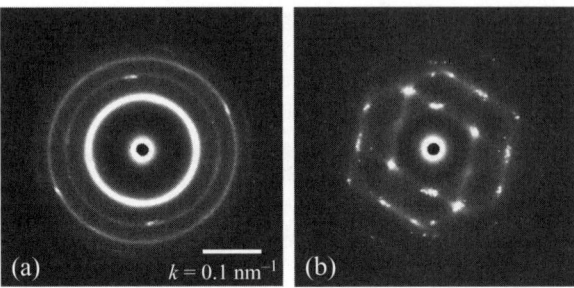

Figure 5.8 Ultra-small-angle X-ray scattering patterns of charged colloidal crystals. Silica particles (diameter = 105 nm), no added salt, $\phi = 0.03$, immobilized by polymer gel. (a) Polycrystal and (b) large crystal. (Courtesy of Dr. Tsutomu Sawada, National Institute of Materials Sciences; Dr. Shigeo Hara, Hamamatsu Photonics Co., Ltd.; and Mr. Yukihiro Sugao, Nagoya City University)

Box 5.2 Miller Index of Crystal Lattice Planes

In crystallography, the lattice planes of crystals are represented by three integers—v_1, v_2, and v_3—which are referred to as the *Miller indices*. A plane orthogonal to a direction (v_1, v_2, v_3) in reciprocal lattice space is donated as $(v_1\ v_2\ v_3)$. The Miller indices of the cubic lattices (the simple, bcc, and fcc lattices) are determined as follows (Kittel, 2005). First, determine the intercepts of the lattice planes along the x, y, and z axes (Figure B5.1). Then, take the reciprocals of the intercepts, and obtain their lowest integer ratio. These values are the Miller indices. In the case illustrated in Figure B5.2(a), the x, y, and z intercepts are 2, 3, and 2, respectively, which yield $(v_1\ v_2\ v_3) = (3\ 2\ 3)$. If the lattice planes are parallel to the x, y, or z axes, the intercepts are considered to be at infinity—that is, the reciprocals are zero. The Miller indices of three lattice planes are shown in Figure B5.1(b). The set of all planes that are equivalent to the $(v_1\ v_2\ v_3)$ plane by symmetry is denoted as $\{v_1\ v_2\ v_3\}$. For example, the planes drawn in Figure B5.1(c) are donated as $\{110\}$ planes.

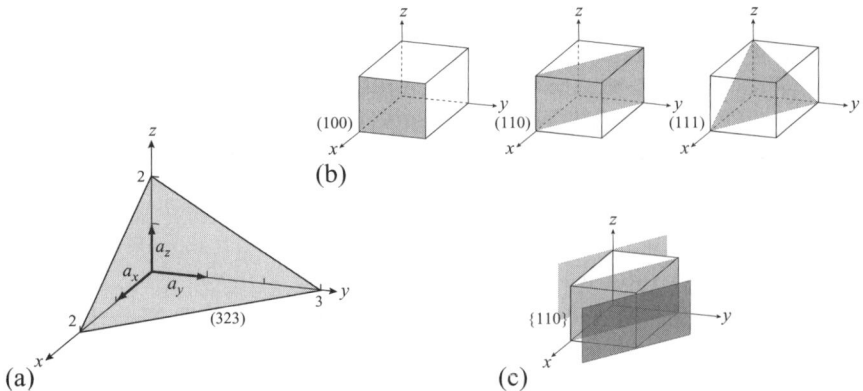

Figure B5.1 Miller indices for various lattice planes of the cubic crystal lattice.

planes $S(\mathbf{k}) \equiv 0$ (Kittel, 2005). For body-centered cubic (bcc) lattices, $S(\mathbf{k}) \equiv 0$ when the value of $v_1 + v_2 + v_3$ is odd. That is, diffractions arise from the (110), (200), (211), ... planes. For fcc lattices, $S(\mathbf{k})$ has nonzero values only when v_1, v_2, and v_3 are all odd or all even. Therefore, the (111), (200), (220), ... planes cause diffraction. As described in Section 5.1.2, diffraction occurs when $\mathbf{G} = \mathbf{k}$. For simplicity, we now consider randomly oriented polycrystals having bcc symmetry. Then, because the diffractions are isotropic, the diffraction conditions are represented as $G = k$. Note that $k = 4\pi \sin\theta/\lambda$, where 2θ is the scattering angle (see Figure 5.4), and λ is the wavelength of X-rays. Therefore, when we measure the scattering intensity I at various θ values and plot it as a function of k, a series of peaks appear in the plot. The Bragg relation is given by $2d \sin\theta = n\lambda$, where d is the corresponding lattice spacing, and n is an integer. Thus, the k values at the peaks are given by $k_m = 2\pi/d$ for first-order diffraction ($n = 1$). The d value for a ($v_1v_2v_3$) lattice is generally expressed as $a_l/\sqrt{v_1^2 + v_2^2 + v_3^2}$, where a_l is the lattice constant (Kittel, 2005). For bcc crystals, the ratio of k_m is represented by the square roots of integers—that is, 1, $\sqrt{2}$, $\sqrt{3}$, ... ($\sqrt{3}$, 2, $2\sqrt{2}$, ... for the fcc lattice). Thus, we can determine the crystal lattice structures from the peak positions. The scattering pattern shown in Figure 5.8(a) shows that the crystals had the bcc lattice symmetry.

We note that the scattering profile is proportional to the product of the form factor $F(k)$ and the structure factor $S(k)$—that is, $I(k) \sim F(k)S(k)$. Here $F(k)$ represents the scattering due to an isolated particle—that is, interference of an incident X-ray inside the particle. Therefore, $F(k)$ is determined solely by the shape and size of a single particle. For a sphere, $F(k)$ is represented by

$$F(k) = 3[\sin(ka) - ka\cos(ka)]/(ka)^3 \qquad (5\text{-}21)$$

(Johnson and Gabriel, 1981). When no interparticle correlation is present (e.g., under sufficiently dilute and high-salt conditions), $S(k)$ is unity regardless of k. Then we can determine the particle size by fitting the form factor.

Spectroscopy

Bragg diffractions from colloidal crystals can also be detected by visible and near-infrared spectroscopy. Figures 5.9(a) and (b) show transmission and reflection spectra, respectively, from colloidal crystals (silica colloid, particle size = 110 nm, $\phi = 0.035$). In Figure 5.9(a), data for the liquid state are also shown by a dashed curve for comparison. Figure 5.10(a) demonstrates the variation in the reflection spectra of a silica colloid (particle size = 230 nm) with changing ϕ (Murai et al., 2012). The spectra were acquired by visible and near-infrared spectroscopy. The first Bragg diffraction peaks (shown by arrows with asterisks in Figure 5.10a) appear in the near-infrared region. They are associated with several higher-order peaks in the visible regime, which are indicated by arrows above the spectra. Note that here we observe the right-angle diffraction—that is, for $\theta = \pi/2$. Thus, k is equal to $2\pi/\lambda$. Here, λ is the wavelength of the incident light beam in the colloid (i.e., $\lambda = \lambda_0/n_r$) where λ_0 is the wavelength in a vacuum, which is almost the same as that in air. Further, n_r

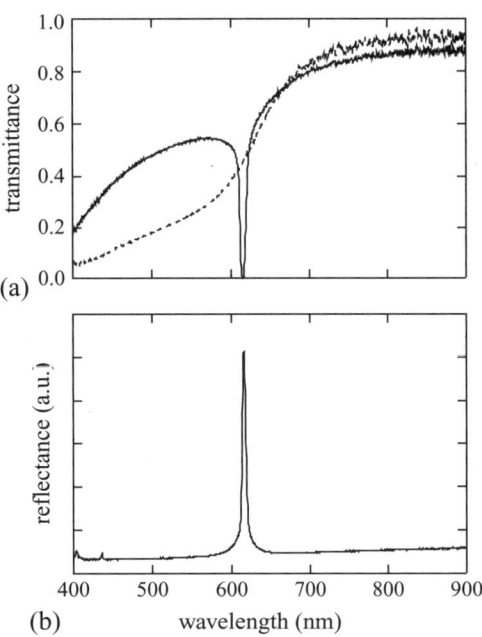

Figure 5.9 (a) Transmission and (b) reflection spectra of charged colloidal crystals of silica particles. The dashed curve in (a) represents data for the liquid state for comparison.

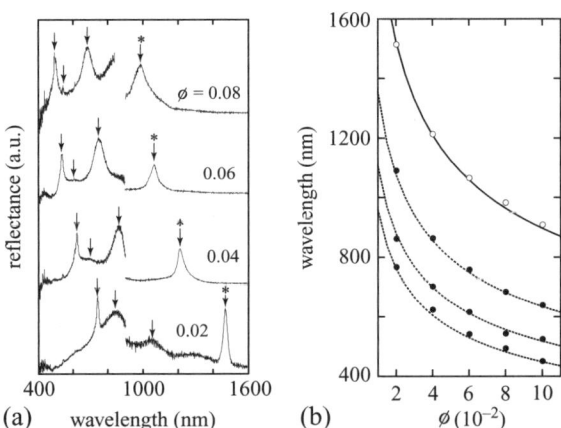

Figure 5.10 (a) Visible and near-infrared reflection spectra of colloidal silica crystals at various values of ϕ. (b) Bragg peak wavelengths plotted against ϕ (curves: calculated; circles: observed).

represents the refractive index of the colloids, which can be approximated for dilute colloids as a volume-averaged value, $n_r = \phi\, n_p + (1-\phi)n_w$, where n_p and n_w are the reflective indices of the particle (1.42 for silica) and water (1.33), respectively.

The Bragg wavelengths λ_B from the (110) plane of the bcc crystal are related to ϕ as follows. The Bragg relation for right-angle diffraction from the (110) plane is given by

$2d_{110} = n\lambda_B/n_r$. Here n is an integer, and d_{110} is the lattice spacing for the (110) planes, which is expressed as $d_{110} = a_l/\sqrt{2}$. On the other hand, the lattice constant a_l is related to ϕ by $a_l = \sqrt[3]{8\pi/(3\phi)}\, a$ for bcc symmetry. From these relationships, we can calculate the λ_B value for (110) diffraction at given values of ϕ. Figure 5.10(b) compares the calculated (solid curve) and observed (open circles) values of λ_B for first-order diffraction ($n = 1$). In scattering measurements, the θ dependence of the diffraction intensity is measured at a constant value of λ_0. On the other hand, in reflection spectroscopy, the diffraction intensity is recorded with varying λ_0 at constant θ ($=\pi/2$). Because λ is inversely proportional to k, the ratios of the diffraction wavelengths in the spectrum are represented as a series of inverse square roots of integers ($= 1, 1/\sqrt{2}, 1/\sqrt{3}$, and $1/\sqrt{4}$). In Figure 5.10(b), the Bragg wavelengths calculated for the isotropic diffractions up to the fourth order (dashed curves) also showed good agreement with the observed values (filled circles).

Microscopy

Microscopy is widely applied for studying the structures of colloids. Conventional microscopic observation is limited to a region close to a coverslip, where the interface between the coverslip and the colloid might strongly affect the structure of the colloids. To eliminate such effects, confocal laser scanning microscopy (CLSM) is frequently used (Wilson, 1990). CLSM produces optical images at selected focal depths. A three-dimensional (3D) image can be obtained by reconstructing the two-dimensional (2D) sliced images. CLSM enables the observation of both individual particles and their higher-order organizations in the interior of the samples, up to several hundred micrometers inside. Sometimes the colloidal particles used are smaller than the resolution limit of the optical microscope (a few hundred nanometers). In that case, the particles imaged by CLSM are the projections of spread images on the focal plane, which correctly demonstrate the positions of the individual particles, but not their sizes. CLSM images of colloidal crystals will be presented in Section 5.3.4.

5.3 Crystallization of Charged Colloids

5.3.1 Charge-Induced Crystallization

The major experimental parameters that govern the interaction magnitude of charged colloids are the particle charge number Z, salt concentration C_s, and particle volume fraction ϕ. The interaction is stronger at higher values of Z and ϕ. On the other hand, salts dissociate into ions, which screen the electrostatic interaction. Therefore, the interaction is weaker at higher C_s. Among these three parameters, C_s and ϕ are easily tunable over a few orders of magnitude. Therefore, most previous experimental studies adopted C_s and ϕ as the experimental parameters (Sood, 1991). However, tuning the Z value has not always been easy. Palberg et al. (1995) examined the Z dependence by using adsorption of an ionic surfactant. Yamanaka et al. (1998) examined the effect of charge on colloidal crystallization by using silica particles of which the Z values can

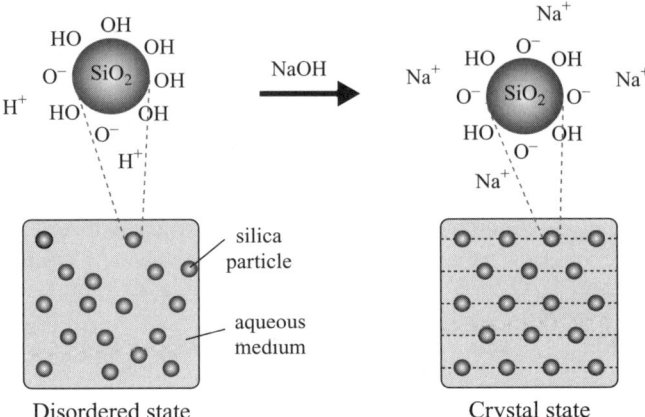

Figure 5.11 Schematic drawing of the tuning of the surface charge number of silica particles and charge-induced crystallization of colloidal silica.

be controlled by varying the pH, as follows. The silica particles are slightly charged in their aqueous dispersions due to self-dissociation of silanol groups on their surfaces ($\equiv$Si–OH $\leftrightarrow$ $\equiv$Si–O$^-$ + H$^+$). Because the silanols are weakly acidic, the Z value increases with the addition of a base, such as sodium hydroxide (NaOH) ($\equiv$Si–OH + Na$^+$OH$^-$ $\rightarrow$ $\equiv$Si–O$^-$ Na$^+$ + H$_2$O; see Figure 5.11). Therefore, under the appropriate conditions, silica colloids undergo charge-induced crystallization with the addition of a base. A crystallization phase diagram of colloidal silica ($\phi = 0.034$) defined by the NaOH and NaCl concentrations (denoted by S and [NaCl], respectively) is shown in Figure 5.12 (Murai et al., 2007). The crystal states were detected by observing the Bragg diffraction. The observed phase boundary is presented in open symbols. The region where [NaCl] is smaller than that at the phase boundary corresponds to the crystal phase.

At sufficiently large [NaOH], where the counterions are solely Na ions, the Z value of the silica particles is calculated as the number of Na ions per particle. However, because of the large electrostatic potential on the particle surface, the counterions are condensed in a region near the particle surface, which reduces the charge number. Thus, the effective charge number Z_e is quite different from the net charge number Z_a; the phase behavior of charged colloids is governed by Z_e, not Z_a. The Z_e values have been estimated by electrical conductivity measurements, surface potential (zeta potential) measurements, and a rheological method. Yamanaka et al. (1998) determined Z_e for PS and silica particles and found that an empirical relationship,

$$\log Z_e = C_1 \log Z_a + C_2, \tag{5-22}$$

holds for both particles (Yamanaka et al., 1998; Yoshida et al., 1999). The values of C_1 and C_2 were approximately 0.5 and 1.2–1.4, respectively.

Robbins and colleagues (1988) reported a theoretical phase diagram based on numerical simulation results using the Yukawa potential. It has been reported that their phase diagram showed a close agreement with the experiments under sufficiently

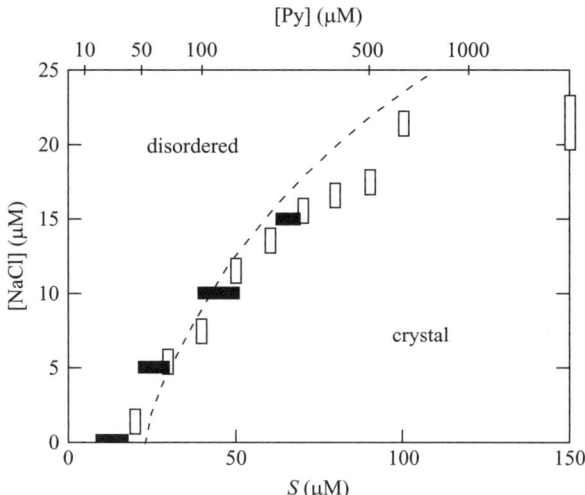

Figure 5.12 Crystallization phase diagram of silica colloids denoted by S and [NaCl]. Open and filled rectangles represent the phase boundaries obtained by using NaOH and Py, respectively. Several values of [Py] are shown on the upper abscissa for reference. The dashed curve is a theoretical phase boundary based on the Yukawa potential.

weak interaction (Sood, 1991). In Figure 5.12, the crystallization phase boundary calculated using their results for the Z_e value is indicated by a dashed curve. The theory agrees very well with the observation under the conditions examined here. We note that a discrepancy was observed between the theory and the experiments at higher Z_e (Yamanaka et al., 1998; Ise and Sogami, 2005).

5.3.2 Unidirectional Crystallization

Colloidal crystals are generally polycrystals composed of small crystal grains. The crystal grains are larger under conditions closer to the crystallization phase boundary. This trend is also common in atomic and molecular crystals, and it is attributed to smaller nucleation and growth rates under conditions closer to the phase boundary. Because the material applications of colloidal crystals—for example, their use in optical materials—depend significantly on their size and quality, controlled crystallization methods have been devised in the field of material science (van Blaaderen, 2004). They include shear annealing of polycrystals confined in submillimeter gaps (Clark et al., 1979; Kanai et al., 2005), epitaxial growth from 2D templates (van Blaaderen et al., 1997b), and electrophoresis (van Blaaderen, 2004). All these methods provide large-area, thin (<100 μm), film-shaped colloidal crystals having good optical quality. However, it is usually difficult to prepare large 3D colloidal crystals.

In atomic and molecular crystals, unidirectional growth under a temperature gradient is used to obtain large single crystals (Dhanaraj et al., 2010). Although the temperature is not an effective parameter for colloidal crystallization, we can expect unidirectional crystal growth under Z, C_s, and ϕ gradients. In addition, by using the temperature dependence of Z, C_s, or ϕ, crystals can be grown under a temperature gradient. Here we describe the unidirectional crystallization under gradients of Z and temperature.

Crystallization Due to Base Diffusion

On the basis of the charge-induced crystallization of silica colloids described in Section 5.3.1, Yamanaka and colleagues examined the unidirectional crystallization of colloidal silica due to diffusion of a base (Yamanaka et al., 2004; Murai et al., 2007). They used the weak base pyridine (Py). In aqueous solutions, Py molecules dissociate only slightly to provide basic species ($py + H_2O \leftrightarrow pyH^+ + OH^-$, where py and pyH^+ denote undissociated and dissociated Py molecules, respectively). When Py is added to silica, negative surface charges are generated on the silica surfaces by the reaction $SiOH + py \rightarrow SiO^- + pyH^+$. For a strong base such as NaOH, unidirectional crystallization was not observed because the NaOH molecules reacted with the silanols almost completely, so the concentration of diffusing species was negligibly small. On the other hand, undissociated Py molecules in the medium were mobile and diffused in the silica colloids.

Figure 5.13(a) illustrates the experimental setup for unidirectional crystallization. Py molecules diffused into colloidal silica from a reservoir of an aqueous Py solution through a semipermeable membrane (sample cell size $= 1 \times 1 \times 4$ cm^3, reservoir volume $= 500$ ml). Photographs of the crystallization process are also shown in Figure 5.13(b) (particle diameter $= 110$ nm, salt-free, $\phi = 0.034$, Py concentration in the reservoir $[Py]_0 = 100$ mM). Columnar crystals having lengths of a few centimeters were formed within one day. The crystal region showed bright Bragg diffraction of visible light. The crystals were larger at a slower growth rate and reached $1 \times 1 \times 3$ cm^3 at the largest. Figure 5.14 shows the crystal growth curves for three values of $[Py]_0$.

The unidirectional crystallization previously described can be regarded as a combination of three processes: (1) diffusion of a base accompanied by the reaction with silica particles, (2) an increase in the Z value of silica, and (3) crystallization of the colloidal silica due to the increase in Z. On the basis of this model, the crystal growth curves were obtained as follows.

The reaction between silica particles and Py is regarded as electrostatic adsorption of Py onto the silica surface. Hereafter, we denote the adsorption amount per particle

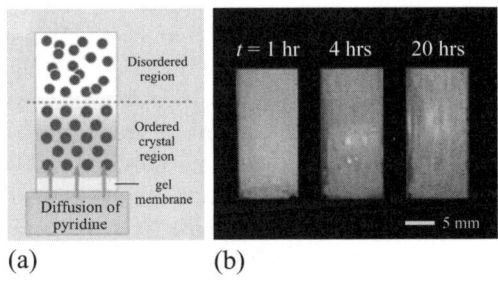

Figure 5.13 (a) Illustration and (b) images of typical crystal growth process of colloidal silica due to diffusion of Py. (For color version of this figure, the reader is referred to the online version of this chapter.)

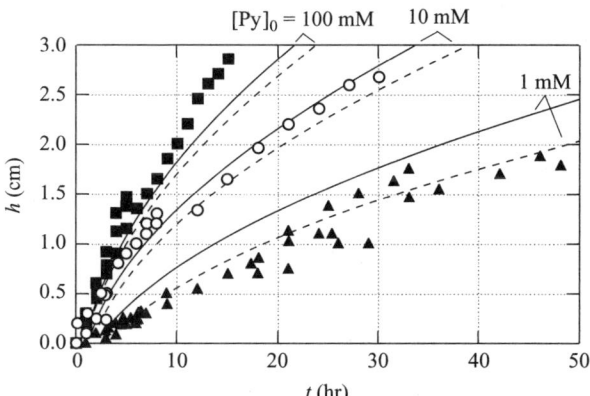

Figure 5.14 Crystal growth curves for colloidal silica due to diffusion of Py at three values of [Py]$_0$. Solid and dotted curves are theoretical growth curves for salt-free conditions and in the presence of 5 μM of salt, respectively.

as S and the concentration of free Py as C. Note that Z_a is proportional to S. The relationship between S and C was determined by performing separate titration experiments. The crystallization phase diagram of the silica+Py system, defined by S and [NaCl], is shown in Figure 5.12 (filled symbols). Because the strong base NaOH reacts almost completely with the silica, we can assume that [NaOH]=S. The phase diagram shows good agreement with that obtained for the silica+NaOH system when the data were plotted using the S values.

We assume instantaneous equilibrium between S and C upon diffusion of Py. Then the adsorption–diffusion equation is given by

$$\partial C/\partial t = [D/(1+\partial S/\partial C)]\partial^2 C/\partial x^2 \qquad (5\text{-}23)$$

where x is the distance from the reservoir, and D is the diffusion coefficient of Py. We numerically solved Eq. (5-23) and obtained $C(x, t)$ and $S(x, t)$. Figures 5.15(a) and (b) show the profiles of $C(x, t)$ and $S(x, t)$, respectively, for [Py]$_0$ = 1 mM. From the crystallization phase diagram (Figure 5.12), we estimated the S value at crystallization, $S^* = 20$ μM. On the basis of the time evolution of S, we calculated the crystal growth curve as the x–t curve that satisfies $S(x, t) = S^*$. The solid curves in Figure 5.14 are the theoretical growth curves for the salt-free condition, which agree well with the observations. The slightly slower growth at low [Py] might be due to the presence of ionic impurities. The case of $C_s = 5$ μM is shown by dotted curves, which are in rather good agreements with the experiments.

Crystallization under Temperature Gradient

Temperature is generally a weaker variable for the crystallization of a charged colloid than Z, ϕ, and C_s. This is because the permittivity of water ε decreases significantly with T. In both κ and the reduced Yukawa potential $U_p/k_B T$, T is introduced as the

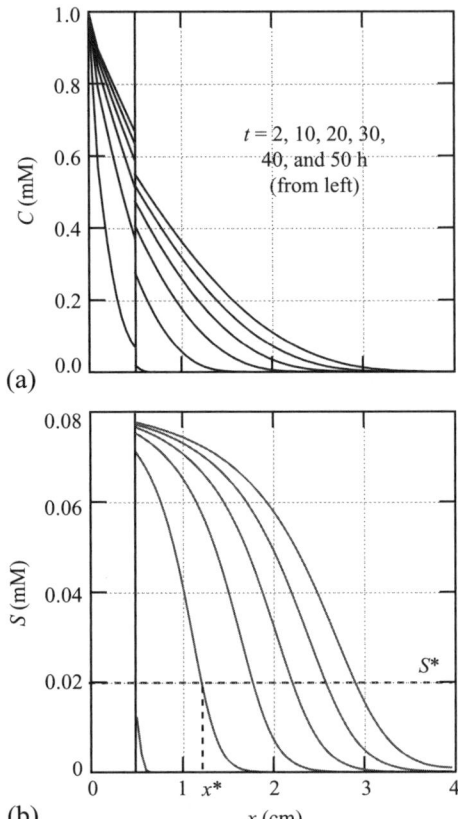

Figure 5.15 Time evolution of profiles of (a) C and (b) S at $[Py]_0 = 1$ mM. Values of t shown in (a) also apply to (b). Solid vertical lines in (a) and (b) indicate the membrane–colloid boundary ($x = 0.5$ cm). In (b), the location of the crystal front ($x = x^*$) is shown when $S^* = 20$ μM at $t = 10$ h.

product εT, which does not change significantly with T. For water, when T is raised from 0 to 100°C, εT decreases slightly (3%). Therefore, the crystallization of a charged colloid does not depend remarkably on T. Toyotama and Yamanaka (2011) reported that charged colloids undergo either freezing or melting transitions with increasing temperature because of a slight change in ε. Freezing occurs in low-Z, low-C_s colloids, whereas melting is observed in high-Z, high-C_s colloids. However, these crystallizations were observed under quite limited experimental conditions. On the other hand, by using the temperature dependence of Z, ϕ, and C_s, one can expect thermally induced crystallization. The dissociation of Py is larger at higher temperatures. Thus, in the presence of Py, the Z value of silica particles increases with T, resulting in crystallization with *increasing* temperature, under appropriate conditions.

Using colloidal silica + Py colloids, Toyotama et al. (2007) examined unidirectional crystallization under a temperature gradient. Figure 5.16(a) shows an

Colloidal Crystals

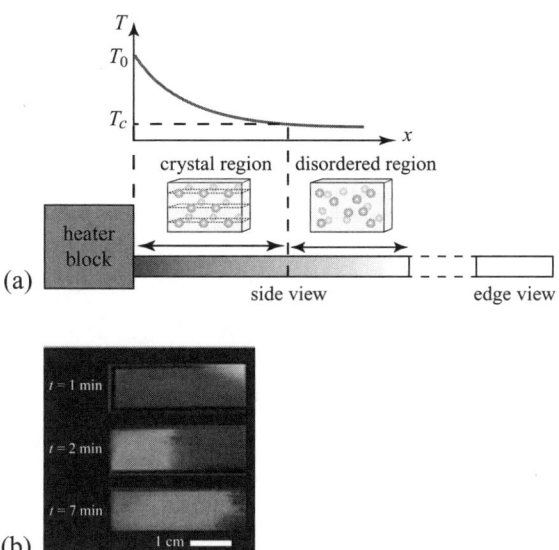

Figure 5.16 (a) Experimental setup for thermally induced unidirectional crystallization. (b) Photographs of crystallization process. (For color version of this figure, the reader is referred to the online version of this chapter.)

illustration of the experimental setup. The colloidal silica sample was introduced into a quartz cell (cell size $= 0.1 \times 1 \times 4.5$ cm^3), and one end of the cell was heated by a heater block. Figure 5.16(b) shows photographs of a typical crystallization process ($\phi = 0.035$, [Py] $= 35$ μM). The heater temperature T_0 was maintained at 60°C. Before heating, the colloid took a disordered liquid state, whereas crystals grew in response to heat conduction. Centimeter-sized crystals were formed within 10 min. Figure 5.17(a) shows the crystal growth curves at three values of T_0. Temperature profiles on the cell surface were measured by a thermocouple array. Figure 5.17(b) shows the change in the temperature profile over time ($T_0 = 60°$C). The sample had a freezing temperature $T_c = 35°$C. The crystal length was calculated at various t assuming instantaneous crystallization. The growth curve (x–t plot) determined from the relation $T(x, t) = T_c$ showed good agreement with the observed crystal growth. This suggests that the present growth can be attributed to a combination of heat conduction and thermally induced crystallization.

Figure 5.18(a) shows transmission spectra of a crystal sample taken every 5 mm by fiber optics in a circular area having a diameter of about 1 mm. The broken curve is the spectrum from the disordered structure for comparison. Sharp dips due to Bragg diffraction of visible light were observed. The half-bandwidth was 5 nm, and the Bragg wavelength exhibited good spatial uniformity (standard deviation of 0.1%). Crystals formed by base diffusion usually have a much wider diffraction peak width (larger than 10 nm) and larger inhomogeneity (a few tens of percent). Heat conduction is a diffusion of thermal energy, which can occur much more rapidly than mass diffusion. Thus, the growth rate for thermally induced crystallization could be much larger

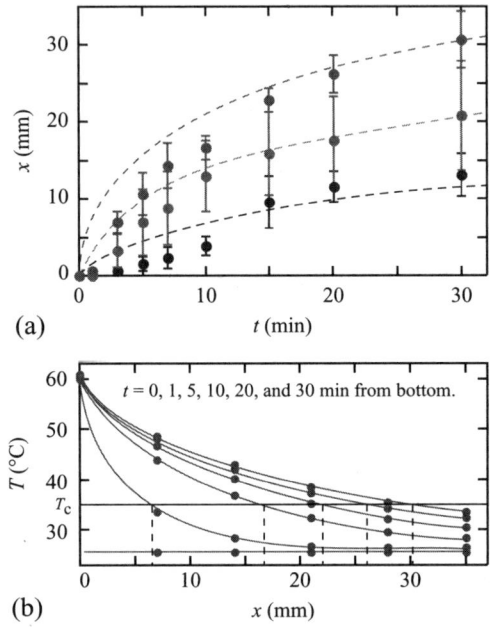

Figure 5.17 (a) Observed (symbols) and theoretical (dashed curves) crystal growth for $T_0=40°C$, $50°C$, and $60°C$. (b) Time evolution of $T(x)$ for $T_0=60°C$. Locations of crystallization fronts at various t are represented by dashed vertical lines.

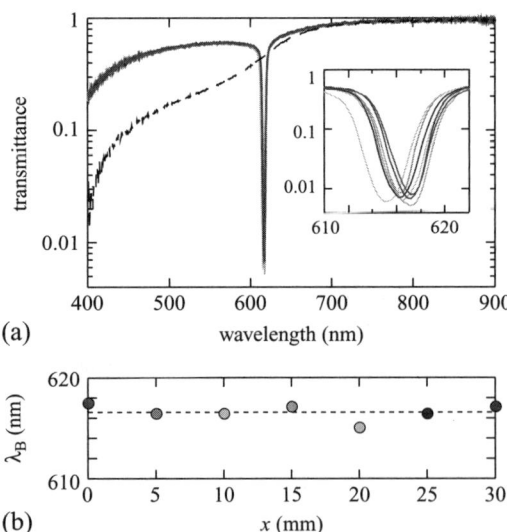

Figure 5.18 (a) Transmission spectra of the crystal at various locations ($x=0$–30 mm; see text). Broken curve is spectrum from disordered sample. Inset shows an enlarged view around the dips. (b) Bragg wavelength at various locations.

than that of Py diffusion (less than a few millimeters per hour). The good uniformity attained in the present crystallization appeared to result from its much faster growth rate.

5.3.3 Gel Immobilization

The colloidal crystals obtained in the studies previously described will be applicable as novel materials such as photonic crystals (Moon and Yang, 2010). However, charged colloidal crystals are easily melted by shear because they are formed by electrostatic interaction in liquid media and are quite fragile. Thus, immobilization of the crystal structures in a polymer gel matrix has been studied (Holtz and Asher, 1997). In many studies, polyacrylamide gels have been used to immobilize colloidal crystals. However, they are not suitable for large crystals because the acrylamide monomer decomposes in aqueous solution to produce small amounts of ionic impurities, which cause the large crystals to melt. Murai et al. (2007) reported immobilization of large colloidal crystals by using N-methylolacrylamide as the gel monomer, which could be satisfactorily deionized by using the ion exchange method. Figure 5.19 shows a photograph of such a large gelled crystal obtained by Py diffusion.

The gelled crystals can be deformed by applying mechanical stress. This deformation causes changes in the Bragg diffraction wavelength of the colloidal crystals fixed in the gel matrix (Figure 5.20a) (Iwayama et al., 2003). In other words, the Bragg wavelengths of the gel-immobilized colloidal crystals can be tuned by applying mechanical stress. The relationship between the gel thickness (reduced by the initial value) and Bragg wavelength is demonstrated in Figure 5.20(b). The plots show good linearity in almost the entire visible light wavelength regime. Gelled colloidal crystals are also attracting attention as novel sensing materials (Holtz and Asher, 1997).

5.3.4 Exclusion of Impurity Particles

Unlike atomic crystals, colloidal crystals are made up of particles of which the size is more or less nonuniform. Sometimes the colloids contain particle aggregates and/or particles of which the sizes differ greatly from the average. They act as "impurities" during colloidal crystallization and lead to the formation of various types of crystal

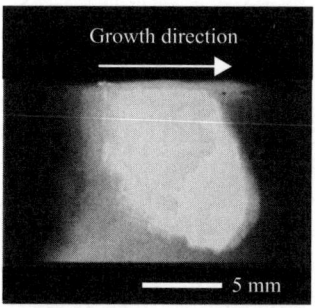

Figure 5.19 Photograph of gel-immobilized colloidal crystal obtained by Py diffusion. (For color version of this figure, the reader is referred to the online version of this chapter.)

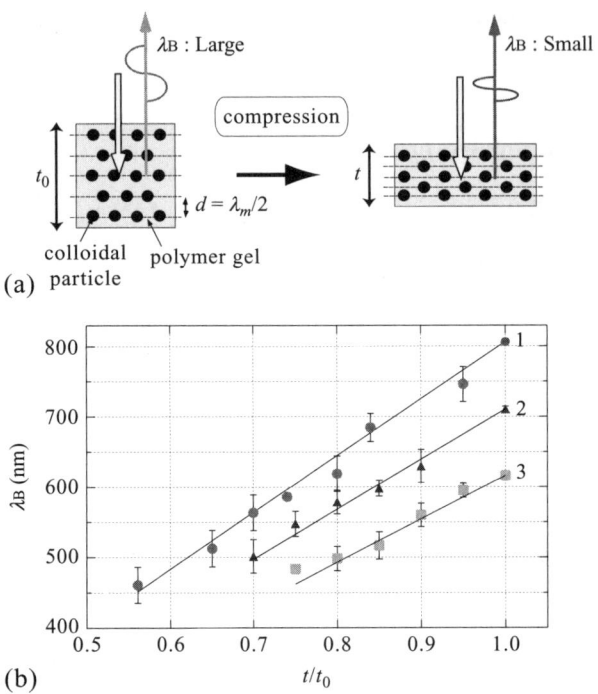

Figure 5.20 (a) Tuning of Bragg wavelength of gel-immobilized colloidal crystal by mechanical compression. (b) Bragg wavelength versus gel thickness (reduced by initial value) for three samples.

defects. Yoshizawa et al. (2011) examined the behavior of these impurity particles of which the size is different from that of the majority of particles.

The sample used was a binary mixture of fluorescent-labeled charged PS particles (diameter = 333 nm, $Z = 1680$, $\phi = 0.0001$) added to colloidal crystals of charged silica particles (diameter = 108 nm, $Z = 220$, $\phi = 0.05$). The charge number of the silica colloid was tuned by the addition of 100 μM Py to form crystals. Figure 5.21(a) shows CLSM images of the binary colloid recorded in the reflection mode ($t = 1260$ min after homogenization of the sample by shaking). The image shows that the sample contained polycrystals composed of crystal grains (domains). Differences in the gray levels of the individual crystal grains is due to differences in the orientation of the crystal lattice planes. Figure 5.21(b) shows a micrograph of the same field of view taken in the fluorescent mode. In this mode, the spatial distribution of the PS particles is easily observed. Figure 5.21(c) shows the superposition of the images in Figures 5.21(a) and (b). Apparently, most of the PS particles are distributed at the crystal grain boundaries. Figure 5.22 is a 3D image ($t = 780$ min), which reveals that the PS particles accumulate at the grain boundaries to form a cellular structure analogous to that observed in metals and froths.

Figures 5.23(a)–(i) are CLSM images showing the time evolution of the PS particle distribution. Individual PS particles can be seen in these images. At $t = 15$ min (a), exclusion of the PS particles at the grain boundaries is already detected, although only

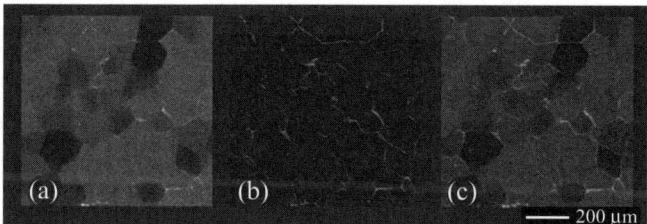

Figure 5.21 CLSM images of binary mixture of silica ($\phi = 0.05$) and fluorescent polystyrene ($\phi = 0.0001$) particles; $t = 1260$ min after homogenization. (a) Reflection and (b) fluorescence images. (c) Superposition of (a) and (b). (For interpretation of the references to color in this figure legend, the reader is referred to the online version of this chapter.)

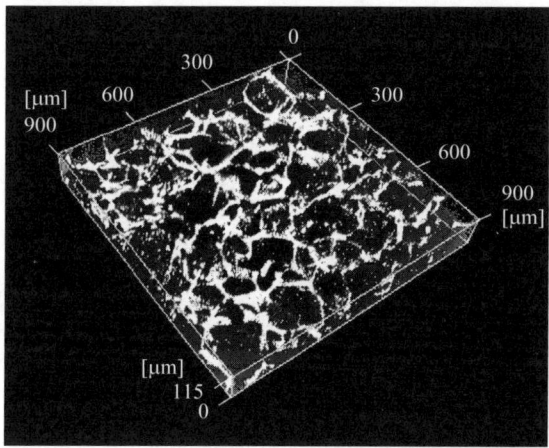

Figure 5.22 Three-dimensional reconstructed image of binary colloid. (For color version of this figure, the reader is referred to the online version of this chapter.)

slightly. Hereafter, we refer to the three types of crystal grains indicated in (a) as nos. I, II, and III. The average grains size in a material is well known to increase over time at high temperature because the free energy of the system is decreased by reducing the total area of the grain boundaries. This process is referred to as grain growth (Dhanaraj et al., 2010). Figures 5.23(b) and (c) show that the grain boundaries start migrating because of grain growth. At $t = 265$ min, grain III increases in size, whereas grain I shrinks and eventually disappears (d). During this process, the PS particles become concentrated and accumulate at the crystal grain boundaries of colloidal silica. Grain II then shrinks with time and eventually vanishes within 435 min (g). We note that the PS particles that are not swept up by the grain boundaries remain near their original positions. This suggests that the observed exclusion of the PS particles is triggered by grain boundary migration.

Yoshizawa and colleagues further studied the exclusion of the impurity particles during unidirectional crystallization. Upon unidirectional crystallization due to Py diffusion, the PS particles sometimes formed submillimeter-sized periodic stripe patterns (Figure 5.24) (Yoshizawa et al., 2012). The mechanism is attributable to a

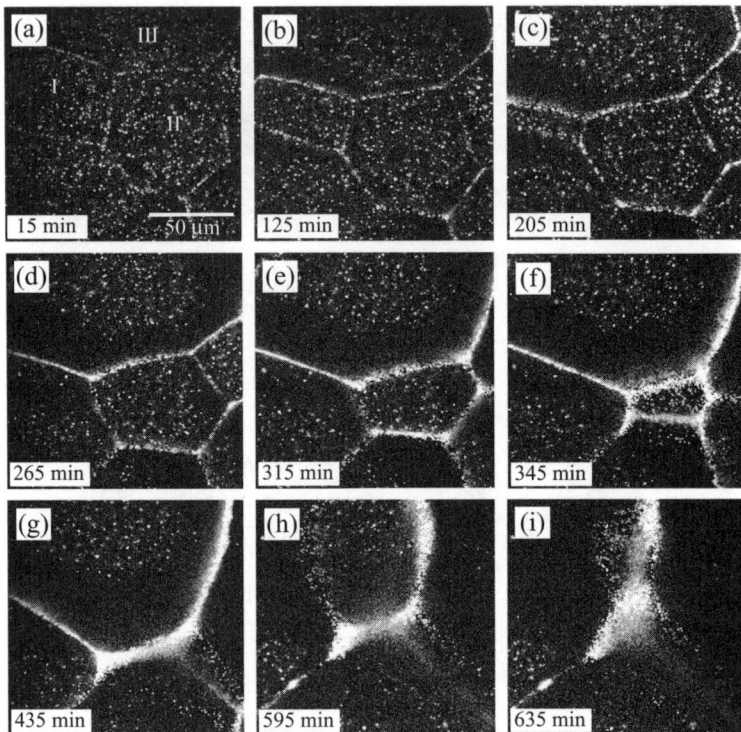

Figure 5.23 (a)–(i) Time-resolved CLSM images of silica + fluorescent PS binary colloid showing the exclusion of the PS particles. (For interpretation of the references to color in this figure legend, the reader is referred to the online version of this chapter.)

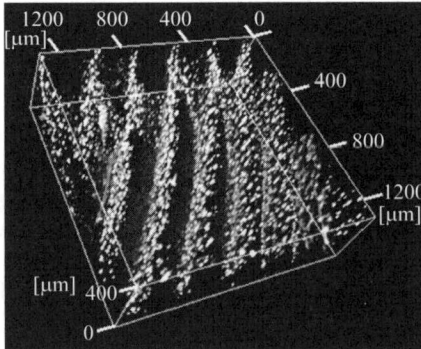

Figure 5.24 Three-dimensional reconstruction of stripe pattern in binary colloid formed during unidirectional crystallization due to Py diffusion. (For color version of this figure, the reader is referred to the online version of this chapter.)

combination of the formation of lamellar crystal grains and the exclusion of PS particles to the crystal grain boundary under sufficiently slow growth. Exclusion was also observed during thermally induced crystallization. In particular, during stepwise unidirectional crystallization achieved by increasing the temperature by 1°C every 10 or

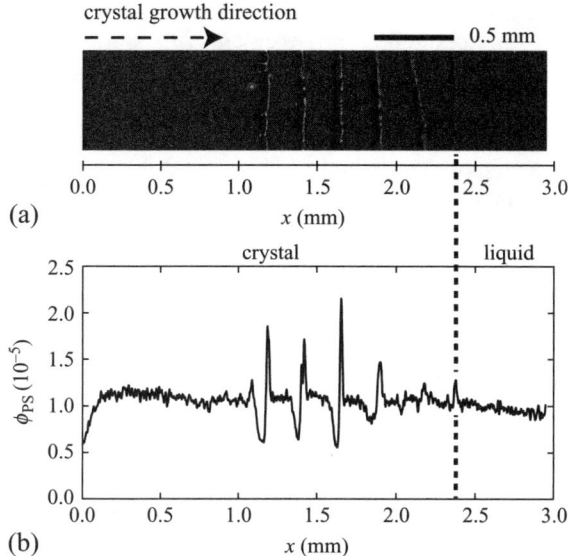

Figure 5.25 (a) CLSM images of stripe patterns observed in binary colloid upon stepwise thermally induced crystallization. (b) Spatial distribution of PS particles in region shown in (a). (For color version of this figure, the reader is referred to the online version of this chapter.)

15 min, the PS particles were arranged in stripe patterns (Figure 5.25) (Sugao et al., 2012). The present stripe patterns appear to be closely analogous to the thermally induced striations that have been found in a number of crystalline materials.

5.4 Current Topics

Here we provide an overview of recent trends in research on structure formation in colloids. Creating novel mesoscale structures is an interesting challenge. The fabrication of complex structures also attracts considerable attention in materials science because of their various novel functions. The desired structures have become much more complex relative to the bcc or fcc crystal lattices. For example, the diamond lattice structure is desirable for photonic crystals (Ozin and Yang, 2001; Colvin, 2001) because it has a perfect photonic band gap. To date, many studies have been made in this field (Zhang et al., 2010; Li et al., 2011). We briefly review recent studies, in particular those on opal crystals and complex colloidal structures.

5.4.1 Opal-Type Crystals

Opal-type colloidal crystals have received considerable attention as photonic materials. They are fabricated mainly by sedimentation (Valsov et al., 2001), 2D deposition, and spin-coating. Because the particles are in contact with each other, the lattice spacing of the opal is determined by the particle diameter. It is impossible in principle to tune the lattice spacing after crystallization. From the opal structure, one can obtain *inverse* opals, which are crystalline arrangements of spherical vacancies (Velev and Kaler, 2000; Velev and Lenhoff, 2000; Stein et al., 2008; Stein and Schrode, 2001; Stein, 2001; Dziomkina and Vansco, 2001). Opals have interior spaces between the individual particles when the samples are dried. Inverse opals are prepared by filling these spaces with a suitable nonvolatile material and then removing the particles by calcination (for organic polymer particles) or acid treatment (for silica particles). The size of the pores and the intervals between them can be tuned by changing the particle size. Inverse opal structures are applicable to photonic crystals, catalysis, and templates having regularly arranged vacancies.

5.4.2 Complex Structures

The arrangements of spherical (charged and hard-sphere) colloidal particles are restricted to the bcc and fcc lattice structures (van Blaaderen and Wiltzius, 1997a). In opals, the hcp structure is also formed. Techniques for the preparation of much more complex structures have been developed recently. Two major approaches have been devised: the use of (1) anisotropic interaction by using anisotropic particles, and (2) multicomponent colloids.

Figure 5.26 shows three types of anisotropic particles. Janus particles have different properties (such as charge, polarity, hydrophilicity, or magnetic properties) on each hemisphere (Chen et al., 2011a, 2012). Patchy particles have patches of regions with different properties on their surface. They are synthesized by chemical vapor deposition using a 2D colloidal crystal structure at the air–water interface, solvent–water interface, or substrate (Li et al., 2008; McGorty et al., 2010). For example, by changing the incident angle of the vapor beam and the number of crystal layers, intricately modified particles can be prepared (Pawar and Kretzschmar, 2008; Zhang et al., 2005).

Furthermore, clusters of particles have been synthesized (Manoharan et al., 2003). Particles combined in a tetrahedron shape have been obtained. Because they show anisotropic interaction, they are expected to act as a chemical unit, like carbon atoms in organic chemistry.

Such particles selectively interact with a certain part having an affinity with other particles, for example, hydrophobic–hydrophobic or cation–anion interactions. Chen et al. (2011b) reported self-assembly of triblock patchy particles to form a "kagome

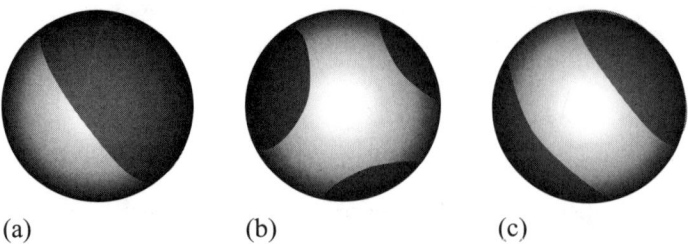

Figure 5.26 Illustrations of (a) Janus, (b) patchy, and (c) tri-block particles.

lattice" structure. Lock-and-key interaction has also been reported (Sacanna et al., 2010). Rycenga and colleagues (2008) reported the assembly of cubic particles. These *colloidal molecules* will open new possibilities for complex structure formation.

Binary colloids may form novel complex structures. The important parameters are the particles sizes and their mixing ratio. They show various phases such as two-phase separation and alloy phases. Ordered 3D superlattices in binary systems have been fabricated (Hachisu and Yoshimura, 1980; Shevchenko et al., 2006; Soto et al., 2002; Chen et al., 2011b; Leunissen et al., 2005). Talapin and colleagues (2009) reported the formation of quasicrystalline-ordered superlattices. Bridging between the particles was studied using DNA molecules (Redl et al., 2003).

Colloidal particles are sometimes modified by fluorescent molecules to facilitate microscopic observation. However, because of quenching of the fluorescent molecules, they are not suitable for long-duration observation. Therefore, syntheses of quantum-dot-modified particles have been developed (Rogach et al., 2000; Yang and Zhang, 2004).

Here we described how the self-assembly of colloidal particles produced various structural patterns. Colloidal crystals were formed in dispersions of hard-sphere and charged colloidal particles, and also in opals. More complex structures were obtainable for novel functional particles that showed anisotropic interactions, and for binary colloids. The methods of colloidal crystal growth will also be useful for anisotropic interaction. One of the ultimate purposes of studies in this field is to construct any desired mesoscale patterns by controlled, designed self-assembly of the constituent particles. Future progress toward this aim is awaited.

Acknowledgments

The studies described in Section 5.3 were supported in part by a Grant-in-Aid from the Ministry of Education, Science and Culture, Japan, and the Project for Promotion of Seeds (JST and JST Plaza Tokai, Japan). Part of the study was performed as the Three-Dimensional Photonic-Crystal (3DPC) Project of the Japan Aerospace Exploration Agency (JAXA). USAXS measurements were performed by Yukihiro Sugao, Nagoya City University, courtesy of Dr. Tsutomu Sawada, National Institute of Materials Sciences, and Dr. Shigeo Hara, Hamamatsu Photonics Co., Ltd. Figure 5.10, Figures 5.12–5.23, and Figures 5.24 and 5.25 are courtesy of Elsevier, the American Chemical Society, and Chemical Society of Japan, respectively.

References

Alder, B.J., Wainwright, T.E., 1957. Phase transition for a hard sphere system. J. Chem. Phys. 27, 1208–1209.

Alexander, S., Chaikin, P.M., Grant, P., Morales, G.J., Pincus, P., Hone, D., 1984. Charge renormalization, osmotic pressure, and bulk modulus of colloidal crystals: Theory. J. Chem. Phys. 80, 5776–5781.

Barrat, J.L., Hansen, J.P., 2003. Basic Concepts for Simple and Complex Liquids. Cambridge University Press, Cambridge.

Bazin, G., Zhu, X.X., 2010. Formation of crystalline colloidal arrays by anionic and cationic polystyrene particles. Soft Matter 6, 4189–4196.

Belloni, L., 2000. Colloidal interactions. J. Phys.: Condens. Matter 12, R549–R587.

Bonse, U., Hart, M., 1966. Small angle X-ray scattering by spherical particles of polystyrene and polyvinyltoluene. Z. Phys. 189, 151–162.

Bosma, G., Pathmamanoharan, C., de Hoog, E.H.A., Kegel, W.K., van Blaaderen, A., Lekerkerker, H.N.W., 2002. J. Colloid Interface Sci. 245, 292–300.

Chandler, D., 1987. Introduction to Modern Statistical Mechanics. Oxford University Press, Oxford.

Chen, Q., Whitmer, J.K., Jiang, S., Luijten, E., Granick, S., 2011a. Supracolloidal reaction kinetics of janus spheres. Science 331, 199–202.

Chen, Q., Bae, S.C., Granick, S., 2011b. Directed self-assembly of a colloidal kagome lattice. Nature 469, 381–384.

Chen, Q., Yan, J., Zhang, J., Bae, S.C., Granick, S., 2012. Janus and multiblock colloidal particles. Langmuir 28, 13555–13561.

Chonde, Y., Krieger, I.M., 1981. Emulsion polymerization of styrene with ionic comonomer in the presence of methanol. J. Appl. Polym. Sci. 26, 1819–1827.

Clark, N.A., Hurd, A.J., Ackerson, B.J., 1979. Single colloidal crystals. Nature 281, 57–60.

Colvin, V.L., 2001. From opals to optics: colloidal photonic crystals. MRS Bull 2001, 637–641.

Dhanaraj, G., Byrappa, K., Prasad, V., Dudley, M. (Eds.), 2010. Handbook of Crystal Growth. Springer, Heidelberg.

Dziomkina, N.V., Vansco, J., 2001. Colloidal crystal assembly on topologically patterned templates. Soft Matter 5, 265–279.

Everett, D.H., 1988. Basic Principles of Colloid Science. Royal Society of Chemistry, London.

Hachisu, S., Yoshimura, S., 1980. Optical demonstration of crystalline superstructures in binary mixtures of latex globules. Nature 283, 188–189.

Hansen, J.P., Löwen, H., 2000. Effective interactions between electric double layers. Ann. Rev. Phys. Chem. 51, 209–242.

Hansen, J.P., McDonald, I.R., 2006. Theory of Simple Liquids, third ed. Elsevier, London.

Holtz, J.H., Asher, S.A., 1997. Polymerized colloidal crystal hydrogel films as intelligent chemical sensing materials. Nature 389, 829–832.

Hunter, R.J., 1981. Zeta Potential in Colloid Science. Academic Press, London.

Iler, R.K., 1979. The Chemistry of Silica, first ed. Wiley-Interscience, New York.

Ise, N., Sogami, I.S., 2005. Structure Formation in Solution. Springer, Berlin.

Israelachivili, J.N., 1992. Intermolecular and Surface Forces, second ed. Academic Press, London.

Iwayama, Y., Yamanaka, J., Takiguchi, Y., Takasaka, M., Ito, K., Shinohara, T., et al., 2003. Optically tunable gelled photonic crystal covering almost the entire visible light wavelength region. Langmuir 19, 977–980.

Johnson Jr., C., Gabriel, D.A., 1981. Laser Light Scattering. Dover Publications, Inc, New York.

Kanai, T., Sawada, T., Toyotama, A., Kitamura, K., 2005. Air-pulse-drive fabrication of photonic crystal films of colloids with high spectral quality. Adv. Funct. Mater. 15, 25–29.

Kittel, C., 2005. Introduction to Solid State Physics, eighth ed. John Wiley and Sons, New York.

Koga, T., Hart, M., Hashimoto, T., 1996. Development of a high-flux- and high-temperature-set-up Bonse-Hart ultra-small-angle X-ray scattering (USAXS) diffractometer. J. Appl. Crystallogr. 29, 318–324.

Leunissen, M., Christova, C.G., Hynninen, A.P., Royall, C.P., Campbell, A.I., Imhof, A., et al., 2005. Ionic colloidal crystals of oppositely charged particles. Nature 437, 235–240.

Li, Y., Cai, W., Duan, G., 2008. Ordered micro/nanostructured arrays based on the monolayer colloidal crystals. Chem. Mater. 20, 615–624.
Li, F., Josephson, D.P., Stein, A., 2011. Colloidal assembly: The road from particles to colloidal molecules and crystals. Angew. Chem. Int. Ed. 50, 360–388.
Luck, W., Klier, M., Wesslau, H., 1963. Über Bragg-Reflexe mit sichtbarem Licht an monodispersen Kunststofflatices. I. Ber. Bunsenges. Phys. Chem. 67, 75–83.
Manoharan, V.N., Elsesser, M.T., Pine, D.J., 2003. Dense packing and symmetry in small clusters of microspheres. Science 301, 483–487.
McGorty, R., Fung, J., Kaz, D., Manoharan, V.N., 2010. Colloidal self-assembly at an interface. Mater. Today (Oxford, U. K.) 13, 34–42.
Moon, J., Yang, S., 2010. Chemical aspects of three-dimensional photonic crystals. Chem. Rev. 110, 547–574.
Murai, M., Yamada, H., Yamanaka, J., Onda, S., Yonese, M., Ito, K., et al., 2007. Unidirectional crystallization of charged colloidal silica due to the diffusion of a base. Langmuir 23, 7510–7517.
Murai, M., Okuzono, T., Yamamoto, M., Toyotama, A., Yamanaka, J., 2012. Gravitational compression dynamics of charged colloidal crystals. J. Colloid Interface Sci. 370, 39–45.
Ozin, G.A., Yang, S.M., 2001. The race for the photonic chip: Colloidal crystal assembly in silicon wafers. Adv. Funct. Mater. 11, 95–104.
Palberg, T., Mönch, W., Bitzer, F., Piazza, R., Bellini, T., 1995. Freezing transition for colloids with adjustable charge: A test of charge renormalization. Phys. Rev. Lett. 74, 4555–4558.
Palberg, T., Evers, M., Garbow, N., Hessinger, D., 1999. Electrophoretic mobility of charged spheres. In: Müller, S.C., Parisi, J., Zimmermann, W. (Eds.), Transport and Structure: Their Competitive Roles in Biophysics and Chemistry. Springer, Berliin, pp. 191–213.
Pawar, A.B., Kretzschmar, I., 2008. Patchy particles by gancing angle deposition. Langmuir 24, 355–358.
Pieranski, P., 1983. Colloidal crystals. Contemp. Phys. 24, 25–73.
Pusey, P.N., van Megen, W., 1986. Phase behavior of concentrated suspensions of nearly hard colloidal spheres. Nature 320, 340–342.
Redl, F.X., Cho, K.S., Murray, C.B., O'Brien, S., 2003. Three-dimensional binary superlattices of magnetic nanocrystals and semiconductor quantum dots. Nature 423, 968–971.
Robbins, M.O., Kremer, K., Grest, G.S., 1988. Phase diagram and dynamics of Yukawa systems. J. Chem. Phys. 88, 3286–3312.
Rogach, A., Susha, A., Caruso, F., Sukhorukov, G., Kornowski, A., Kershaw, S., et al., 2000. Nano- and microengineering: Three-dimensional colloidal photonic crystals prepared from submicrometer-sized polystyrene latex spheres pre-coated with luminescent polyelectrolyte/nanocrystal shells. Adv. Mater. 12, 333–337.
Russel, W.B., Saville, D.A., Schowalter, W.R., 1989. Colloidal Dispersions. Cambridge University Press, Cambridge.
Rycenga, M., McLellan, J.M., Xia, Y., 2008. Controlling the assembly of silver nanocubes through selective functionalization of their faces. Adv. Mater. 20, 2416–2420.
Sacanna, S., Irvine, W.T., Chaikin, P.M., 2010. Lock and key colloids. Nature 464, 575–578.
Shevchenko, E.V., Talapin, D.V., Kotov, N.A., O'Brien, S., Murray, C.B., 2006. Structural diversity in binary nanoparticle superlattices. Nature 439, 55–59.
Sood, A.K., 1991. Structural ordering in colloidal suspensions. Solid State Phys 45, 1–73.
Soto, C.M., Srinivasan, A., Ratna, B.R., 2002. Controlled assembly of mesoscale structures using DNA as molecular bridges. J. Am. Chem. Soc. 124, 8508–8509.
Stein, A., 2001. Sphere templating methods for periodic porous solids. Microporous Mesoporous Mater. 44–45, 227–239.

Stein, A., Schrode, R.C., 2001. Colloidal crystal templating of three-dimensional ordered macroporous solid: materials for photonics and beyond. Curr. Opin. Solid State Mater. Sci. 2001, 553–564.

Stein, A., Li, F., Denny, N.R., 2008. Morphological control in colloidal crystal templating of inverse opals, hierarchical structures, and shaped particles. Chem. Mater. 20, 649–666.

Sugao, Y., Yoshizawa, K., Toyotama, A., Okuzono, T., Yamanaka, J., 2012. Striation pattern of impurity particle in charged colloidal crystals formed by stepwise thermally induced crystallization. Chem, Lett. 41, 1163–1165.

Talapin, D.V., Shevchenko, E.V., Bodnarchuk, M.I., Ye, X., Chen, J., 2009. Quasicrystalline order in self-assembled binary nanoparticle superlattices. Nature 461, 964–967.

Toyotama, A., Yamanaka, J., Yonese, M., Sawada, T., Uchida, F., 2007. Thermally driven unidirectional crystallization of charged colloidal silica. J. Am. Chem. Soc. 129, 3044–3045.

Toyotama, A., Yamanaka, J., 2011. Heating-induced freezing and melting transitions in charged colloids. Langmuir 27, 1569–1572.

Valsov, Y.A., Bo, X.Z., Sturm, J.C., Norris, D.J., 2001. On-chip natural assembly of silicon photonic bandgap crystals. Nature 414, 289–293.

van Blaaderen, A., Wiltzius, P., 1997a. Growing large, well-oriented colloidal crystals. Adv. Mater. 9, 833–835.

van Blaaderen, A., Ruel, R., Wiltzius, P., 1997b. Template-directed colloidal crystallization. Nature 385, 321–324.

van Blaaderen, A., 2004. Colloids under external control. MRS Bull. 29, 85–90.

van Roij, R., Dijkstra, M., Hansen, J.P., 1999. Phase diagram of charge-stabilized colloidal suspensions: van der Waals instability without attractive forces. Phys. Rev. E 59, 2010–2025.

Velev, O.D., Kaler, E.W., 2000. Structured porous materials via colloidal crystal templating: From inorganic oxides to metals. Adv. Mater. 12, 531–534.

Velev, O.D., Lenhoff, A.M., 2000. Colloidal crystals as templates for porous materials. Curr. Opin. Colloid Interface Sci. 5, 56–63.

Verwey, E.J.W., Overbeek, J.Th.G., 1948. Theory of the Stability of Lyophobic Colloids. Elsevier, New York.

Wilson, T. (Ed.), 1990. Confocal Microscopy. Academic Press, London.

Yamanaka, J., Yoshida, H., Koga, T., Ise, N., Hashimoto, T., 1998. Reentrant solid-liquid transition in ionic colloidal dispersions by varying particle charge density. Phys. Rev. Lett. 80, 5806–5809.

Yamanaka, J., Murai, M., Iwayama, Y., Yonese, M., Ito, K., Sawada, T., 2004. One-directional crystal growth in charged colloidal silica dispersions driven by diffusion of base. J. Am. Chem. Soc. 126, 7156–7157.

Yang, X., Zhang, Y., 2004. Encapsulation of quantum nanodots in polystyrene and silica micro-/nanoparticles. Langmuir 20, 6071–6073.

Yoshida, H., Yamanaka, J., Koga, T., Koga, T., Ise, N., Hashimoto, T., 1999. Transitions between ordered and disordered phase and their coexistence in dilute ionic colloidal dispersions. Langmuir 15, 2684–2702.

Yoshizawa, K., Okuzono, T., Koga, T., Taniji, T., Yamanaka, J., 2011. Exclusion of impurity particles during grain growth in charged colloidal crystals. Langmuir 27, 13420–13427.

Yoshizawa, K., Onda, S., Sawada, T., Yamanaka, J., 2012. Formation of stripe patterns in charged colloids during unidirectional crystallization in the presence of impurity particles. Chem. Lett. 41, 322–324.

Zhang, J., Li, Y., Zhang, X., Yang, B., 2010. Colloidal self-assembly meets nanofabrication: From two-dimensional colloidal crystals to nanostructure arrays. Adv. Mater. 22, 4249–4269.

Zhang, G., Wang, D., Möhwald, H., 2005. Decoration of microspheres with gold nanodots—Giving colloidal spheres valences. Angew. Chem. Int. Ed. 44, 7767–7770.

6 Structural Color in Nature
Basic Observations and Analysis

Shinya Yoshioka

Chapter Contents
6.1 Introduction 199
    6.1.1 Multiscale Systems and Structural Color in Nature 199
6.2 Basic Observations 201
    6.2.1 How to Illuminate 201
    6.2.2 *Morpho* Butterfly 202
    6.2.3 Jewel Beetle 204
    6.2.4 Liquid Immersion Experiment 206
6.3 Optical Characterization 207
    6.3.1 Angle-Dependent Reflection 207
    6.3.2 θ–2θ Scan Measurement 212
    6.3.3 Integrated Optical Properties 215
    6.3.4 Refractive Index Value 217
6.4 Analysis I 218
    6.4.1 Fresnel's Equations 218
    6.4.2 Single Thin-Layer Interference 221
    6.4.3 Multilayer Interference 227
    6.4.4 Analysis of Jewel Beetle's Iridescence 234
6.5 Analysis II 237
    6.5.1 Fraunhofer Diffraction 237
    6.5.2 Diffraction Grating 240
    6.5.3 Analysis of *Morpho* Butterfly's Structural Color 242
    6.5.4 Role of Pigment 246
6.6 Concluding Remarks 248

6.1 Introduction

6.1.1 Multiscale Systems and Structural Color in Nature

Nature exhibits many examples of symmetrical or regular spatial patterns, such as a hexagonal snow crystal, the colorful eye pattern of a peacock feather, and the regularly arranged black patches of a leopard. These examples pose the following fundamental questions: What are the physical principles that govern the formation processes? What physical factors determine the symmetry and spatial period? These are problems of pattern formation under nonequilibrium physical conditions. Moreover, some natural examples contain hierarchical or multiscale structures with different sizes, in which the structures are organized at each scale. As an example, let us briefly observe a peacock feather.

A bird feather generally consists of a main shaft with two types of structures branching out from it (Figure 6.1). One type comprises primary branches of a main shaft called feather barbs, and the other consists of secondary smaller branches projecting from a barb, which are called barbules (Figures 6.1b and c). Barbs branch off of the main shaft with a constant submillimeter separation. Similarly, but at a much smaller size, barbules periodically project from the barb with a constant separation of several dozens of microns. In addition, another smaller periodic structure exists beneath the barbule surface. In the cross-section of the barbule, a rectangular arrangement of small granules appears at a size comparable to the wavelength of light (Figure 6.1d). When the barbule is longitudinally sectioned, the granules are observed to be elongated, as shown in Figure 6.1(e). This periodic arrangement of the granules causes optical interference and results in the brilliant color of the peacock feather (Yoshioka and Kinoshita, 2002; Zi et al., 2003); the wavelength of the reflected light is determined mainly by the lattice constant. Therefore, the colorful eye pattern of the peacock feather is realized by controlling the submicron arrangement of granules over several centimeter planes of the feather, which consists of numerous barbs with

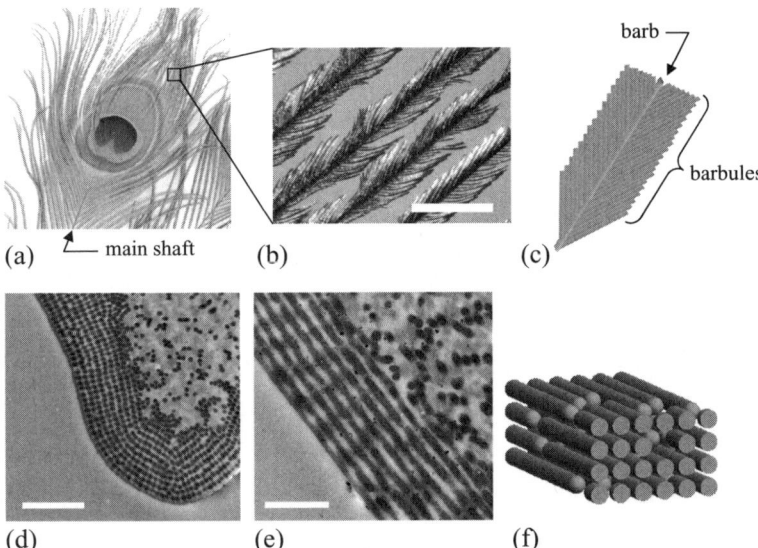

Figure 6.1 Peacock feather. (a) Photograph of the upper tail covert, (b) optical micrograph showing several barbs and numerous barbules projecting from the barbs, and (c) schematic illustration of a barb with many barbules. (d, e) Transmission electron micrographs of the cross-section and longitudinal section of the barbule, respectively. (f) Schematic illustration of the color-causing microstructure of the peacock feather. Scale bars: (b) 500 μm, (d) 2 μm, and (e) 1 μm. (For color version of this figure, the reader is referred to the online version of this chapter.)

barbules. It is quite interesting to consider how developmental processes are controlled in such a multiscale structure.

In this chapter, we treat optical phenomena that originate in submicron microstructures that are found in nature. In not only the peacock feather but also diverse species of animals, submicron structures produce brilliant colors, called *structural colors*. The physical mechanisms of the coloration are directly related to optical processes such as interference, diffraction, and scattering, which become discernible when the sizes of the structures become comparable to the wavelength of light. The efficiencies of these optical processes depend fundamentally on the wavelength. Thus, they can cause wavelength-selective reflection that, in turn, produces brilliant color. Structural color exhibits interesting optical properties (e.g., iridescence, high reflectance, strong polarization effects) that are rarely obtained by the conventional coloration mechanism (optical absorption due to pigment). This is one reason that structural color has recently received considerable attention, although it has a very long research history; Newton (1704) already described the similarities between the iridescent color of the peacock feather and thin platelets in his book. Research in this field is progressing rapidly in attempts to apply structural color in the textile, car-painting, and cosmetic industries (e.g., Saito, 2011).

Several scientific monographs and review papers have already been published on structural color in natural systems (e.g., Parker, 2000; Vukusic and Sambles, 2003; Berthier, 2007; Kinoshita, 2008). Thus, this chapter is devoted to a description of basic observations and analysis, mainly for those who are newly interested in this topic. For this purpose, the chapter covers two insect species heavily rather than introducing many examples; one is the *Morpho* butterfly, which has very brilliant blue wings, and the other is a jewel beetle that shows a beautiful iridescence. These two species are among the most typical examples of structural color in nature, yet they have contrasting reflection mechanisms.

This chapter is organized as follows. The next section describes basic observations of the two species using optical and electron microscopes. Next, several optical methods are introduced for the quantitative characterization of structural color. They are followed by theoretical treatments of fundamental optical processes that are necessary for the analysis of the coloration mechanisms. Finally, a few examples of structural color are mentioned, in which not only submicron structures but also other factors are essential to the coloration. As this book has the motto "to see, to try, to think," I hope readers obtain specimens and try doing basic experiments. Thus, some experimental details are described for convenience.

6.2 Basic Observations

6.2.1 How to Illuminate

When structurally colored samples are observed, it is important to pay attention to the directions of illumination and observation because the appearance can differ greatly depending on these directions. One such phenomenon, iridescence, is defined as a color change depending on the observation and/or illumination angles. There are generally two extreme types of illumination: directional illumination, in which the illuminating light comes from a certain direction, and ambient illumination, in which light is omnidirectional with equal intensity. Both types have benefits depending on the situation. In fact, the two types are employed in a complementary manner in the following observations of the *Morpho* butterfly and jewel beetle.

6.2.2 Morpho *Butterfly*

Morpho butterflies belong to the genus *Morpho* and are classified into several dozen species, although the classification has not been firmly established. However, a common type of microstructure is observed in many *Morpho* species. In this chapter, we treat mainly such commonly observed structural and optical features. A few species-specific factors contributing to the coloration are briefly mentioned in Section 6.5.4.

Figures 6.2(a)–(c) show photographs of a *Morpho* butterfly taken from three different directions under nearly directional illumination; lamps on the ceiling were turned off except for one lamp, which was located directly above the butterfly specimen. Under this illumination, a shadow clearly appears behind the wing. The wing looks brilliantly blue when it is observed from the nearly normal direction (Figure 6.2a). However, the brilliant color vanishes almost completely when it is observed from an oblique direction perpendicular to the body of the butterfly, as shown in Figure 6.2(c); the blue part becomes very dark, whereas the white stripe remains bright. Conversely, the blue color looks as bright as from the normal direction when the direction is oblique along the butterfly's body (Figure 6.2b). This large change in the intensity depending on the oblique direction is one of the important

Figure 6.2 *Morpho cypris* butterfly observed from (a) the normal direction, (b) an oblique direction along the body of the butterfly, and (c) an oblique direction perpendicular to the body. (d) Optical micrograph showing the arrangement of the wing scales at the border between the blue part and white stripe. (e) High-magnification close-up photograph of the blue scale. Scale bars: (d) 1 mm and (e) 10 μm. (For interpretation of the references to color in this figure legend, the reader is referred to the online version of this chapter.)

reflection properties of the blue *Morpho* wing. On the other hand, the white part looks bright regardless of the observation direction. The difference in the observation-direction dependence implies that the white stripe has a different reflection mechanism from the blue part.

Next, let us use an optical microscope to observe the wing structures in more detail. The butterfly wing generally consists of the wing membrane and numerous tiny scales covering both sides of it. The scales usually have a flat rectangular shape with typical dimensions of 150 µm in length, 80 µm in width, and several microns in thickness. They are regularly arranged just like roof tiles, as shown in Figure 6.2(d). In general, each scale has one color, and the arrangement of many scales with different colors forms the pattern of the wing as a mosaic. In Figure 6.2(d), the blue and white scales are clearly photographed at the border of the white stripe on the wing of an *M. cypris* butterfly. (Here, we abbreviate the scientific name of the butterfly *Morpho cypris* as *M. cypris* following a convention. Similar abbreviations are used throughout the text.)

A high-power objective lens with a 100 × magnification reveals very thin blue lines on the blue scale, as shown in Figure 6.2(e). The spacing between adjacent lines is approximately 1 µm. This suggests that there are linelike structures that strongly reflect blue light. The spatial resolution of an optical microscope is close to a fundamental limit because of the wave nature of light. It is necessary to use an electron microscope to examine more detailed microstructures inside the scale.

Figure 6.3(a) shows the scale arrangement observed by a scanning electron microscope. Observation under high magnification confirms that many lines exist on the surface of the scale with a spacing of approximately 0.8 µm (Figure 6.3b). This type of structure, called *ridges*, is generally found in a butterfly wing scale (Ghiradella, 1991).

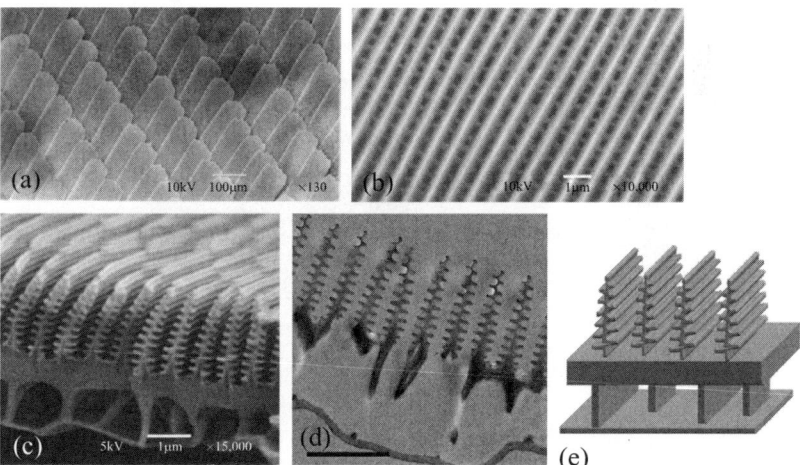

Figure 6.3 Scanning and transmission electron micrographs of the *Morpho* butterfly. (a) Scale arrangement, (b) part of the scale showing many ridges separated by a constant distance, (c) cross-section of a scale observed by scanning electron microscope, (d) cross-section of a scale observed by transmission electron microscope (scale bar: 2 µm), and (e) schematic illustration of the microstructure of the *Morpho* scale. Species: (a)–(c) *M. cypris* and (d) *M. rhetenor*.

An important structural characteristic of the *Morpho* scale can be found in the cross-section of the ridges; treelike complicated microstructures are observed in the cross-section, as shown in Figures 6.3(c) and (d). Each "tree" corresponds to one ridge. The height of the tree is approximately 2.5 µm, and about 10 branches project from the trunk to both sides. A careful inspection reveals that the branches on both sides have alternate heights. The vertical distance between the adjacent branches is observed to be 150–240 nm, and the thickness of a branch is approximately 50–90 nm (Kambe et al., 2011). The treelike structures exist on a type of scale structures, called *pillars*, that stand on the bottom plate of the scale. A schematic illustration of the *Morpho* scale's microstructures is shown in Figure 6.3(e). It is confirmed later in the chapter that the treelike structure is the essential origin of the scale's blue color. In fact, the brown scales on the ventral side, for example, do not have such an elaborate tree-like structure.

6.2.3 Jewel Beetle

Jewel beetles are insects that belong to the family *Buprestidae*, which constitutes the largest group of beetles. Although not all the species are brilliantly colored, some exhibit strong metallic reflection. A Japanese jewel beetle, *Chrysochroa fulgidissima*, is an example having a green elytron with dark red stripes. As shown in Figure 6.4(a), it exhibits a beautiful iridescent color changing from green to blue as the observation angle increases. This species is particularly famous because it was used as a decorative element in the tamamushi-no-zushi at Horyou Temple, Japan (Hariyama et al., 2005). Although the decoration was crafted more than 1000 years ago, the green color of the elytra still remains at least in part, clearly demonstrating the anti-fading characteristic of the structural color.

When the jewel beetle is observed under directional illumination, only a small part of the elytron looks much brighter than the other parts. This indicates that the reflection from the elytron is more or less similar to that of a mirror ball, which consists of small specular surfaces with different orientations. In this case, ambient illumination

Figure 6.4 Iridescence of the jewel beetle. (a) Photographs taken from several different directions under ambient illumination. The number indicates the observation angle, which is defined as the angle between the direction of the observation and surface normal of the jewel beetle. Note that the constant color of the red stripe is due to the wide wavelength sensitivity of the digital camera, which extends to the infrared region where the spectral peak is located under normal incidence. (b) Experimental setup for obtaining ambient illumination. See the text for details. (For interpretation of the references to color in this figure legend, the reader is referred to the online version of this chapter.)

is more convenient for examining the change in color with the observation angle. The photographs shown in Figure 6.4(a) were taken under this type of illumination condition using the experimental setup shown in Figure 6.4(b). Two ring-type fluorescent lamps, which were taken from commercial ceiling lighting, were used as the light source. They were stacked with a spacing of approximately 15 cm and placed on a white plate made of Teflon. The two lamps were covered with a white cloth so that light was multiply reflected inside. The sample was placed around the center axis of the two lamps and photographed from a small aperture in the surrounding white cloth.

Iridescence is clearly visible in Figure 6.4(a); the jewel beetle looks green when it is observed at a small observation angle, and the color changes to blue when observed from an oblique direction. Although the red stripes do not seem to change color, this is due to the color expression of the digital camera. Its sensitivity extends to the near-infrared region; therefore, the stripes appear red even when the wavelength of reflection is beyond the visible range of human vision. The abdominal side reportedly shows a similar but slightly shifted color change from red to green (Kinoshita and Yoshioka, 2005).

The metallic glittering appearance suggests that the elytron has a smooth surface similar to a polished mirror. However, many deep irregular bumps exist on the surface, as shown in Figures 6.5(a) and (b). They scatter light and make it difficult to perform quantitative optical measurements. However, the density of the bumps depends markedly on their position: fewer of them appear near the thorax, whereas the density is very high in the peripheral region. A similar tendency in the surface bump distribution is observed in other jewel beetle species, for example, *Chrysochroa rajah* and *C. fulminans*. The part of the elytron with a small bump density is used in the following observations and measurements.

Under higher magnification, hexagonal tiles of approximately 10 μm in size are observed on the surface. There are small pits at the vertexes, as shown in

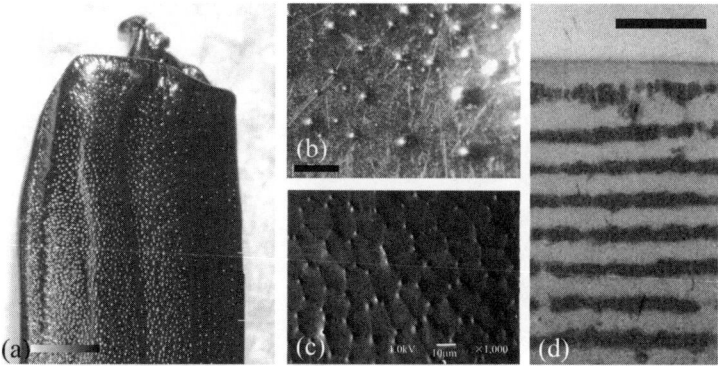

Figure 6.5 Surface of jewel beetle's elytron. (a, b) Optical micrographs showing numerous bumps on the surface, (c) electron micrograph showing small hexagonal patterns approximately 10 μm in size, and (d) transmission electron micrograph of the cross-section of the elytron. In (d), alternately stacked electron-dense and electron-sparse layers are clearly observed. Scale bars: (a) 2 mm, (b) 200 μm, (c) 10 μm, and (d) 500 nm. (For color version of this figure, the reader is referred to the online version of this chapter.)

Figure 6.5(c). The surface inside the hexagons looks quite smooth, and no elaborate structures that seem to cause optical interference are found. Conversely, transmission electron microscopy of the cross-section reveals that electron-dense and electron-sparse layers are stacked alternately beneath the surface, as shown in Figure 6.5(d). It has been experimentally clarified that the electron-dense layers have a higher refractive index value than the electron-sparse layers (Yoshioka and Kinoshita, 2011). The observation of the stacked layers suggests that multilayer interference is the origin of the wavelength-selective reflection. Although the interfaces between the layers are not perfectly flat, especially those of the top dark layer, the effective thickness of each layer can be determined by image analysis (Yoshioka et al., 2012). The average thicknesses of the electron-sparse and electron-dense layers are estimated to be 100 nm and 79 nm, respectively, which are appropriate for optical interference at the wavelengths of visible light. However, as described in Section 6.4.4, a detailed analysis reveals that a simple periodic multilayer structure is not an appropriate model for the jewel beetle's structural color. Instead, the layers near the elytron surface are modified to play an important role in the multilayer interference.

6.2.4 Liquid Immersion Experiment

One of the striking characteristics of structural color is the color change due to liquid immersion. Strictly speaking, ordinary colors originating from pigments are also affected by liquid because the change in the refractive index of the surrounding medium inevitably alters the magnitudes of scattering and reflection and their spectral shape. However, when optical interference is involved in the coloration mechanism, the effect of the refractive index change appears to be dramatic. Figure 6.6(a) compares the *Morpho* wing color in three different media: air and two liquids. The blue wing instantaneously changes to green when it is immersed in methanol but looks brown in toluene.

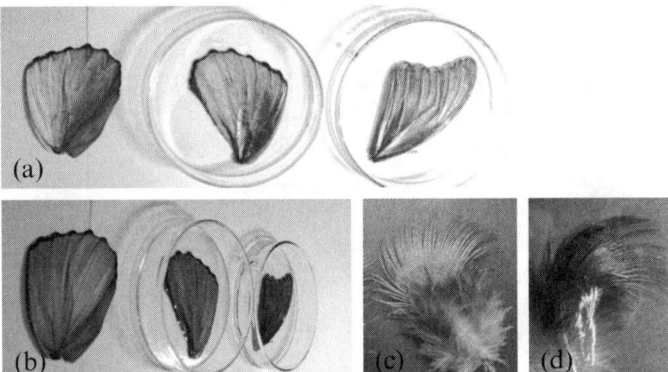

Figure 6.6 Color change due to liquid immersion. (a, b) Wing of *M. didius* butterfly observed from two different directions. The wings are in air (left), methanol (center), and toluene (right). (c, d) Peacock feather in air and liquid methanol, respectively. (For color version of this figure, the reader is referred to the online version of this chapter.)

The color change toward longer wavelengths is explained qualitatively by the modification in the interference condition. Replacement of the air included in the microstructure with a liquid having a larger refractive index than air lengthens the optical path length; consequently, the interference condition is satisfied at a longer wavelength of light. When the methanol-immersed wing is observed from an oblique direction, the color changes to blue, as shown in Figure 6.6(b). This iridescence indicates that optical interference is certainly involved in the coloration mechanism. Conversely, the brown color in liquid toluene originates from the dark pigment contained inside the wing structures; as in many butterfly species, the ventral side of the *Morpho* wing is covered with dark-pigmented black or brown scales, which probably serve as cryptic coloration because this side is exposed when the butterfly is at rest with its wings closed. Because the wing is made of cuticle, the refractive index of which is close to that of liquid toluene (1.50), reflection and scattering are almost completely suppressed by immersion in the liquid toluene, and the optical absorption due to the pigment becomes conspicuous.

The color change due to liquid immersion is not restricted to the butterfly wing; the peacock feather also shows a similar change, as shown in Figures 6.6(c) and (d), although it is not as dramatic as in the *Morpho* butterfly. In contrast, the jewel beetle does not show a large color change in response to liquid immersion unlike the *Morpho* butterfly. The reason is differences in the color-causing microstructures: liquid does not easily penetrate the multilayer microstructure of the jewel beetle because it consists of two different types of materials, unlike the *Morpho* scales, which contain air inside the treelike ridge structures. However, it has been reported that long-term liquid immersion over several months causes a gradual color change in the jewel beetle's elytron (Adachi, 2007).

6.3 Optical Characterization

6.3.1 Angle-Dependent Reflection

Simple observations under directional illumination revealed that the blue color of the *Morpho* wing changes in intensity depending largely on the observing direction (Figure 6.2). Another simple experiment, the screen projection method, gives us a more intuitive understanding of this optical property. A schematic illustration of the experimental setup is shown in Figure 6.7. In this method, the sample is illuminated through a hole in a white screen so that the reflected light is projected onto it. The spatial pattern appearing on the screen intuitively expresses the angular dependence of the reflection. Figure 6.8(a) shows that a narrow blue band appears on the screen when a small part of the wing of a *Morpho* butterfly is illuminated. This indicates that the reflection is neither diffuse nor specular, but anisotropic. The large intensity change depending on the observation direction is naturally explained by this reflection pattern; the wing looks brilliant when it is observed from a direction that is included in the anisotropic reflection band, whereas it appears very dark when observed from the other directions.

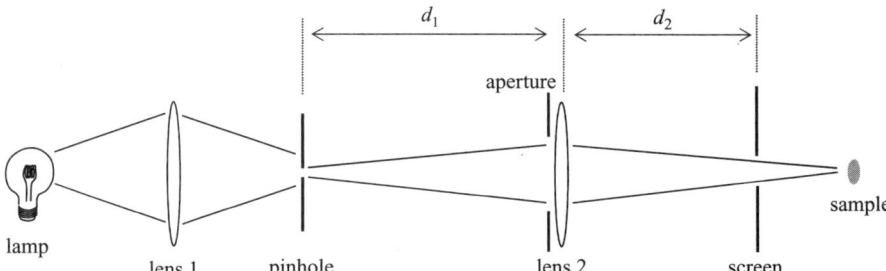

Figure 6.7 Example of optical setup for the screen projection method. The illuminating light is first focused on a pinhole, and the image is projected on the sample by a second lens. An aperture is used to restrict the angular range of the illumination. The optical system is designed so that the focused spot is small enough to illuminate a part of the sample with a uniform color. For example, when the focal length f of the second lens is 50 mm, a pair of distances, $d_1 = 75$ and $d_2 = 150$ mm, satisfy the well-known formula for imaging by a lens: $1/d_1 + 1/d_2 = 1/f$. In this case, the lateral magnification M of the image is equal to $d_2/d_1 = 2$. Thus, when the pinhole diameter r is assumed to be 1 mm, the calculated spot size l is $l = rd_2/d_1 = 2$ mm, which is usually small enough to illuminate part of a butterfly wing. An ordinary light source such as a tungsten lamp is bright enough to observe a highly reflecting sample such as a *Morpho* butterfly wing. However, when a small sample is examined (e.g., a 100 μm single scale, as shown in Figure 6.9), a high-luminance xenon lamp is preferred because a correspondingly small pinhole is necessary.

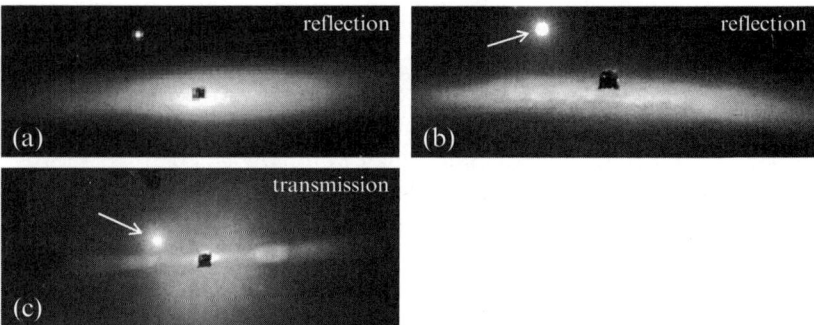

Figure 6.8 Results of observations using the screen projection method. (a) Reflection pattern from a wing of *M. rhetenor* butterfly. (b) Reflection and (c) transmission pattern from a single scale of *M. sulkowskyi*. Scale appears as bright spot indicated by white arrow in the photographs. The experimental geometry is shown in Figure 6.9. (For color version of this figure, the reader is referred to the online version of this chapter.)

It is possible to examine a single wing scale by attaching it to the tip of a needle and illuminating it with tightly focused white light, as shown in Figure 6.9 (Yoshioka and Kinoshita, 2006a). Although it is much easier to handle a piece of the wing than a tiny single scale, single-scale examination has at least three advantages. First, the reflection originating purely from the scale microstructure can be examined by completely avoiding the effects of multiple reflections that could occur on the wing membrane

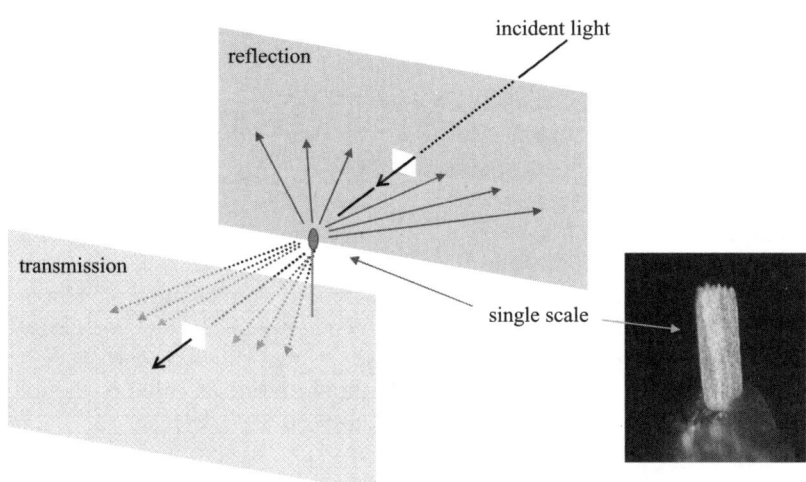

Figure 6.9 Geometry of single-scale examination. The reflection and transmission patterns from a single *Morpho* scale are projected on two white screens placed in front of and behind the sample. The sample is illuminated by a focused spot of white light emitted from a xenon lamp using an optical system similar to that shown in Figure 6.7. The incident light comes through a hole in the screen on the reflection side. The screen on the transmission side also has a hole so as to avoid an excessively bright spot of directly transmitted light, which would make it difficult to observe the diffracted light. Image on the right shows a single scale attached to the tip of a needle. The width of the scale is approximately 80 μm. (For color version of this figure, the reader is referred to the online version of this chapter.)

and the scales on both sides of it. Second, color variations specific to the scale can be investigated; butterfly wing patterns generally consist of a mosaic of differently colored scales. The reflection pattern from a single *Morpho* scale, shown in Figure 6.8(b), looks slightly narrower than that of the wing. This indicates that the thickened reflection band of the wing results from a distribution in the angle of the scale face. In addition, it is immediately understood from the experimental geometry that the narrow reflection band is nearly perpendicular to the longer side of the scale (and to the ridges).

The third advantage of single-scale examination is that the transmission pattern can be observed. As shown in Figure 6.8(c), a spectrally dispersed color band appears in both the left and right sides of the direction of the directly transmitted light. This pattern clearly indicates that the periodically spaced ridges work similarly to a ruled diffraction grating. It is more convenient to use a monochromatic laser beam instead of focused white light for quantitative analysis because the diffraction angle is directly related to the ridge spacing and the wavelength. Such an experiment is described in Box 6.1.

Careful inspection of the transmission pattern (Figure 6.8c) reveals that a diffuse circular pattern appears around the square hole made on the screen. This scattering pattern probably originates in one type of scale structure called *pillars*, which connect the treelike ridges and the bottom plate of the scale (see Figure 6.3e). Thus, the screen projection method is quite convenient for getting an idea of what types of optical

Box 6.1 Diffraction Ring

The ridges on a butterfly wing scale are usually separated by a few microns. The separation is so regular that the ridges act just like a ruled diffraction grating in transmission (but usually not in reflection). Using this feature, we can observe an interesting diffraction pattern, as shown in Figure B6.1, by performing a simple experiment.

The scales were randomly and densely scattered on a glass slide; a small brush was used to take the scales off the wing membrane, and touching the glass slide with the brush left many scales on it. As shown in Figure B6.1(c), the glass slide was held vertically using a clip, and two laser beams with different wavelengths were used as light sources. A ringlike diffraction pattern appeared on the screen. The radius of the ring differed depending on the wavelength (compare Figures B6.1a and b).

Each singe scale works as a diffraction grating. However, because they are randomly oriented, the superposition of the diffraction spots from different scales appears as a ring pattern. In Section 6.5.2, the interference condition for the diffraction grating is used to estimate the ridge spacing from the diffraction angle.

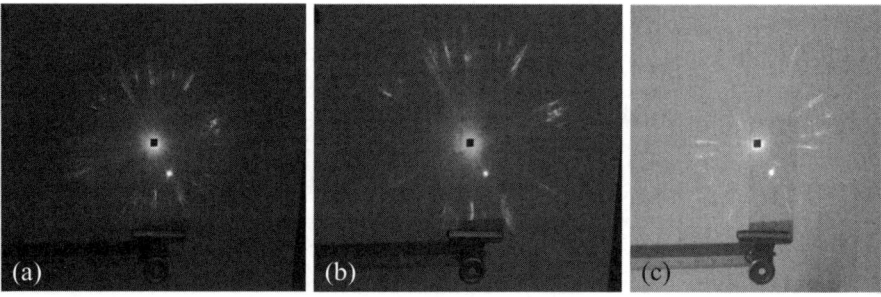

Figure B6.1 Diffraction ring observed for the scales of *M. sulkowskyi*. The light sources are (a) a green laser pointer with a wavelength of 532 nm, (b) an He–Ne gas laser with a wavelength of 633 nm, and (c) both these lasers. (For interpretation of the references to color in this figure legend, the reader is referred to the online version of this chapter.)

processes occur inside the sample. Summarizing the preceding observations, we noticed seemingly contradictory optical properties in the *Morpho* butterfly: the single scale acts as a diffraction grating in transmission, whereas on the reflection side, only a blue narrow band is observed without any apparent diffraction spots.

After the reflection pattern is intuitively checked by the screen projection method, the next step is quantitative characterization—that is, measurement of the angle-resolved optical spectrum. One type of experimental setup is shown in Figure 6.10. White light illuminates the sample through a narrow aperture, and an optical fiber pointing at the sample rotates around it to guide the reflected light into a spectrometer

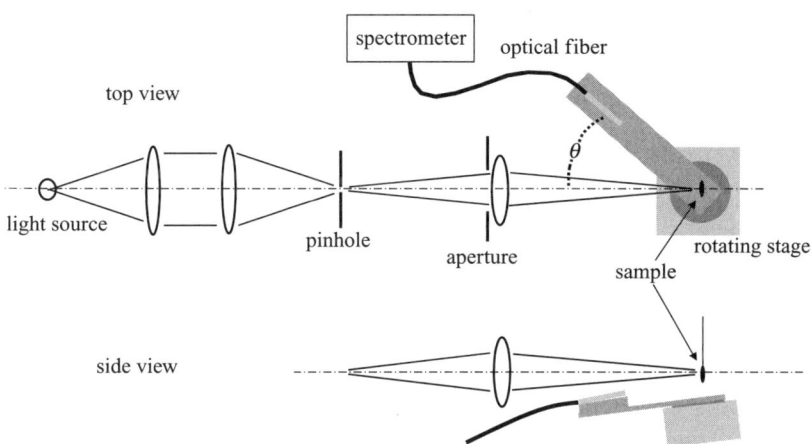

Figure 6.10 Experimental diagram for measurement of the angle-resolved reflection spectrum. As in the system shown in Figure 6.7, white light emitted from a source is focused on a small spot on the sample, which is placed on the axis of a rotating stage. The optical fiber rotates around the sample to detect the reflected light as a function of the angle. The spectrum is measured by a fiber-optic spectrometer. Note that the plane of the fiber rotation is not exactly parallel to the horizontal plane, but tilted by approximately 10°, as shown in the side view.

as a function of the angle between the incident and reflection directions. There are many variations of this type of angle-resolved experiment. For example, the incident light can first be monochromatized with a monochromator, and a photomultiplier can be used as a detector instead of a spectrometer. Figure 6.11(a) shows the results of such measurements of the *Morpho* wing under normal incidence, which confirm quantitatively that the reflected light is widely spread in an angle. The reflectance tends to reach a maximum around the direction of the incident light for long wavelengths of

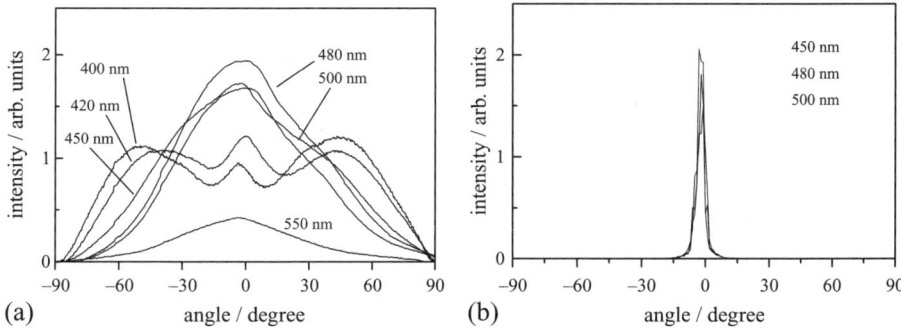

Figure 6.11 Angular dependence of reflected light intensity at several wavelengths. (a) Wing of *M. didius* without cover scales (see Section 6.5.4 for the types of wing scale). The detecting fiber is rotated in the plane containing the narrow reflection band. ((a) Reproduced from Kinoshita et al., 2002a.) (b) Angular dependence when the detection plane is perpendicular to the narrow reflection band (*M. cypris*).

light, whereas it shows three peaks at approximately 0 and ±50° at shorter wavelengths. Consequently, the reflectance is higher at large angles for light with short wavelengths. In these observations, the plane of the fiber rotation includes the narrow reflection band observed by the screen projection method.

Conversely, when the sample is rotated by 90° within the plane of the wing, the angle dependence exhibits a sharp peak around the normal direction with a full width at half-maximum (FWHM) of less than 10°, as shown in Figure 6.11(b). This is because the plane of the fiber rotation becomes perpendicular to the narrow reflection band. These two measurements quantitatively confirm the anisotropic reflection of the *Morpho* butterfly wing.

The angle-resolved experiment is quite important for optical characterization of structural color. This is because optical interference, which is usually involved in the coloration mechanism, is expected to hold some relation between the angle and wavelength of the reflected light; this should be tested experimentally. Thus, methodological research is still in progress to address the disadvantages of the two methods previously described. The screen projection method is not efficient for examination at highly oblique directions, and it is difficult to quantitatively determine the reflectance value. In addition, the measurement becomes difficult for weakly reflecting samples because the secondary scattered light from the screen is detected. Conversely, in the fiber rotation method, the measurement is obviously limited to the plane in which the fiber rotates. A complicated system is necessary if measurements are extended into two dimensions.

Stavenga et al. (2009) developed a new method of observing the angle dependence called imaging scatterometry, which overcomes the preceding disadvantages. An ellipsoidal mirror is employed as the central component of the scatterometer; the sample is placed at the primary focal point of the ellipse and illuminated from a hole at the center of the mirror. Hemispherically scattered light is focused on the secondary focal point of the ellipse, and the narrow cone of scattered light is collected by a lens and imaged on a two-dimensional (2D) photodetector such as a CCD camera. Several studies using this optical setup have already examined the structural colors of, for example, beetles (Stavenga et al., 2011a; Pouya et al., 2011), butterflies (Wilts et al., 2009; Trzeciak et al., 2012), and bird feathers (Stavenga et al., 2011b). Kambe and colleagues (2011) reported a differently extended experimental system, in which both the sample and optical fiber are rotated. They demonstrated that a 2D map of the reflectance (i.e., the reflectance as a function of both the incident and reflection angles) is quite useful for characterizing the reflection properties of structural color.

6.3.2 θ–2θ *Scan Measurement*

An application of the screen projection method reveals that the jewel beetle, in contrast to the *Morpho* butterfly, shows nearly specular reflection when a small part (several dozen square microns) of the elytron is illuminated; the reflected light spot appearing on the screen has nearly the same size as the specular reflection, as shown in Figure 6.12(a). However, the slightly ambiguous edge of the circular spot indicates

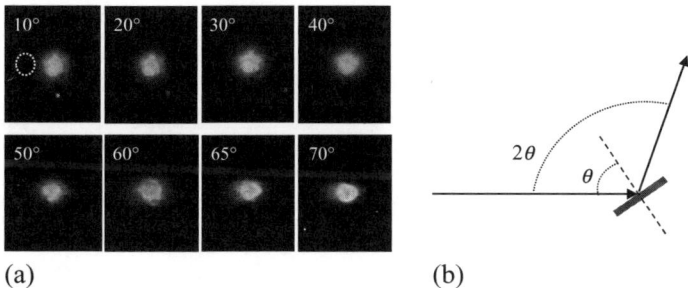

Figure 6.12 Observation of reflected light under θ–2θ geometry. (a) Photographs of the reflected light spot appearing on the white screen. A small part of the elytron of the jewel beetle is illuminated by unpolarized light, and the screen is placed at the position of the optical fiber shown in Figure 6.10 to project the reflected light. The incidence angles are given at the upper left. The spot appears yellow or yellow–green at smaller angles and gradually changes to blue with increasing angles. However, it becomes more or less white at large angles. The dotted circle in the 10° image shows the estimated reflection spot when the reflection is assumed to be specular. ((a) Reproduced from Yoshioka et al., 2012.) (b) Geometry of specular reflection. The incident light is reflected in the 2θ direction when the incidence angle θ is defined as the angle between the surface normal and direction of the incident light. (For interpretation of the references to color in this figure legend, the reader is referred to the online version of this chapter.)

that the specularity is not perfect, and the surface is slightly irregular. The iridescent color is clearly confirmed by the gradual color change with changes in the angle: the light spot appears yellow–green or green at small incident angles but changes to blue with increasing incident angle. However, careful inspection reveals that the spot appears white rather than blue at large incident angles: the vividness of the green color observed at small angles is much stronger than that of the blue color at large angles. These observations indicate that the angle dependence of the reflectance spectrum is not merely a shift in the spectral peak.

When a reflection is specular, an experimental geometry called the θ–2θ scan is often employed to characterize the angle dependence. That is because, as shown in Figure 6.12(b), the light is reflected along the 2θ direction when the incidence angle θ is defined as the angle between the surface normal of the sample and the incident direction. Artificial multilayered optical components such as dielectric thin-film mirrors and band-pass filters are usually characterized by the θ–2θ scan measurement, which entails measurement of the reflection spectrum along the 2θ direction while the incident angle is varied. It is important to note whether or not the surface is specular, especially when a natural sample is examined. In the study of the jewel beetle reported by Yoshioka and colleagues (2012), a small portion of several dozen microns square in size was illuminated to minimize the effects of structural irregularities such as bumps and scratches on the surface and also the inhomogeneity in the layer thicknesses. The angle- and polarization-dependent reflectance spectra obtained by the small-spot θ–2θ scan measurement are shown in Figure 6.13.

Figure 6.13(a) shows the spectra under s-polarization, in which the electric vector is perpendicular to the plane of reflection. The reflectance peak is found to be located

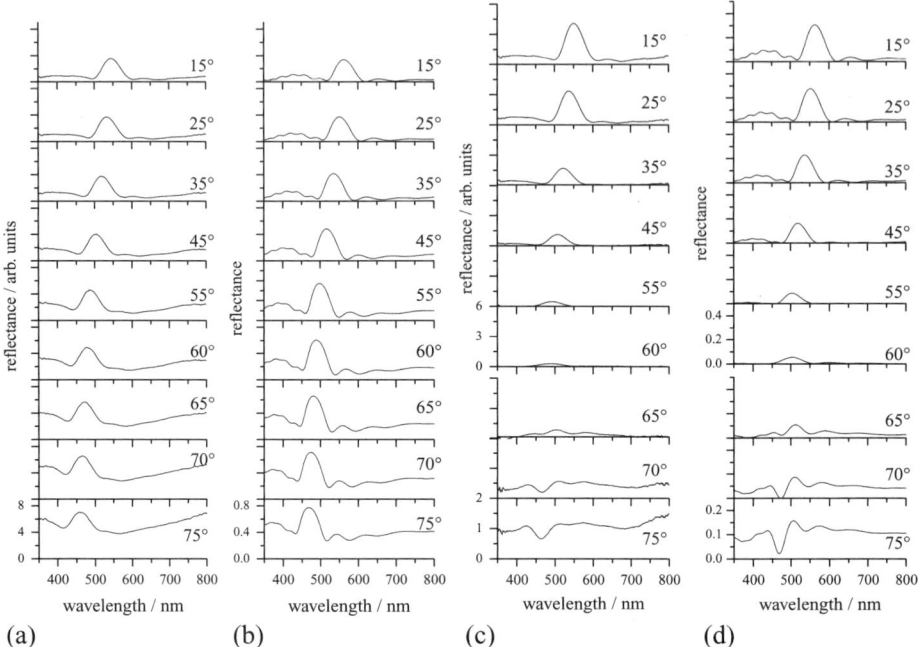

Figure 6.13 Angle-dependent reflectance spectra of the jewel beetle under (a, b) *s*-polarization and (c, d) *p*-polarization. (a, c) Experimental results of a small-spot θ–2θ scan measurement for *s*- and *p*-polarization, respectively; (b, d) theoretical calculation for the multilayer system using the complex wavelength-dependent refractive index values shown in Figures 6.15(b) and (c). In the reflectance calculation, the multilayer structure is assumed to consist of 16 alternating low- and high-index layers. The effective thicknesses of the layers were determined by image analysis to be 141, 84, 138, 72, 101, 71, 101, 72, 105, 79, 101, 80, 94, 93, 104, and 51 nm, from the top to the bottom layer. The substrate is assumed to be made of low-index material. The incident angles are given at the upper right corners. Note that the vertical scale in the lowest three plots in (c) and (d) is magnified. (Reproduced from Yoshioka et al., 2012.)

at approximately 545 nm at a small incident angle of 15°, and the FWHM is approximately 55 nm. The peak gradually shifts to shorter wavelengths with increasing incident angle. However, the reflectance increases over the entire wavelength range examined in this measurement. The appearance of this background-like component explains the weakness of the blue color, although the origin of this component should be identified by analysis.

Different behavior is observed under *p*-polarization, in which the electric vector is in the plane of reflection. As shown in Figure 6.13(c), with an increasing angle, the reflectance gradually decreases, and the peak gradually shifts to shorter wavelengths. However, a dip instead of a peak appears in the background when the angle becomes very large. The dip retains the wavelength shift; it is located at a slightly shorter wavelength in the spectrum observed at the incidence angle of 75° than that at 70°.

Summarizing the preceding measurements, we can say that the important spectral characteristics of the jewel beetle are (1) the wavelength shift of the reflectance peak, (2) the appearance of the background-like component, and (3) the change of the peak into a dip for p-polarization at large angles. These features are analyzed in detail in Section 6.4.4.

6.3.3 Integrated Optical Properties

Another important method of optical characterization is measurements of the total reflectance and transmittance by an integrating sphere. This optical apparatus consists of a spherical cavity, the inner surface of which is covered with a diffuse white material that has nearly 100% reflectance. The integrating sphere usually has several ports, at one of which is placed a photodetector such as a photomultiplier tube. Figure 6.14 (a) shows an optical setup for determining the integrated reflectance; light reflected from the sample is multiply scattered inside the spherical cavity and eventually captured by the photodetector. Namely, reflection is integrated over a hemispherical angular range. Note that not all the reflected light is detected because some portion of the light inevitably escapes through the open ports. However, the detected light intensity is proportional to the magnitude of the total reflection, and dividing the spectrum by that of a white standard such as a $BaSO_4$ plate, which is assumed to have 100% reflectance, yields a quantitative reflectance value. The effects of the port size on the reflectance and dual reflection from the sample are discussed by Yoshioka and Kinoshita (2006a). The integrated transmittance can be similarly measured by modifying the experimental geometry as shown in Figure 6.14(b).

After both the integrated reflectance R and integrated transmittance T are determined, the quantity $1 - R - T$ can be attributed to absorptance because the sum of these three quantities becomes unity when the fluorescence is considered negligible. Examples of such integrated optical properties are shown in Figures 6.14(c) and (d) for

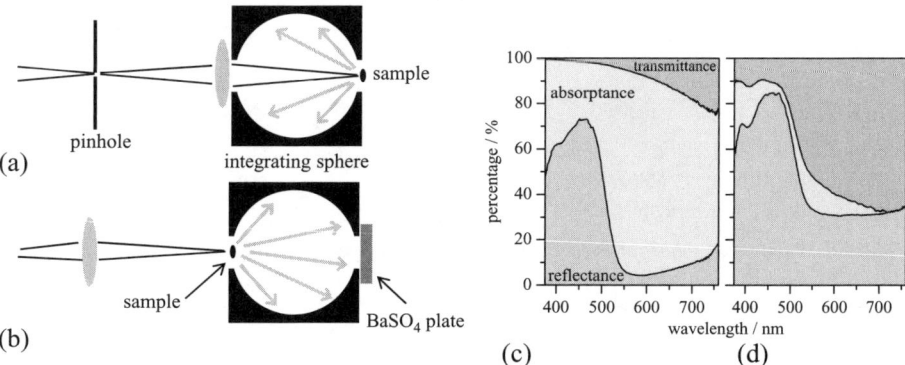

Figure 6.14 Optical setup for measurement of integrated (a) reflectance and (b) transmittance. Experimentally determined optical properties of the (c) blue and (d) white areas of the intact dorsal wing of *M. cypris*. Reflectance, transmittance, and absorptance are expressed as three different gray levels.

the blue and white parts of the wing of *M. cypris*, respectively (Yoshioka and Kinoshita, 2006b). As expected from the wing color, the blue part exhibits high reflectance at wavelengths below 500 nm. A maximum reflectance of more than 70% is observed at approximately 450 nm. Interestingly, despite the large difference in color, the white stripe also shows a strong reflectance of more than 80% at shorter wavelengths. The most prominent difference between the two areas appears in the magnitude of the absorptance, indicating the difference in the amount of pigmentation. Another difference is found in the reflectance at wavelengths longer than 550 nm; the blue area has a minimum reflectance of only 5% at approximately 580 nm, whereas the white one has more than 30%. Because the reflectance determines what we see, this quantitative difference at longer wavelengths should explain the difference in color. The origin of the distinctive whiteness of the stripe is addressed in Section 6.5.4.

Conversely, the integrated reflectance of the jewel beetle is just 20–30%, as shown in Figure 6.15(a), although the elytron surface appears to glitter like a metal. This indicates that the metallic appearance is related more directly to the specular characteristics of the reflection than to the magnitude of the reflectance. The wavelength of the spectral peak is consistent with the green color, although it is distributed in a wavelength range between 545 nm and 570 nm. This distribution in the peak wavelength and peak value implies that the multilayer structure of the jewel beetle is not exactly uniform but differs slightly with the sample and position of the elytron, as is generally

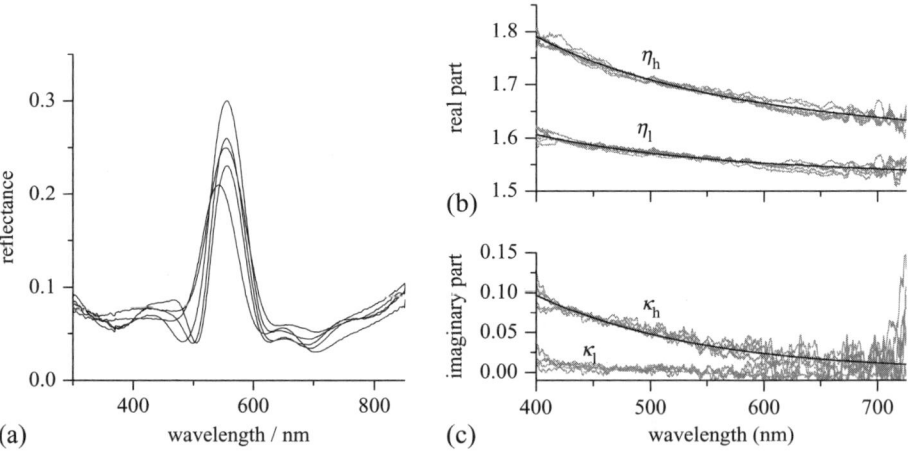

Figure 6.15 (a) Integrated reflectance spectra of several jewel beetles (green region). (b) Real and (c) imaginary parts of the wavelength-dependent refractive index values of the two types of material that comprise the jewel beetle's multilayer system. The index values are denoted by $\hat{n}_h = \eta_h + i\kappa_h$ and $\hat{n}_l = \eta_l + i\kappa_l$ for the electron-dense (high-index) and electron-sparse (low-index) layers, respectively. Six gray curves were determined by using six pairs of reflectance and transmittance spectra for different parts of the semi-frontal thin section. See text for the details on the index determination. Black curves were drawn according to a model formula that approximates the experimentally determined index values. (Reproduced from Yoshioka and Kinoshita, 2011.)

expected for natural samples. When a multilayer reflector consists of materials having a large difference in refractive index (e.g., tantalum pentoxide and magnesium fluoride, which have refractive indices of 2.2 and 1.4, respectively), the reflectance reaches nearly unity at the peak wavelength. Thus, the moderate reflectance of the jewel beetle indicates that the difference in refractive index between the two types of materials that comprise the elytron's multilayer structure is not very large.

6.3.4 Refractive Index Value

The refractive index is one of the essential physical parameters that govern optical phenomena. Many research efforts have been devoted to determining this index to quantitatively analyze optical spectra. For the cuticle of the *Morpho* butterfly wing scale, a few values have been reported, for example, 1.55 (Mason, 1927; Land, 1972) and 1.56 + 0.06i (Vukusic et al., 1999). Leertouwer and colleagues (2011) recently reported careful measurements for a transparent scale of a *Papilio* butterfly (*Graphium sarpedon*) using Jamin–Lebedeff interference microscopy. They determined the cuticle's refractive index n_c to be 1.572, 1.552, and 1.541 at wavelengths of 400 nm, 500 nm, and 600 nm, respectively. The slight wavelength dependence is described well by Cauchy's equation,

$$n_c(\lambda) = A + \frac{B}{\lambda^2}, \qquad (6\text{-}1)$$

with the parameters $A = 1.517$ and $B = 8.8 \times 10^{-3}$ μm². Therefore, the previously reported index values for the butterfly wing agree well and are distributed only slightly at approximately 1.55, although Berthier and colleagues (2003) reported an exceptional value. They discussed the polarization-dependent effective refractive index values, which are much higher than 1.55.

Conversely, for the two types of materials that comprise the beetle's multilayer reflector, various combinations have been reported so far, for example, 1.4 and 1.73 (Bernard and Miller, 1968), 1.5 and 1.6 (Mossakowski, 1979), 1.5 and 2.0 (Schultz and Rankin, 1985a), and 1.55 + 0.14i and 1.68 + 0.03i (Noyes et al., 2007). The distribution in the reported index values is attributed to difficulties arising from a few inherent features of the beetles' elytra as natural multilayer systems. First, unlike the butterfly wing, it is not possible to obtain one type of layer-constituent material as a bulk form in air. Therefore, the convenient index-matching method using a series of liquids with different index values is not applicable to the jewel beetle. Second, surface irregularities and variations in the layer thicknesses make it difficult to directly apply standard optical techniques for thin-film characterization, such as ellipsometry, because these techniques require a large, specular, homogeneous surface such as an artificial optical component.

Yoshioka and Kinoshita (2011) developed a new experimental procedure for independently determining the complex refractive index values of two types of layer-constituent materials in natural multilayer systems. The procedure consists of two experimental steps. First, the multilayer structure is thin-sectioned nearly parallel to the layers (semi-frontal thin sectioning). Because natural multilayer systems are not perfectly flat, a semi-frontal section is naturally obtained without intentionally tilting the sample during sectioning. Although the individual layers in the multilayer system are approximately 100 nm thick, they appear as broad stripes in the semi-frontal thin section, the widths of which are expanded to more than a few microns. Second, a microscope equipped with a spectrometer, which is generally called a microspectrophotometer, is employed to quantitatively measure the reflection (R) and transmission (T) spectra for a small part of the section corresponding to individual layers. Using the two experimental values R and T, the real and imaginary parts of the refractive index values can be independently estimated as a function of the wavelength.

As shown in Figures 6.15(b) and (c), this method was successfully applied to the jewel beetle, and the complex refractive index value and wavelength dependence were determined. The real part for the high-index layer η_h depends greatly on the wavelength; it increases from 1.65 to 1.8 with decreasing wavelength in the examined spectral range. The wavelength dependence is approximated well by Eq. (6-1) with the parameters $A = 1.56$ and $B = 3.6 \times 10^{-2}$ µm^2. Conversely, for the low-index layer, the real part η_l shows a slight increase from 1.55 to 1.6 with decreasing wavelength with the parameters $A = 1.51$ and $B = 1.53 \times 10^{-2}$ µm^2. As expected from the moderate reflectance value obtained with the integrated sphere, these results confirm that the difference in the refractive index is not very large but rather small—for example, 0.12 at a wavelength of 550 nm. Section 6.4.4 shows that this pair of refractive index values yields reflectance spectra that are consistent with the experimental results. Note that the imaginary part for the high-index layer κ_h also depends on the wavelength, increasing it to approximately 0.1 for the shortest wavelength. In contrast, the low-index layer has a negligibly small imaginary part.

6.4 Analysis I

Single-thin-layer and multilayer interference phenomena are among the fundamental optical processes that cause structural color. The goal of this section is to provide mathematical methods for calculating the reflectance spectrum of such systems. We begin with Fresnel's equations, which describe the reflectance and transmittance at a plane boundary between different media. Next, they are applied to the treatment of single-thin-layer interference. A system with varying thickness is considered as an extension of the theory to analysis of a natural example. This is followed by a description of multilayer interference and its application to an analysis of the jewel beetle's iridescence.

6.4.1 Fresnel's Equations

Consider that a light wave traveling in a medium with a refractive index n_i impinges on a different medium with a refractive index n_j, as illustrated in Figure 6.16(a). Part of the incident light is reflected at the interface, and the rest enters the medium with the

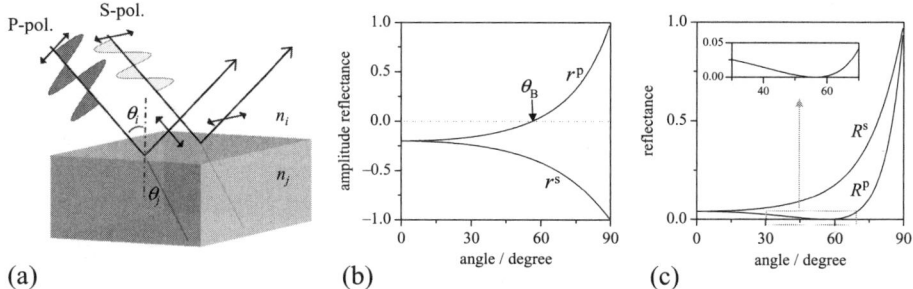

Figure 6.16 Reflectance at the interface between two media with refractive indices n_i and n_j. (a) Geometry and definitions of two linear polarizations. Electric vectors are illustrated by double-headed arrows. (b, c) Amplitude and energy reflectance, respectively, as a function of the incidence angle. Incident medium is assumed to be air ($n_i = 0$), and $n_j = 1.5$ is assumed for the medium in the transmission region. Inset in (c) shows magnified reflectance for p-polarization so as to clearly show Brewster's angle, at which r_p becomes 0.

propagation direction refracted. The amplitude reflectance r and transmittance t are defined as the ratio of the electric fields to that of the incident wave. Namely, the relations $E_r = rE_0$ and $E_t = tE_0$ hold at the interface, where E_0, E_r, and E_t denote the electric fields of the incident, reflected, and transmitted light waves, respectively. By using Maxwell's equations under the conditions that the tangential components of the electric and magnetic fields are continuous at the interface, the following Fresnel equations are obtained:

$$r_{i,j}^s = \frac{n_i \cos\theta_i - n_j \cos\theta_j}{n_i \cos\theta_i + n_j \cos\theta_j} = -\frac{\sin(\theta_i - \theta_j)}{\sin(\theta_i + \theta_j)}, \qquad (6\text{-}2)$$

$$t_{i,j}^s = \frac{2n_i \cos\theta_i}{n_i \cos\theta_i + n_j \cos\theta_j}, \qquad (6\text{-}3)$$

$$r_{i,j}^p = \frac{n_i \cos\theta_j - n_j \cos\theta_i}{n_i \cos\theta_j + n_j \cos\theta_i} = -\frac{\tan(\theta_i - \theta_j)}{\tan(\theta_i + \theta_j)}, \qquad (6\text{-}4)$$

$$t_{i,j}^p = \frac{2n_i \cos\theta_j}{n_i \cos\theta_j + n_j \cos\theta_i}, \qquad (6\text{-}5)$$

where subscripts i and j indicate the media in the incident and transmission regions, respectively, and superscripts s and p indicate two polarizations; s- and p-polarized waves have electric fields perpendicular and parallel, respectively, to the reflection plane, which is defined as the plane containing the two directions of the incident and reflected light. The incidence and refraction angles are denoted by θ_i and θ_j, respectively, and the relation called Snell's law holds for these two angles, $n_i \sin\theta_i = n_j \sin\theta_j$. One example is shown in Figure 6.16(b), where the amplitude reflectance is plotted against the incidence angle for the interface between air and a glass material ($n_i=1$ and $n_j=1.5$). At normal incidence, the two polarization states become physically the same, and their common amplitude reflectance is estimated as $r^s=r^p=(1-1.5)/(1+1.5)=-0.2$. The minus sign indicates that the reflected wave is opposite in phase to the incident wave. Note that a different definition of the positive direction in the electric field is sometimes adopted for p-polarization, which results in a different sign for r^p. When the incidence angle deviates from $0°$, the two polarizations show different angular dependence; r^s decreases monotonically to -1 at $\theta_i=90°$, whereas r^p increases and the sign reverses for an incidence angle larger than Brewster's angle θ_B, where r^p becomes zero. The right side of Eq. (6-4) indicates that the relation $\theta_B+\theta_j=\pi/2$ holds at Brewster's angle because of the condition that the denominator diverges. By substituting $\theta_j=\pi/2-\theta_B$ into Snell's law, the following equation is easily obtained:

$$\tan\theta_B = \frac{n_j}{n_i}. \tag{6-6}$$

When $n_i=1$ and $n_j=1.5$, the angle θ_B is calculated to be $56.3°$. The angle θ_B approaches $45°$ as the refractive index difference becomes small.

Some useful relations between the amplitude reflectance and transmittance are expressed as

$$t_{i,j}t_{j,i} + r_{i,j}^2 = 1, \tag{6-7}$$

$$r_{i,j} + r_{j,i} = 0, \tag{6-8}$$

which can be confirmed by direct calculations using Eqs. (6-2)–(6-5). However, Stokes (1849) derived them from a careful consideration of the time-reversed optical process that should be physically realized. Let us consider reflection and transmission of the light wave shown in Figure 6.17(a). In the corresponding time-reversed process, shown in Figure 6.17(b), two light waves are incident on the interface; one has amplitude $t_{i,j}E_0$ at the interface propagating upward (wave 1) and the other has

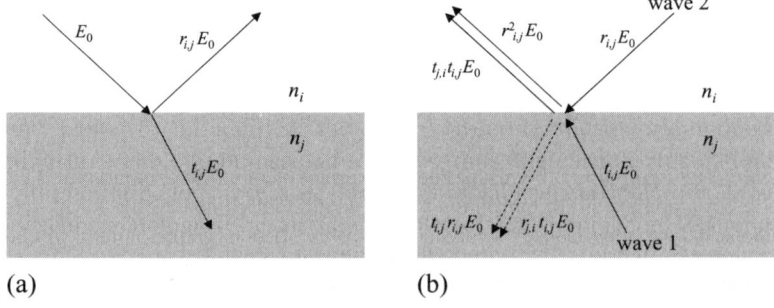

Figure 6.17 (a) Geometry of reflection and transmission at an interface and (b) the corresponding time-reversed process for the derivation of Eqs. (6-7) and (6-8) (the Stokes relations).

amplitude $r_{i,j}E_0$ propagating downward (wave 2). The superposition of transmission of wave 1 and the reflection of wave 2 becomes $t_{i,j}t_{j,i}E_0 + r_{i,j}^2 E_0$ at the interface, which should be equal to E_0. This condition is reduced to Eq. (6-7). Similarly, Eq. (6-8) can be obtained from the condition that the reflection of wave 1 and the transmission of wave 2 are superimposed to become zero.

The energy reflectance R is defined as the ratio of the energy flux carried by the reflected light to that of the incident light. By calculating the Poynting vectors, R is found to be the absolute squared value of the amplitude reflectance—that is, $R = |r|^2$. As shown in Figure 6.16(c), the energy reflectance is just 4% under normal incidence for the interface between air and glass. Conversely, the energy transmittance T differs from $|t|^2$ and is estimated to be $|t|^2 n_j \cos \theta_j / n_i \cos \theta_i$ owing to refraction. The relation $R + T = 1$ is satisfied as the law of energy conservation.

6.4.2 Single Thin-Layer Interference

Interference Conditions

We consider optical interference of a single thin layer of thickness d and refractive index n_1. As shown in Figure 6.18, the system is divided into three parts having refractive indices of n_0, n_1, and n_2. The incident light is partially reflected at the upper surface, and the rest is transmitted with refraction of its propagation direction. The transmitted light is reflected back at the bottom surface, and reflection occurs infinitely many times inside the layer. However, it is often a good approximation to consider only the two major waves shown in Figure 6.18(a), which are reflected from the top and bottom surfaces. This two-wave approximation is valid when the amplitude of the light wave decreases rapidly as reflection is repeated within the layer. This

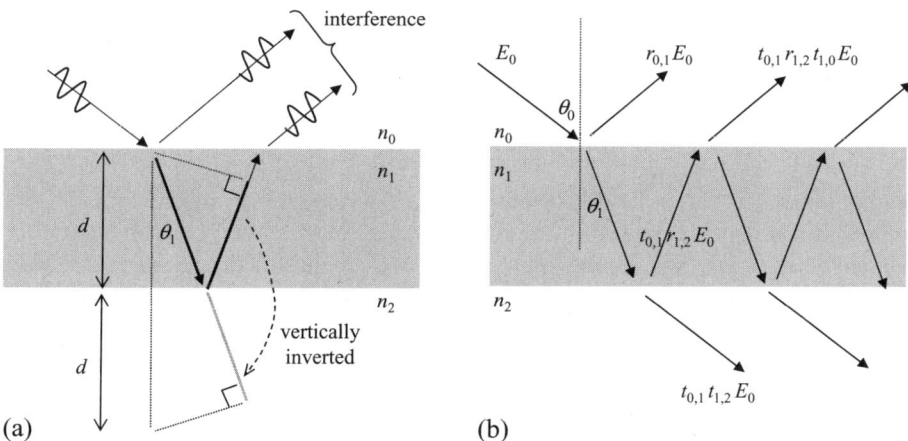

Figure 6.18 Single-thin-layer interference. (a) Illustration of geometrical estimation of the optical path length difference between two reflection paths, one reflected at the top surface and the other at the bottom surface. (b) Multiple reflections within the layer. Amplitudes of several reflection paths are given.

approximation yields a useful interference condition as follows. Geometrical estimation reveals that the two waves have a phase difference ϕ_d because of the difference in the optical path length,

$$\phi_d = 2k_0 n_1 d \cos \theta_1, \tag{6-9}$$

where $k_0 = 2\pi/\lambda$ is the wavenumber, and λ is the wavelength of the light wave in a vacuum. Another phase difference ϕ_r originates from the phase of the amplitude reflectance,

$$\phi_r = \arg(r_{0,1}) - \arg(r_{1,2}), \tag{6-10}$$

where $\arg(x)$ is the argument of a complex value x, and $r_{i,j}$ denotes the amplitude reflectance when a light wave impinges on the medium having a refractive index n_j from a medium having a refractive index n_i. When the refractive indices are real, the argument is 0 or π depending on their magnitudes. However, it generally takes a value ranging from 0 to 2π when the refractive indices (and the amplitude reflectance) are complex. The sum of these two phase differences governs the type of optical interference; when it is equal to 2π multiplied by an integer m—that is,

$$\phi_d + \phi_r = 2m\pi, \tag{6-11}$$

the interference is constructive (i.e., reflection is enhanced and transmission is decreased). In many cases, the materials can be considered optically transparent. Then, the phase difference ϕ_r becomes 0 or π.

Corresponding to these two values, there are two common cases of thin-film interference: in one, called the soap bubble case, the refractive index of the thin layer is larger than those of the other two media—that is, $n_1 > n_0$ and $n_1 > n_2$. In this case, because the phase is reversed only for the reflection at the upper surface, ϕ_r is equal to π, and the preceding interference condition is reduced to

$$2n_1 d \cos \theta_1 = \left(m - \frac{1}{2}\right)\lambda_m, \tag{6-12}$$

where λ_m denotes the wavelength of mth-order optical interference. Conversely, in the second case, called the anti-reflective coating case, the substrate has the largest refractive index—that is, $n_0 < n_1$ and $n_1 < n_2$. Equation (6-11) becomes

$$2n_1 d \cos \theta_1 = m\lambda_m.$$

The preceding conditions are convenient for estimating the wavelengths of the reflectance peak. When the materials have complex refractive indices, ϕ_r generally takes a value between 0 and 2π. Even in such a case, the interference condition given in the form of Eq. (6-11) is still valid.

When the wavelengths of the minimum reflectance are considered, the condition for destructive interference is useful; it is similarly obtained by replacing the right side of Eq. (6-11) with $(2m-1)\pi$. In this case, the interference decreases reflection and enhances transmission.

Reflectance Spectrum

Reflectance and transmittance spectra of thin-film interference can be calculated by rigorously considering the multiple reflections inside the layer, as illustrated in Figure 6.18(b). The total electric field E_r of the reflected light is obtained as an infinite series,

$$E_r = r_{0,1}E_0 + t_{0,1}r_{1,2}\left(1 + r_{1,0}r_{1,2}e^{i\phi_d} + (r_{1,0}r_{1,2})^2 e^{2i\phi_d} + \cdots\right)t_{1,0}e^{i\phi_d}E_0,$$

$$r_t \equiv \frac{E_r}{E_0} = r_{0,1} + \frac{t_{0,1}r_{1,2}t_{1,0}e^{i\phi_d}}{1 - r_{1,0}r_{1,2}e^{i\phi_d}} = \frac{r_{0,1} + r_{1,2}e^{i\phi_d}}{1 + r_{0,1}r_{1,2}e^{i\phi_d}}, \quad (6\text{-}13)$$

where r_t is defined as the total amplitude reflectance. Equations (6-7) and (6-8) are used to derive the right side of Eq. (6-13). Then, the calculated energy reflectance R_t is

$$R_t = \left|\frac{r_{0,1} + r_{1,2}e^{i\phi_d}}{1 + r_{0,1}r_{1,2}e^{i\phi_d}}\right|^2, \quad (6\text{-}14)$$

which is reduced to

$$R_t = \frac{r_{0,1}^2 + r_{1,2}^2 + 2r_{0,1}r_{1,2}\cos\phi_d}{1 + r_{0,1}^2 r_{1,2}^2 + 2r_{0,1}r_{1,2}\cos\phi_d} \quad (6\text{-}15)$$

when all the refractive indices (and the amplitude reflectance) are real. The total amplitude transmittance t_t is similarly obtained by calculating the infinite series as

$$t_t \equiv \frac{E_t}{E_0} = t_{0,1}t_{1,2}e^{\frac{i\phi_d}{2}}\sum_{n=0}^{\infty}\left(r_{1,0}r_{1,2}e^{i\phi_d}\right)^n = \frac{t_{0,1}t_{1,2}e^{i\phi_d/2}}{1 + r_{0,1}r_{1,2}e^{i\phi_d}}. \quad (6\text{-}16)$$

The phase factor $\exp[i\phi_d/2]$ in the numerator comes from the optical path length of one-way propagation through the thin layer. An example of a spectrum is shown in Figure 6.19(b), assuming $n_0 = n_2 = 1$, $n_1 = 1.5$, $\theta_0 = 45°$, and $d = 650$ nm. These values were chosen to model the structural color of the rock dove's neck feather, which is the most typical example of thin-layer interference as a coloration mechanism (Yin et al., 2006; Yoshioka et al., 2007). The outer cortex of the feather barbule consists of a single layer of keratin, as shown in Figure 6.19(d). In fact, moderate agreement is obtained between the theory and the experimental results shown in Figure 6.19(a).

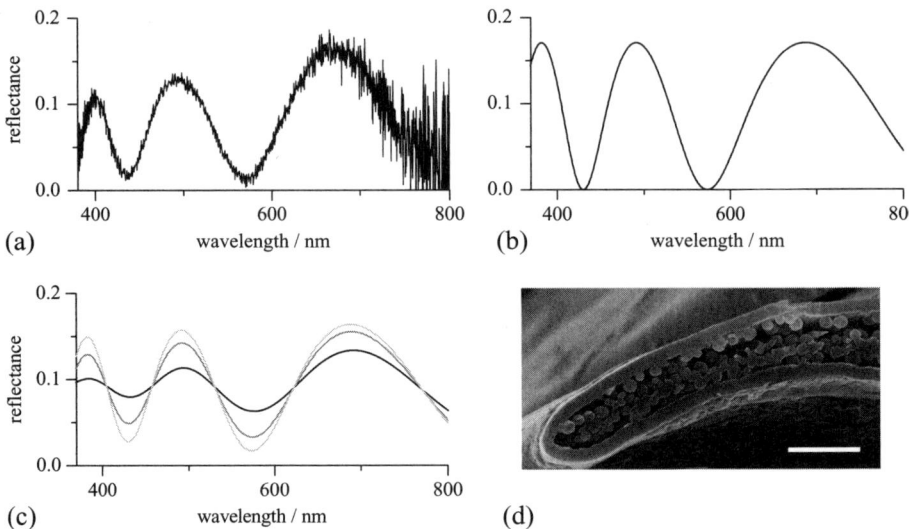

Figure 6.19 Reflectance spectra of single-thin-layer interference. (a) Experimentally obtained spectrum for a single barbule of rock dove's neck feather under 45° incidence of unpolarized light; (b) theoretically calculated spectrum with $n_0 = n_2 = 1$, $n_1 = 1.5$, $\theta_0 = 45°$, and $d = 650$ nm, for unpolarized light; and (c) reflectance spectra of three systems with varying thickness calculated using Eqs. (6-19) and (6-20). The standard deviation σ_d of the thickness distribution is assumed to be 20 (light gray curve), 30 (gray curve), and 50 nm (black curve). (d) Cross-section of the barbule of the rock dove's neck feather observed by scanning electron microscope. Scale bar: (d) 3 μm.

The condition for constructive interference given by Eq. (6-12) is satisfied at the wavelengths $\lambda_m = 382$ nm, 491 nm, and 688 nm with the preceding parameters for the interference orders $m = 5, 4,$ and 3, respectively, producing several reflectance peaks in the wavelength range of visible light. These spectral characteristics cause a peculiar iridescent effect called *two-color iridescence* (Yoshioka et al., 2007) owing to the match between the several reflectance peaks and the peaks of the color-matching functions of human vision, which correspond roughly to the sensitivity curves for the three primary colors (e.g., Srinivasarao, 1999).

When the reflectance spectrum is measured in a broader wavelength range using an integrating sphere, more peaks with different orders are clearly observed, as shown in Figure 6.20(a). When the spectrum is plotted against the wavenumber (Figure 6.20b), the peaks are separated by a constant distance, which is estimated to be $1/(n_1 d \cos \theta_1)$ from Eq. (6-12), assuming that the refractive index is independent of the wavelength. The high reflectance in the infrared region is thought to originate from multiple scattering inside the complicated feather structure because the reflectance was measured for the whole feather shown in Figure 6.20(c); the melanin pigment inside the feather barbule reduces the multiple scattering in the visible wavelength range due to optical absorption, but this effect becomes negligible in the infrared region, where the pigment becomes transparent (Nakamura et al., 2008).

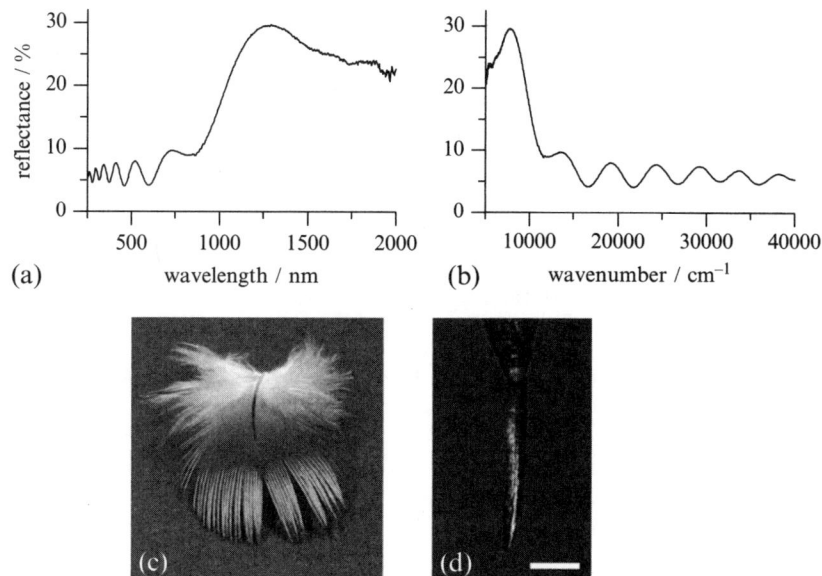

Figure 6.20 Reflectance spectra of rock dove's neck feather determined using an integrating sphere. The same data are plotted against the (a) wavelength and (b) wavenumber. (c, d) Photographs of rock dove's neck feather and a single barbule attached to the tip of a needle, respectively. Scale bar: (d) 100 μm. (For color version of this figure, the reader is referred to the online version of this chapter.)

Interference in Thin Layer with Varying Thickness

The experimental spectrum shown in Figure 6.19(a) was measured for a single barbule attached to the tip of a needle, as shown in Figure 6.20(d), not for the whole feather. Thus, the effects of multiple reflection within the feather structure were absent. However, a careful comparison of the experimental and theoretical spectra reveals that they do not agree perfectly (Figures 6-19a and b). One discrepancy is the fact that in the experimental spectrum, the reflectance peak gradually decreases as the wavelength becomes shorter and the minima do not reach zero, whereas in theory the peaks have the same height and the minima are zero. The discrepancy might originate in structural irregularities in the thin layer, such as surface roughness and variations in the thickness of the layer. The surface roughness has already been discussed by Nakamura and colleagues (2008). Therefore, we consider the effects of the variations in thickness on the spectral line shape. The variations are taken into account as the statistical average of the reflectance spectrum. For this purpose, Eq. (6-15) is rewritten as

$$R_t(\phi_d) = 1 + b\frac{1}{1 - a\cos\phi_d} = 1 + b\sum_{n=0}^{\infty} a^n \cos^n \phi_d, \tag{6-17}$$

where a and b are defined as

$$a = -\frac{2r_{0,1}r_{1,2}}{1+r_{0,1}^2 r_{1,2}^2},$$

$$b = \frac{r_{0,1}^2 + r_{1,2}^2 - r_{0,1}^2 r_{1,2}^2 - 1}{1+r_{0,1}^2 r_{1,2}^2}.$$

Because the relation $|a| < 1$ is easily confirmed, the preceding expansion is sure to converge. According to a mathematical formula, the nth power of $\cos \phi_d$ can be expanded as a series of $\cos m\phi_d$ using binomical coefficients,

$$\cos^n \phi_d = \begin{cases} \dfrac{1}{2^{n-1}} \displaystyle\sum_{m=0}^{(n-1)/2} \binom{n}{m} \cos(n-2m)\phi_d, & \ldots \text{ for odd } n \\[2ex] \dfrac{1}{2^n}\binom{n}{n/2} + \dfrac{1}{2^{n-1}} \displaystyle\sum_{m=0}^{\frac{n}{2}-1} \binom{n}{m} \cos(n-2m)\phi_d, & \ldots \text{ for even } n \end{cases},$$

which is rewritten by newly defining the expansion coefficient c_m^n as

$$\cos^n \phi_d = \sum_{m=0}^{n} c_m^n \cos m\phi_d. \tag{6-18}$$

Taking the statistical average of Eq. (6-17), we obtain the following expression using Eq. (6-18):

$$\langle R_t(\phi_d) \rangle = 1 + b \sum_{n=0}^{\infty} \sum_{m=0}^{n} a^n c_m^n \langle \cos m\phi_d \rangle, \tag{6-19}$$

where $\langle \ldots \rangle$ denotes the statistical average. Because the variation in the layer thickness affects the reflectance only through the phase ϕ_d according to Eq. (6-9), the distribution function of the phase variation $f(\delta\phi)$ is introduced, where the phase variation $\delta\phi$ is defined as $\delta\phi \equiv \phi_d - \phi_{\bar{d}}$, and $\phi_{\bar{d}}$ is the mean phase related to the mean thickness $\bar{d}$ by $\phi_{\bar{d}} = 2k_0 n_1 \bar{d} \cos\theta_1$. Assuming a Gaussian distribution function,

$$f(\delta\phi) = \frac{1}{\sqrt{2\pi}\sigma_\phi} \exp\left(-\frac{(\delta\phi)^2}{2\sigma_\phi^2}\right),$$

where σ_ϕ is the standard deviation related to that in the thickness σ_d by $\sigma_\phi = 2k_0 n_1 \sigma_d \cos\theta_1$, we can perform the statistical average as

$$\langle \cos m\phi_d \rangle = \frac{1}{\sqrt{2\pi}\sigma_\phi} \int \cos m(\bar{\phi} + \delta\phi) \exp\left(-\frac{(\delta\phi)^2}{2\sigma_\phi^2}\right) d(\delta\phi)$$

$$= \cos m\bar{\phi} \exp\left(-\frac{m^2 \sigma_\phi^2}{2}\right). \tag{6-20}$$

Thus, the averaged reflectance is analytically expressed as the infinite series of Eq. (6-19) with its coefficient Eq. (6-20). However, because the coefficients contain an exponential factor, $\exp\left(-m^2\sigma_\phi^2/2\right)$, higher-order harmonics do not contribute greatly to the series.

Figure 6.19(c) shows reflectance spectra for a few systems with varying thickness. The amplitude of the reflectance oscillation decreases as the standard deviation σ_d becomes larger. In addition, the peak value decreases as the wavelength becomes shorter, in agreement with the experimental results. That is because the phase becomes more sensitive to the thickness variation at shorter wavelengths. The varying-thickness system with $\sigma_d = 20$ nm produces the reflectance spectrum that is most similar to the experimental results among the three systems shown in Figure 6.19(c). However, the agreement is still not perfect, indicating that some additional factors contribute to the reflection mechanisms.

6.4.3 Multilayer Interference

Interference Conditions

When a stack of thin layers is considered, light is reflected at each interface, and the optical interference becomes quite complicated because there are many reflection paths. Thus, the reflectance and transmittance spectra of the multilayer system can have various shapes depending on the design. For example, when the layers are periodically stacked, the reflectance can reach nearly 100% in a narrow wavelength range, and the system acts as a laser-reflecting mirror. In another design, the transmittance becomes high for wavelengths longer than a certain edge wavelength, and the system acts as a long-wavelength pass filter. Such multilayer structures are now widely applied in many types of optical components. Here we consider a periodic multilayer system that causes wavelength-selective reflection, which is among the primary origins of structural color.

Let us assume that the multilayer stack consists of pairs of two types of layers, which are distinguished by subscripts a and b, as shown in Figure 6.21(a). Each type of layer has the same thickness d_j and refractive index n_j ($j = a$ or b). The interference condition for this periodic multilayer system can be derived by considering the phase difference between two light waves reflected at interfaces that are separated by one period (e.g., interfaces i_2 and i_4), as labeled in Figure 6.21(a). The path length difference of the two waves can be estimated similarly as that for single-thin-layer interference, and the following condition is obtained for constructive interference,

$$2(n_a d_a \cos\theta_a + n_b d_b \cos\theta_b) = m\lambda_m, \tag{6-21}$$

where the refraction angles inside the two type of layers are denoted by θ_a and θ_b. These angles are related to the incidence angle θ_0 by Snell's law, $n_a \sin\theta_a = n_b \sin\theta_b = n_0 \sin\theta_0$, where n_0 is the refractive index of the incident medium. The integer m in Eq. (6-21) is called the order of interference.

This interference condition is quite useful for estimating the wavelength of the reflection, and it usually gives correct values. However, it is important to remember that in exceptional cases the condition does not work correctly: in one case, a

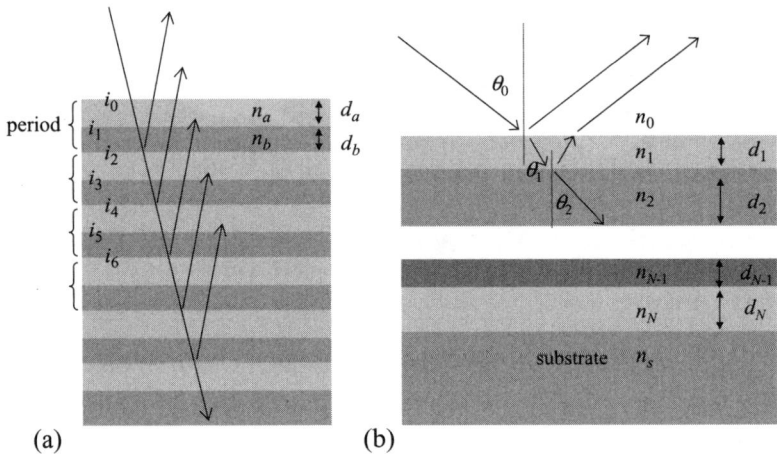

Figure 6.21 Multilayer system. (a) A periodic system consisting of pairs of two types of layers having refractive indices n_a and n_b and thicknesses d_a and d_b. Interfaces between different layers are labeled i_m, where $m = 1, 2, 3, \ldots$ from the top to the bottom. (b) A generic multilayer system consisting of a total of N layers having refractive index values n_j, thicknesses d_j, and refraction angles θ_j, where $j = 1 \sim N$. The incident region and substrate have refractive indices of n_0 and n_s, respectively.

reflectance peak is not located at the wavelength predicted by Eq. (6-21) but is shifted, and in another case, a reflectance peak does not exist. These exceptions are not surprising because the interference condition is derived simply by considering the optical path length difference between only two waves. Thus, it is important to rigorously calculate the reflectance spectrum for the multilayer stack in question. One simple method is described in the next subsection.

Reflectance Spectrum

To treat multilayer interference exactly, we have to consider all the reflection paths that can occur within the system. Many methods have been used, for example, several types of the transfer matrix method (e.g., Yeh, 1988; Born and Wolf, 1975), Huxley's method (Huxley, 1968), and the recursive method (e.g., Noyes et al., 2007). Here we describe the last one, in which the formula for single-thin-layer interference is recursively employed in the calculation. A multilayer structure is assumed to consist of a total of N layers with refractive index n_i and thickness d_i ($i = 1 \sim N$) that are stacked on the substrate with a refractive index n_s as shown in Figure 6.21(b). The calculation starts by considering the thin-layer interface of the bottom (Nth) layer. Consider the situation shown in Figure 6.22(a), in which the reflectance ρ_N of the thin layer sandwiched by media having index values n_{N-1} and n_s is calculated directly using Eq. (6-13) as

$$\rho_N = \frac{r_{N-1,N} + r_{N,s} e^{i\phi_N}}{1 + r_{N-1,N} r_{N,s} e^{i\phi_N}}, \tag{6-22}$$

Structural Color in Nature

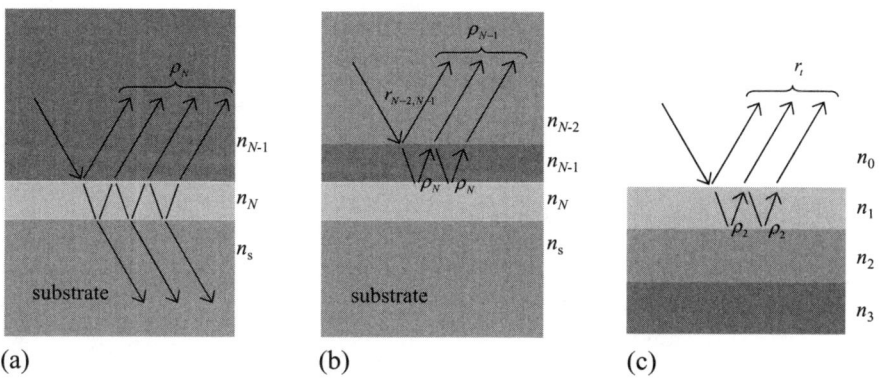

Figure 6.22 Recursive method for calculating the reflectance of the multilayer system: (a) first step, (b) second step, and (c) last step. See text for the definition of ρ_j.

where ϕ_N is the phase difference similar to that defined in Eq. (6-9) as $\phi_N = 2k_0 n_N d_N \cos\theta_N$ with the refractive index n_N, refraction angle θ_N, and thickness d_N. In the next step, the next layer up from the bottom is included, as shown in Figure 6.22(b). The reflectance of this two-layer system, denoted by ρ_{N-1}, can be calculated by replacing the amplitude reflectance of the bottom surface $r_{1,2}$ in Eq. (6-13) with ρ_N to take the multiple reflections inside the Nth layer into account. Namely, the reflectance is given as

$$\rho_{N-1} = \frac{r_{N-2,N-1} + \rho_N e^{i\phi_{N-1}}}{1 + r_{N-2,N-1} \rho_N e^{i\phi_{N-1}}}. \tag{6-23}$$

Repeating this procedure by recursively calculating the reflectance until the top layer is included in the calculation gives the reflectance value for the entire multilayer system.

Figures 6.23(a)–(c) show some examples of spectra of multilayer systems called quarter-wave stacks, where the thicknesses of the two types of layers, d_a and d_b,

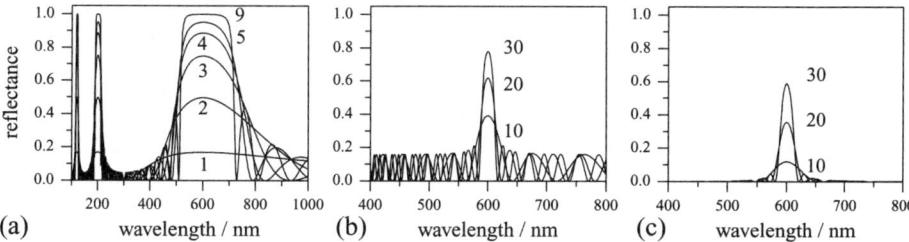

Figure 6.23 Reflectance spectra of several periodic multilayer systems. (a) A quarter-wave stack with parameters that satisfy the condition $n_a d_a = n_b d_b = \lambda_1/4$ with $\lambda_1 = 600$ nm. The refractive indices of the two types of layers are $n_a = 1.55$ and $n_b = 1$, and those of the incident and transmission regions are assumed to be $n_0 = 1$ and $n_s = 1$, respectively. (b) A similar system with a low–refractive index contrast. The parameters are $n_a = 1.55$, $n_b = 1.5$, and $n_0 = n_s = 1$. (c) Same multilayer stack as (b) but immersed in a medium having a refractive index of 1.5 ($= n_0 = n_s$). The numbers of high-index layers N_a are given near the spectra. Normal incidence is assumed in all the preceding calculations.

are determined such that the optical thickness of each layer becomes one-quarter of the wavelength λ_1 at which the reflectance is expected to become high under normal incidence. Namely, $n_a d_a = n_b d_b = \lambda_1/4$. Under this assumption, the interference condition of Eq. (6-21) is satisfied at λ_1 with $m = 1$, which ensures that waves reflected at the interfaces of different periods interfere constructively. In addition, the waves reflected at the middle interfaces within a period $(i_1, i_3, i_5, \ldots$ in Figure 6.21a) also interfere constructively at λ_1 because the roundtrip of the half-period has an optical path length of $\lambda_1/2$, and the corresponding phase delay π compensates for the phase inversion that occurs at the reflection at the middle interface. Thus, the quarter-wavelength stack is generally expected to have a very high reflectance.

Figure 6.23(a) shows several cases for different layer numbers with parameters $n_a = 1.55$, $n_b = 1$, and $\lambda_1 = 600$ nm. The peak reflectance increases rapidly as the layer number N_a, which is the number of high-index (type a) layers, increases. The reflectance spectrum for a system with $N_a = 9$ (a total of 17 layers) has a flat-topped shape, and the reflectance becomes nearly 1 in a wavelength range including λ_1. The reflectance is expected to reach 1 if the layer number is increased infinitely. Huxley (1968) reported an analytical expression for estimating this wavelength range,

$$\cos^2 \psi \leq r_{a,b}^2, \tag{6-24}$$

where ψ is defined as $2\pi n_a d_a/\lambda = \pi \lambda_1/(2\lambda)$, and $r_{a,b}$ is the amplitude reflectance given by Fresnel's equations (Eqs. 6-2 or 6-4 depending on the polarization).

Calculations using the preceding refractive index values reveal that the condition is satisfied in a wavelength range $\lambda_l \leq \lambda \leq \lambda_u$, where $\lambda_l = 527$ nm, and $\lambda_u = 696$ nm. The reflectance spectrum for $N_a = 9$ seems to have a slightly wider reflection band than this estimated wavelength range. However, almost complete agreement with the theory is obtained for spectra of a system with a large layer number (e.g., $N_a = 100$). Such a multilayer system is considered to represent a one-dimensional (1D) photonic crystal that has a photonic band gap in that wavelength range. Thus, light having a wavelength in the gap cannot propagate inside the multilayer system and is completely reflected.

The condition given by Eq. (6-24) implies that the reflection band becomes narrower when the amplitude reflectance $r_{a,b}$ is smaller—that is, when the refractive index difference is smaller. This is confirmed in the spectra shown in Figure 6.23(b), in which the refractive index values $n_a = 1.55$ and $n_b = 1.5$ are assumed in the calculation. Because the amplitude reflectance $r_{a,b}$ is small, many more layers are necessary for the reflectance peak to become as high as those in the spectra of the systems with a large refractive index difference shown in Figure 6.23(a). In addition, an oscillatory spectral feature is noticeable in the entire spectral range except for a narrow reflection band at approximately 600 nm. This oscillation is understood as optical interference of a single thick layer: the multilayer system behaves as a single thick layer overall because the reflections at the top and bottom surfaces are much stronger than the internal reflections owing to the large refractive index difference. This type of oscillatory spectrum is often called Fabry–Perot interference because the mechanism is quite similar to that of a Fabry–Perot interferometer. One way to suppress this oscillation in the spectrum is to place the multilayer system in a medium having a refractive index similar to n_a or n_b. This is theoretically demonstrated in Figure 6.23(c), where the system is surrounded by a medium with a refractive index of 1.5.

Let us check the higher-order interference peaks in the spectra shown in Figure 6.23 (a). Three remarkable peaks appear at wavelengths of 600 nm, 200 nm, and 125 nm. It is confirmed from the interference condition (Eq. 6-21) that these peaks correspond to the interference orders $m = 1, 3,$ and 5, respectively. However, the peaks are not observed at wavelengths of 300 nm and 150 nm, which are predicted for $m = 2$ and 4, respectively. In these cases, the satisfaction of the interference conditions does not necessarily ensure the presence of reflectance peaks. The disappearance of the even-order peaks is explained by destructive interference between the waves reflected at interfaces $i_1, i_3, i_5, \ldots$ and those reflected at $i_2, i_4, i_6, \ldots$ (see Figure 6.21a), which will be described in the next section in more detail.

Ideal and Nonideal Multilayer Systems

The quarter-wave stack is sometimes called an ideal multilayer structure because all the reflected waves can interfere constructively at a wavelength λ_1 owing to the strict condition $n_a d_a = n_b d_b$. Here, let us loosen this condition and consider more general cases in which the optical thicknesses of the two types of layers are not necessarily equal to each other (i.e., $n_a d_a \neq n_b d_b$), while keeping the relation $2(n_a d_a + n_b d_b) = \lambda_1$ with $\lambda_1 = 600$ nm. This type of stack is called a nonideal periodic multilayer structure (Land, 1972). To express the degree to which is it nonideal, the parameter p is defined as

$$p \equiv \frac{4 n_a d_a}{\lambda_1}, \tag{6-25}$$

where p takes a value ranging from 0 to 2. The ratio of the optical thicknesses of the two types of layers is expressed as $n_a d_a : n_b d_b = p : 2 - p$, and the ideal stack corresponds to the case with $p = 1$. Figures 6.24(a) and (b) show reflectance spectra for

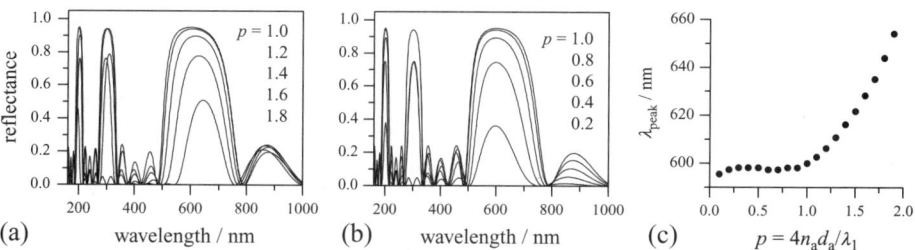

Figure 6.24 (a, b) Reflectance spectra of various nonideal multilayer structures. The parameter p ($= 4 n_a d_a / \lambda_1$) is given to the right of the spectra. The case with $p = 1$ corresponds to the ideal multilayer structure (the quarter-wave stack). Refractive indices $n_a = 1.55$ and $n_b = 1$, layer number $N_a = 5$ (a total of nine layers), and normal incidence are assumed in the calculations. (c) The wavelength of the reflectance peak plotted against various values of p for nonideal periodic multilayer systems that satisfy the interference condition at $\lambda_1 = 600$ nm (see text for details).

various p values. The reflectance at the peak around λ_1 decreases as p deviates from 1 because interference with reflections at the middle interfaces becomes not completely constructive. In addition, the reflectance does not take the maximum value at $\lambda_1 = 600$ nm, as shown in Figure 6.24(c); the reflectance peak shifts toward longer wavelengths as p increases from 1, whereas the shift is not very large for $p < 1$.

In some of the spectra shown in Figures 6.24(a) and (b), the second-order peak is clearly visible, although the magnitude differs greatly depending on p. Similarly, the appearances of the higher-order peaks are affected by p. The conditions at which a peak disappears can be derived from the conditions for destructive interference between the waves reflected at interfaces $i_1, i_3, i_5, \ldots$ and those reflected at $i_2, i_4, i_6, \ldots$. When the roundtrip in the a-type layers equals the optical thickness of the mth-order wavelength λ_m multiplied by an integer l, the phase inversion due to the reflection causes destructive interference, and consequently the disappearance of the mth-order peak. This condition is expressed as

$$2 n_a d_a = l \lambda_m, \qquad (6\text{-}26)$$

where the integer l lies in the range $0 < l < m$. This condition is equivalent to

$$p = \frac{2l}{m}. \qquad (6\text{-}27)$$

For example, when the second-order interference $m = 2$ is considered, the reflectance peak vanishes for the stack with $p = 1$ because the condition is satisfied with $l = 1$. This is the case of the quarter-wave stack and is consistent with the spectra shown in Figure 6.23(a). Similarly, the third-order peak ($m = 3$) disappears in systems with $p = 2/3$ and $4/3$ obtained for $l = 1$ and $l = 2$, respectively. Figure 6.25 shows several examples of reflectance spectra where various higher-order peaks are not observed. They confirm that the preceding condition correctly predicts the disappearances of the higher-order peaks in the reflectance spectra.

There is a natural example of a highly nonideal multilayer structure that produces coloration. The Madagascan sunset moth, *Chrysiridia rhipheus* shown in Figure 6.26 (a), has wings covered with variously colored scales. Electron microscopy revealed that the scales contain a multilayer structure consisting of several cuticle layers with air spacers (Lippert and Gentil, 1959). The scales in the green part of the hind dorsal wing have the thickest cuticle layers of all the variously colored regions (Yoshioka et al., 2008); the thicknesses of the cuticle and air layers were determined to be $d_{cut} = 270$ nm and $d_{air} = 130$ nm, respectively, from the electron micrograph shown in Figure 6.26(b). Assuming the refractive index of the cuticle $n_{cut} = 1.55$, the interference condition of Eq. (6-21) predicts that the reflectance peaks are located at wavelengths of 1097 nm, 549 nm, and 366 nm, which correspond to the interference orders $m = 1, 2,$ and 3, respectively. In fact, the three peaks are clearly observed experimentally in the reflectance spectrum (Figure 6.26c). Thus, the green color of the wing is produced mainly by second-order interference. The parameter p for this multilayer system is calculated to be 1.5 from the optical thicknesses 270×1.55 and 130×1.

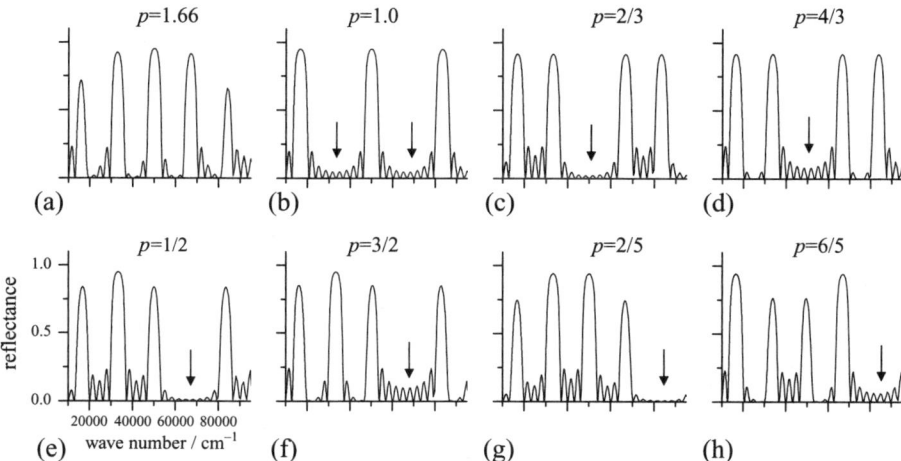

Figure 6.25 Reflectance spectra of multilayer systems with various values of nonideal parameter p plotted against the wavenumber. The value of p is given in each graph. (a) All five orders of interference peaks are seen, for example, at $p = 1.66$. In the other plots, the second-, third-, fourth-, and fifth-order peaks vanish in (b), (c) and (d), (e) and (f), and (g) and (h), respectively. The condition expressed by Eq. (6-27) is satisfied at (b) $p = 1$ with $m = 2$ and $l = 1$, and with $m = 4$ and $l = 2$; at (c) $p = 2/3$ with $m = 3$ and $l = 1$; at (d) $p = 4/3$ with $m = 3$ and $l = 2$; at (e) $p = 1/2$ with $m = 4$ and $l = 1$; at (f) $p = 3/2$ with $m = 4$ and $l = 3$; at (g) $p = 2/5$ with $m = 5$ and $l = 1$; and at (h) $p = 6/5$ with $m = 5$ and $l = 3$. The refractive index values and layer numbers are the same as in the systems considered in Figure 6.24.

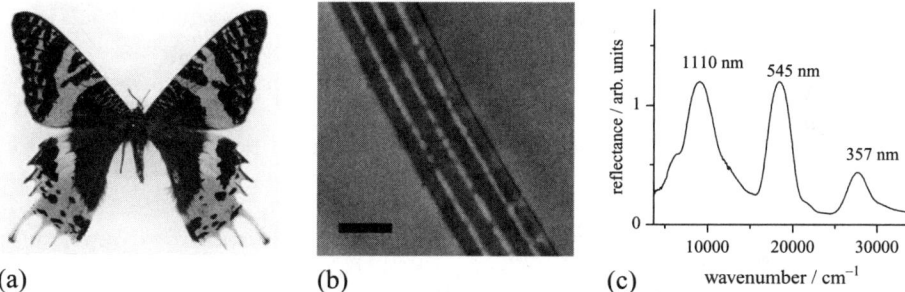

Figure 6.26 Madagascan sunset moth, *Chrysiridia rhipheus*. (a) Dorsal side of wing, (b) transmission electron micrographs of cross-section of a scale in the green region of the hind dorsal wing, and (c) reflectance spectrum obtained using a broad-band spectrometer. Scale bar: (b) 1 μm. (Reproduced from Yoshioka et al., 2008.) (For interpretation of the references to color in this figure legend, the reader is referred to the online version of this chapter.)

This value does not satisfy the condition of Eq. (6-27) for $m = 1, 2$, and 3, so the three corresponding peaks do not disappear, although the fourth-order peak is expected to do so. This is theoretically confirmed in Figure 6.25(f).

Interestingly, second-order interference produces the green color in this moth because higher-order interference requires more precise control of the layer thicknesses when the microstructures develop in the larval stage. The insect eye is generally not believed to be sensitive to infrared light. This may imply that the third-order peak in the ultraviolet region is important because the ultraviolet is usually visible to insects; both the second- and third-order reflection peaks produce the color of the wing as a mixture of these two wavelengths. Thus, the green part of the moth wing might look totally different to insect eyes.

6.4.4 Analysis of Jewel Beetle's Iridescence

It is interesting to see how well the multilayer interference theory explains the structural color of the jewel beetle. For a detailed analysis, the thicknesses of the layers were determined by image analysis of an electron micrograph (Yoshioka et al., 2012) as schematically depicted in Figure 6.27(a). The refractive indices of the

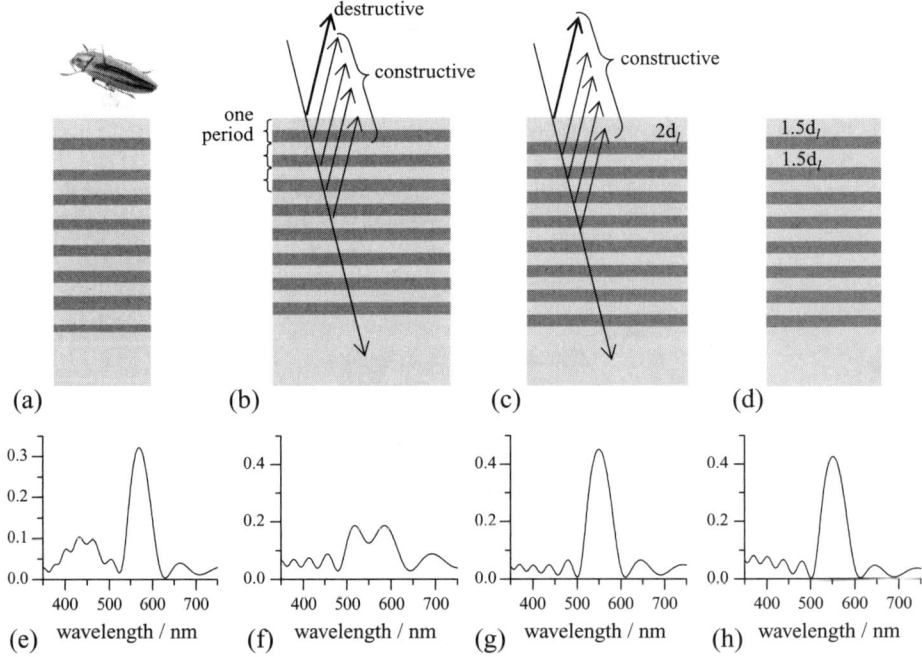

Figure 6.27 Several multilayer system models for the jewel beetle. Reflectance spectra of models (a)–(d) are shown in (e)–(h), respectively. Light and dark gray layers correspond to low– and high–refractive index layers, respectively. Thicknesses of the layers are depicted such that they are proportional to the optical thicknesses (the product of the refractive index and physical thickness). In calculating the reflectance, for simplicity, we assume real and constant refractive indices, $n_l = 1.56$ and $n_h = 1.68$, for the low- and high-index layers, respectively. (a) The model obtained from image analysis of an electron micrograph; (b) quarter-wave stack model in which layer thicknesses are denoted by $d_l = 88$ nm and $d_h = 82$ nm for the low– and high–refractive index layers, respectively, which are determined such that the interference condition is satisfied at a wavelength of 550 nm; (c) a modified model obtained from (b) where the first low-index layer has a double-layer thickness of $2d_l$; and (d) another modified model where both the first and third layers have a thickness of $1.5\, d_l$.

high- and low-index layers were experimentally determined as shown in Figures 6.15 (b) and (c) by semi-frontal thin sectioning. Using these parameter values, the recursive method was applied to calculate the angle- and polarization-dependent reflectance spectra. The results, shown in Figures 6.13(b) and (d), roughly agree with the experimental results because they reproduce the following three major spectral characteristics:

1. The gradual shift in the reflectance peaks as the incidence angle increases.
2. The appearance of the background-like component at large incidence angles.
3. The spectral dip instead of a peak at incidence angles greater than 65° for p-polarization.

We employed different combinations of layer thicknesses obtained from other electron micrographs and confirmed that the calculated spectra always reproduce these major characteristics. This agreement indicates consistency among the structural observations, determined refractive index values, and spectral measurements.

Spectral feature (1) is generally observed for multilayer interference, and it can be directly understood in terms of the interference condition, which is satisfied at a shorter wavelength for a larger incidence angle; the cosine factor associated with the refraction angle becomes smaller as the incidence angle increases. In fact, for the jewel beetle, the condition reportedly reproduces the angle dependence of the peak wavelength well (Yoshioka and Kinoshita, 2011). To more intuitively understand the physical origins of spectral characteristics (2) and (3), we consider several simple multilayer system models, as shown in Figures 6.27(b)–(d). In the following calculations, for simplicity, we assume real and constant refractive indices, $n_l = 1.56$ and $n_h = 1.68$, for the low- and high-index layers, respectively, which were chosen from the real part at a wavelength of 550 nm. It has been confirmed that this approximation is adequate for considering the wavelength range around the main peak; the multilayer system model shown in Figure 6.27(a) with these real constant refractive index values has a spectral shape quite similar to that calculated with the wavelength-dependent complex refractive index values.

First, we consider the quarter-wave stack model shown in Figure 6.27(b). The thicknesses of the low- and high-refractive index layers, d_l and d_h, respectively, are determined such that the interference condition under normal incidence is satisfied at $\lambda_1 = 550$ nm. Although the quarter-wave stack generally has a high reflectance at λ_1, as described in Section 6.4.3, the theoretical calculation reveals that the reflectance spectrum does not have a clear peak at 550 nm, as shown in Figure 6.27(f). Instead, a dip appears at this wavelength. Thus, the simple quarter-wave stack is not an appropriate model for the jewel beetle. The appearance of the dip can be interpreted as an additional effect of the top surface; because light is incident from a lower–refractive index medium (air) to a higher–refractive index material (the n_l layer), the phase of the reflected light is inverted, but this phase inversion does not occur for light incident from the n_h layer to the n_l layer inside the multilayer structure. In other words, the phases of the reflected light waves are $\pi, 2\pi, 2\pi, 4\pi, 4\pi, 6\pi, 6\pi, \ldots$ for reflection from the top surface toward the inner interfaces. These phases are calculated at the wavelength λ_1, and the phase of the incident light is assumed to be 0 at the top surface.

Therefore, destructive interference occurs between the reflection at the top surface and other light waves reflected at the inner interfaces.

This destructive interference with the top surface strongly affects the reflectance spectrum; the reason is that the top surface contributes the most to the overall reflection because it has the largest refractive index difference in the multilayer system. A simple calculation using Fresnel's equation shows that the amplitude reflectance $r_{a,l}$ at the interface between air and the n_l layer is approximately six times larger than $r_{l,h}$, which is the amplitude reflectance between the n_l and n_h layers under normal incidence. Even when multiple reflections within the multilayer system are considered, the contribution of the reflection at the top surface to the total amplitude reflectance is estimated to be approximately 40% (Yoshioka et al., 2012). Thus, the destructive interference caused by the phase inversion at the top surface induces the dip at λ_1 in the reflectance spectrum, even though the interference condition is satisfied at that wavelength.

Such destructive interference can be prevented in several simple ways. One way is to use a high–refractive index material to construct the first layer. However, this is not the case with the jewel beetle's elytron because the electron-sparse (low-index) material forms the first layer (Figure 6.5d). Another simple way is to double the thickness of the first layer, as shown in Figure 6.27(c). The thickened layer makes the light waves reflected at the inner interfaces travel a longer path, and the phase is delayed by π at λ_1. Consequently, the effect of the phase inversion is canceled, and interference with the top surface becomes constructive. As a result, the reflectance spectrum of such a system has a distinctive peak at 550 nm, as shown in Figure 6.27(g).

However, the first layer of the jewel beetle's multilayer system does not seem to have a doubled thickness. Instead, both the first and third layers are observed to be thicker than the other low–refractive index layers. If these two layers are assumed to have a thickness of 1.5 d_l, as shown in Figure 6.27(d), the reflectance spectrum has a distinctive peak at 550 nm (Figure 6.27h). This is because the additional thickness is equal to a total of d_l for the two layers, and this causes a phase delay of π. We examined various thickness pairs for the first and third layers while keeping an additional phase delay of π (Yoshioka et al., 2012). We found that the reflectance peak is always produced regardless of the additional thicknesses of the two layers. Thus, we can conclude that the total phase delay is important for forming the peak in the reflectance spectrum.

Now that we have understood that the reflection from the top surface makes the dominant contribution to the overall multilayer interference, the origins of spectral characteristics (2) and (3) are interpreted without difficulty. Fresnel's equations show that the amplitude reflectance increases monotonically for s-polarization, as shown in Figure 6.16(b). Thus, the incident light is reflected merely by the top surface regardless of the wavelength for a large incidence angle. Because this reflection process does not depend on the wavelength as long as the wavelength dependence of the refractive index is ignored, it causes the background-like component in the reflectance spectrum. For p-polarization, the phase of the reflected light is reversed when the incidence angle is larger than Brewster's angle. Thus, the interference becomes destructive even for the phase-adjusted multilayer system, causing the dip in the spectrum of the background-like component.

It is often thought that a multilayer structure should be periodic to produce saturated structural color through constructive interference at a specific wavelength of light. However, the analysis of the jewel beetle's structural color clearly demonstrates that this is not always the case; rather, modification is necessary in some cases. These structural characteristics pose a deeper question about the formation process of the elytron's multilayer system. The cuticle layers of several structurally colored tiger beetles are reportedly deposited before ecdysis, and melanization or sclerotization subsequent to ecdysis completes the developmental process of the multilayer reflector (Schultz and Rankin, 1985b). It is interesting to consider how the layer thicknesses, particularly those of the layers near the surface, are effectively controlled with subwavelength accuracy during development.

6.5 Analysis II

When light propagation is obstructed by an object, a shadow appears behind it. However, the edge of the shadow is not perfectly sharp but is more or less ambiguous. This smearing is due to a wave characteristic called diffraction. In this section, we describe the basic theory of diffraction. After a few examples, the theory is applied to the analysis of the *Morpho* butterfly's structural color.

6.5.1 Fraunhofer Diffraction

Let us consider that a light wave emitted from a source at a point Q impinges on a screen with a small slit, as depicted in Figure 6.28(a). Secondary spherical waves are assumed to be emitted from each point within the slit. In the 19th century, Kirchhoff gave a mathematical formula that describes diffraction by making several reasonable assumptions regarding the boundary conditions on the screen (Born and Wolf, 1975). According to his theory, the amplitude u_P of the diffracted wave at a point P is described by the formula

$$u_P = -\frac{ikA}{4\pi} \int \frac{e^{ik(r+r_0)}}{rr_0} [\cos(\mathbf{n},\mathbf{r}) + \cos(\mathbf{n},\mathbf{r}_0)] dS, \qquad (6\text{-}28)$$

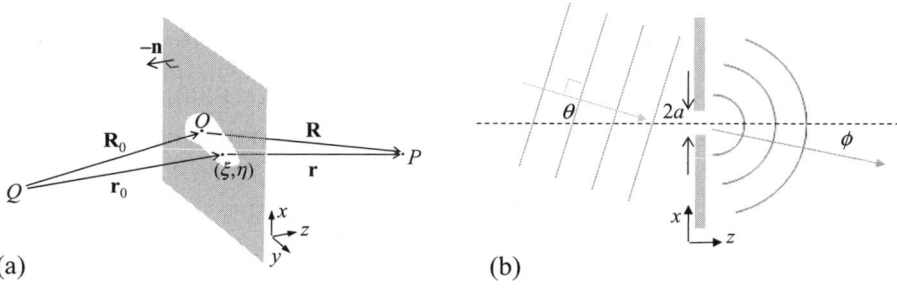

Figure 6.28 Geometry of diffraction. (a) Light wave emitted from point Q is diffracted by a small slit on the x–y plane. The field amplitude is observed at point P. Vector $\mathbf{n}$ is the unit vector normal to the x–y plane (note that $-\mathbf{n}$ is illustrated). Coordinates (ξ,η) describe a position on the slit. (b) Diffraction from a single slit with a width $2a$. Angle ϕ expresses the direction of diffraction.

where light is treated as a scalar wave having a wavelength λ and a wavenumber $k = 2\pi/\lambda$, and cos(**a**,**b**) denotes the cosine of the angle between two vectors **a** and **b**. The amplitude of the light source is denoted by A, and the vector **n** is a unit vector normal to the screen, which is assumed to be in a plane with $z=0$. The vectors **r**, **r**$_0$, **R**, and **R**$_0$ with lengths r, r_0, R, and R_0, respectively, are defined as shown in Figure 6.28(a). The integral is performed over the slit S on the screen. Assuming that the light source $Q(x_0, y_0, z_0)$ and observation point $P(x, y, z)$ are located very far from the screen, we can make the following approximations:

$$r_0 = \sqrt{(x_0 - \xi)^2 + (y_0 - \eta)^2 + z_0^2} \cong R_0\left(1 - \frac{x_0\xi + y_0\eta}{R_0^2}\right), \quad (6\text{-}29)$$

and similarly,

$$r \cong R\left(1 - \frac{x\xi + y\eta}{R^2}\right), \quad (6\text{-}30)$$

where $R_0 = \sqrt{x_0^2 + y_0^2 + z_0^2}$ and $R = \sqrt{x^2 + y^2 + z^2}$, and higher-order terms of ξ/R_0, η/R_0, ξ/R, and η/R are ignored in the expansion of the square root. The coordinates (ξ,η) denote a point on the slit. Then, Eq. (6-28) is rewritten as

$$u_P(\alpha, \beta) = C \iint d\xi d\eta \, \exp[-ik\{(\alpha - \alpha_0)\xi + (\beta - \beta_0)\eta\}], \quad (6\text{-}31)$$

where $\alpha_0 = -x_0/R_0$, $\alpha = x/R$, $\beta_0 = -y_0/R_0$, and $\beta = y/R$ are newly defined, and a constant C includes a slowly varying factor $[\cos(\mathbf{n}, \mathbf{r}) + \cos(\mathbf{n}, \mathbf{r}_0))/(RR_0)]$ that is taken outside the integral as an approximation.

The treatment of diffraction using the far-field approximations of Eqs. (6-29) and (6-30) is called the Fraunhofer diffraction. These equations indicate that the incident light wave propagating into a direction $\mathbf{k}_0 = k\left(\alpha_0, \beta_0, \sqrt{1 - \alpha_0^2 - \beta_0^2}\right)$ is diffracted into a direction $\mathbf{k} = k\left(\alpha, \beta, \sqrt{1 - \alpha^2 - \beta^2}\right)$ with amplitude u_P.

It is not difficult to experimentally observe the characteristic diffraction pattern from a narrow slit; as depicted in Figure 6.29(a), a narrow slit can be prepared by using two razor blades, and a laser pointer works as a light source. Figure 6.29(b) shows the experimentally observed diffraction patterns with various slit widths. When the slit is much wider than the beam width of the laser, only a circular spot appears on the screen. However, as the slit becomes narrower, the diffracted light spreads in a wider angular range with an oscillating light intensity.

The mathematical expression of the diffraction intensity pattern is calculated by integrating Eq. (6-31). The coordinates shown in Figure 6.28(b) are assumed—that is, the slit is assumed to have a width $2a$—and the center is located at the origin of the x axis. In addition, the slit is assumed to be infinitely long in the y direction,

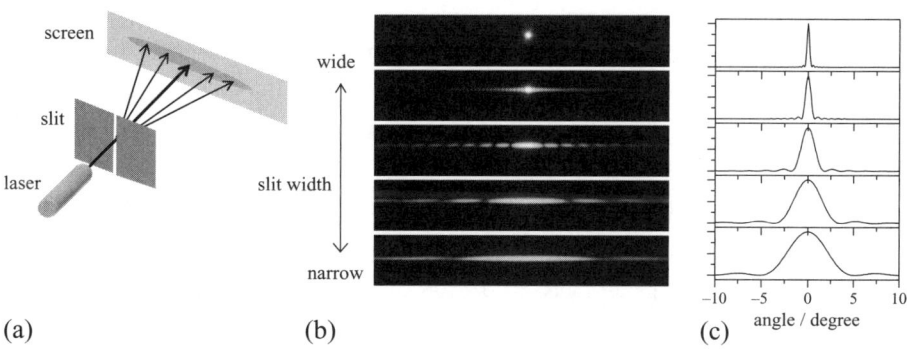

Figure 6.29 Diffraction from a single slit. (a) Experimental geometry, (b) diffraction patterns experimentally observed on a screen, and (c) theoretically calculated diffracted light intensity for a wavelength of 633 nm with various slit widths. Slit width $2a$ is assumed to be 100 μm, 50 μm, 20 μm, 10 μm, and 7 μm, from top to bottom.

and the light source is on the x–z plane—that is, $\beta_0 = 0$. Under these conditions, it is sufficient to consider diffraction only in the x–z plane, and the diffracted wave amplitude u_s is obtained from Eq. (6-31) as

$$u_s \propto \int_{-a}^{a} \exp[-ik\{(\alpha - \alpha_0)\xi\}]d\xi = 2a\frac{\sin[k(\alpha - \alpha_0)a]}{k(\alpha - \alpha_0)a}. \quad (6\text{-}32)$$

Through the factors α_0 and α, u_s depends on the propagation directions. By newly defining two angles θ and ϕ as $\sin\theta = \alpha_0$ and $\sin\phi = \alpha$, as shown in Figure 6.28(b), the normalized diffraction pattern $I(\phi) = |u_s(\phi)|^2/|u_s(0)|^2$ can be plotted for various slit widths, as shown in Figure 6.29(c), assuming $\theta = 0$. The results theoretically confirm that the diffraction pattern spreads in a wider angular range as the slit becomes narrower. The first intensity minimum appears for $k\alpha a = \pi$, which is equivalent to $\sin\phi = \lambda/2a$. Thus, we can understand that the angular spread due to diffraction becomes negligible when the slit is much wider than the wavelength of the light. This corresponds to the common observation that the propagation of a laser beam is straight.

Box 6.2 Designing Diffraction Pattern

Fraunhofer diffraction, which is formulated in Eq. (6-31), has the same mathematical form as the 2D Fourier transformation except for a constant. Thus, the inverse Fourier transformation can be used to calculate a mask that causes a designed diffraction pattern. Here, we present one simple way to prepare such a mask.

The inverse Fourier transformation is calculated after a diffraction pattern is designed as a 2D array of dots, and the power spectrum can be used as the mask pattern. However, it was empirically found that a mask with a binarized intensity produces a more distinctive diffraction pattern than one with a continuous

Continued

Box 6.2 Designing Diffraction Pattern—cont'd

grayscale. One example is shown in Figure B6.2(a). Using this pattern as a unit tile, 10 × 10 tiles are repeatedly arranged using computer software and printed on A3-size copy paper, for example, as shown Figure B6.2(b). Finally, the paper is photographed with a camera using a reversal film. The developed film acts as an intensity mask that causes diffraction.

Figure B6.2(c) shows an observed diffraction pattern for a mask irradiated by a laser pointer. The pattern essentially has inversion symmetry with respect to the origin, which is the direction of the transmitted laser. That is because the mask prepared in this way modulates only the amplitude of the incident light, but not the phase.

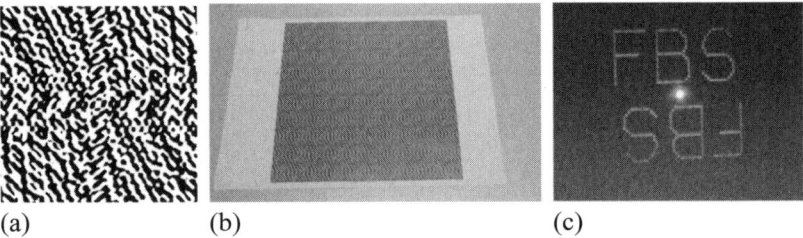

(a)   (b)   (c)

Figure B6.2 Designed Fraunhofer diffraction pattern. (a) Binarized power spectrum of the 2D inverse Fourier transformation, (b) an example of copy paper where the mask patterns are repeatedly arranged, and (c) diffraction pattern observed for a photographic film used as a mask illuminated with a laser pointer. FBS stands for the Graduate School of Frontier BioSciences, which is the author's institution.

6.5.2 Diffraction Grating

A compact disc or DVD is probably the most familiar example of the diffraction grating. Below the protective layer, the disc contains ridges separated by a few microns, as shown in Figure 6.30(b), and they cause rainbow-like dispersed color under white-light illumination. Diffraction by a ruled grating can be similarly treated to the case of a single slit. Let us consider a diffraction grating that consists of N regularly spaced thin slits along the y axis with a width $2a$, as shown in Figure 6.30(a). When the center of the nth slit is denoted by $\xi_n = nd$, the diffracted light amplitude u_N is given as

$$u_N = \sum_{n=1}^{N} e^{-ik(\alpha-\alpha_0)\xi_n} \int_{-a}^{a} e^{-ik(\alpha-\alpha_0)\xi'} d\xi' = \frac{1 - e^{-iNk(\alpha-\alpha_0)d}}{1 - e^{-ik(\alpha-\alpha_0)d}} u_s, \qquad (6\text{-}33)$$

where u_s is given by Eq. (6-32). Equation (6-33) consists of the product of two terms: one comes from the arrangement of the N slits, and the other corresponds to the diffraction from a single slit. Figure 6.31(a) shows the calculated diffraction patterns for

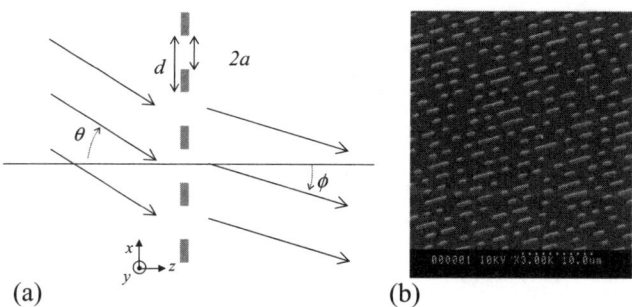

(a) (b)

Figure 6.30 (a) Geometry for multislit diffraction and (b) scanning electron microscopy image of the surface of a compact disc. The protective layer of the disc was removed before the observation.

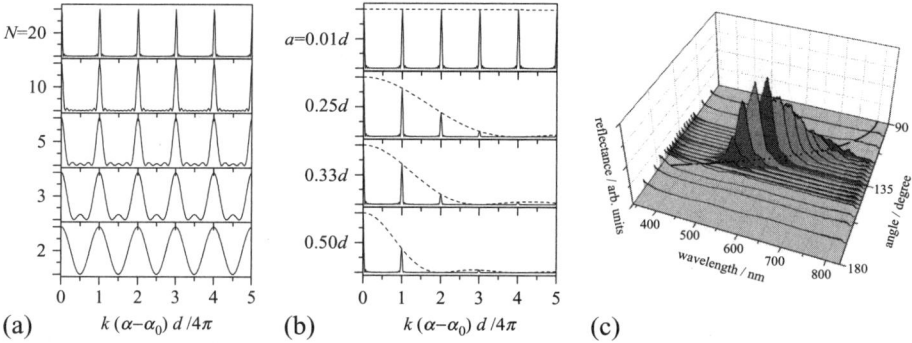

Figure 6.31 Normalized diffraction intensity pattern from multiple slits calculated by Eq. (6-33). (a) Diffraction patterns with a very narrow slit $2a = 0.01d$ for various slit numbers. Slit number N is given on the left. (b) Diffraction patterns with $N = 20$ for several slit widths of $2a$. Broken curves show the normalized diffraction factor of a single slit $|u_s|^2$. (c) Experimentally obtained transmission spectra from a single *Morpho* scale. The angle 180° indicates the direction of directly transmitted light, whereas 90° corresponds to the lateral direction. Solid-dotted curve indicates the theoretically calculated relation between the wavelength and diffraction angle for a grating with $d = 0.8$ μm. (Reproduced from Kinoshita et al., 2002a.)

a very narrow slit with $2a = 0.01d$ for various slit numbers N. The pattern with $N = 2$ resembles a sinusoidal curve because it corresponds to Young's well-known double-slit experiment. However, the peaks become sharper as N increases. They are located at the position where the denominator of Eq. (6-33) become zero—that is, $kd(\sin \phi - \sin \theta) = 2m\pi$, where m is an integer. This relation is reduced to the interference condition for the diffraction grating,

$$d(\sin \phi - \sin \theta) = m\lambda. \tag{6-34}$$

When N becomes very large, the diffraction peak becomes very sharp. Thus, a ruled diffraction grating is commonly employed in a spectrometer that separates white light into different directions depending on the wavelength. When the slit width $2a$ is comparable to the distance d, the factor associated with the single-slit diffraction becomes more important, as shown in Figure 6.31(b). The factor $|u_s|^2$ behaves like an envelope of many sharp peaks. As a special case, the l th-order diffraction peak vanishes when the slit width satisfies $a=d/l$, where l is an integer. The reason is that the phase factor in the factor u_s becomes $k(\alpha - \alpha_0)a = 2l\pi/d \times d/l = 2\pi$.

A single *Morpho* scale works as a diffraction grating in transmission (Figure 6.8c). Figure 6.31(c) shows the transmission spectrum from a single scale as a function of the angle. The peak wavelength is reproduced well by Eq. (6-34) with the parameter $d = 0.8$ μm, which is consistent with scanning electron microscopy observations.

6.5.3 Analysis of Morpho Butterfly's Structural Color

The brilliant blue wing of the *Morpho* butterfly has attracted a considerable amount of scientific interest, and it has become a representative example of structural color in nature. Spectroscopic measurements have confirmed that the wing has a high reflectance in a short wavelength range (Figure 6.14c). To explain this feature, a periodic multilayer model has long been considered, since Lord Rayleigh (1917) described a mathematical treatment of optical interference in periodically stacked thin films. This model successfully explains the highly efficient wavelength-selective reflection in the blue color; a multilayer structure that consists of approximately 10 cuticle layers with air spacing has nearly 100% reflectance, if a quarter-wave stack is assumed (Figure 6.23a). However, this model fails to explain the angular dependence of the reflection; the reflected light from the *Morpho* wing is spread broadly in one plane perpendicular to the ridges of the scales, although the multilayer model predicts specular reflection. Conversely, a multislit diffraction grating model seems consistent with the observation of the dispersed color in transmission. However, the diffraction grating model does not explain the wavelength-selective reflection; more importantly, diffraction spots are not observed on the reflective side.

To explain these contradictory optical characteristics, Kinoshita and colleagues (2002a) proposed a simple analytical model called the discrete multilayer model. This model successfully explains the wavelength-selective reflection with an anisotropic reflection pattern. The essence of this model is to consider both structural regularity and irregularity within the scale. In addition, the diffraction theory is applied to analytically predict the angular dependence of the reflected light. Recent advances in computer technology make it possible to directly treat a complicated microstructure to calculate the reflectance by numerically solving Maxwell's equations. For example, the finite-difference time-domain method has been employed in many recent studies (e.g., Plattner, 2004; Lee and Smith, 2009; Saito et al., 2011). However, an analytical model is still valuable for intuitively understanding the fundamental optical processes that occur in the microstructure. Here, we describe the discrete multilayer model according to the paper by Kinoshita *and* colleagues (2002b) with some modifications and newly presented results.

As shown in Figure 6.32(a), the treelike microstructure within a ridge is modeled as a stack of N narrow platelets of width a and length l, and they are assumed to be

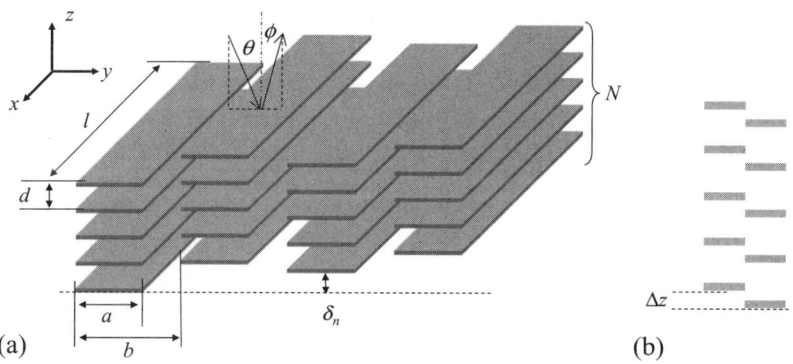

Figure 6.32 Discrete multilayer model. (a) Treelike microstructure within the *Morpho* scale modeled as N thin platelets having a width a and length l that are vertically stacked with a constant separation d. The distance between adjacent stacks is denoted by b. The height irregularity of the ridges is described by a factor δ_n in the nth ridge. (b) Modified model illustrated by a cross-section of the ridges (y–z plane). Alternating heights of the branches projecting to both sides are introduced. The height difference between the left and right sides is denoted by the factor Δz.

infinitely thin. They are stacked with a regular interval d, and the distance between adjacent stacks is denoted by b. The height irregularity in the nth ridge is described as δ_n, which is the deviation from the mean height. This irregularity is introduced on the basis of careful inspections of scanning and transmission electron micrographs; observations of the scale microstructure in the longitudinal section reveals that the branches of the treelike microstructure are not parallel to the bottom plate of the scale but are slanted (Ghiradella, 1991). As a result, the upper edges of the branches appear on the tops of the ridges, and their positions are randomly distributed (Kinoshita et al., 2002a). Therefore, the branches on different ridges are not at the same height. In fact, the correlation in the height variation has been investigated, and the correlation length was estimated to be less than the ridge separation (Kinoshita et al., 2002b). This height distribution is one reason that a flat multilayer structure is not an appropriate model for the treelike microstructure of the scales.

To consider reflection in this model, diffraction theory is employed with the following assumptions: the multiple reflections between the platelets are ignored, and the amplitude of the incident light is assumed to be unaffected. Then, the amplitude of the reflected light wave E is estimated by integrating the second wave emitted from the platelets by applying Eq. (6-31),

$$E = \sum_n \sum_{j=0}^{N-1} \int_{-l/2}^{l/2} dx \int_{-a/2}^{a/2} dy \exp\left[-i(k_x - k_{0,x})x\right] \exp\left[-i(k_y - k_{0,y})(bn + y)\right]$$

$$\times \exp\left[-i(k_z - k_{0,z})(jd + \delta_n)\right],$$

where the wave vectors of the incident and reflected light are denoted by $\mathbf{k}_0 = (k_{0,x}, k_{0,y}, k_{0,z})$ and $\mathbf{k} = (k_x, k_y, k_z)$, respectively. Summation and integration can be performed separately to yield

$$E = 4al \frac{\sin\left[(k_x - k_{0,x})l\right]}{(k_x - k_{0,x})l} \frac{\sin\left[(k_y - k_{0,y})a\right]}{(k_y - k_{0,y})a} \frac{1 - e^{-iN(k_z - k_{0,z})d}}{1 - e^{-i(k_z - k_{0,z})d}}$$
$$\times \sum_n \exp\left[-i\{(k_z - k_{0,z})\delta_n + (k_y - k_{0,y})bn\}\right].$$

When many ridges are illuminated in an experiment, the intensity of the reflected light is observed as the statistical average of the height variation,

$$I = \langle |E|^2 \rangle$$
$$= (4al)^2 \left|\frac{\sin\left[(k_x - k_{0,x})l\right]}{(k_x - k_{0,x})l}\right|^2 \left|\frac{\sin\left[(k_y - k_{0,y})a\right]}{(k_y - k_{0,y})a}\right|^2 \left|\frac{1 - e^{-iN(k_z - k_{0,z})d}}{1 - e^{-i(k_z - k_{0,z})d}}\right|^2 F_R,$$

(6-35)

where the factor F_R is introduced as

$$F_R \equiv \left\langle \left|\sum_n \exp\left[-i\{(k_z - k_{0,z})\delta_n + (k_y - k_{0,y})bn\}\right]\right|^2 \right\rangle. \tag{6-36}$$

If the randomness in height is not considered (i.e., $\delta_n = 0$), F_R becomes a factor corresponding to the multislit diffraction grating (absolute value of Eq. 6-33 squared). However, when the randomness causes a phase variation ranging from $-\pi$ to π, F_R is reduced to the number of ridges N_R because the interference terms vanish in the statistical average. This happens when the randomness is distributed over ± 120 nm, which is estimated for normal incidence and normal reflection ($k_z - k_{0,z} = 2k_{0,z}$) and a wavelength of 480 nm, at which the reflectance takes a maximum value. In fact, this estimation is consistent with the electron micrographs of the cross-section, where the branches of the trees cannot be smoothly expanded to those of the adjacent ridges.

Owing to this irregularity, the reflection intensities are simply expressed as the products of the three terms except for a constant. Their meanings are interpreted as follows: the first and second terms in Eq. (6-35) correspond to diffraction from a single platelet and express the angular dependence along the x and y directions, respectively, and the third term expresses the interference effect of the stack of N platelets in one ridge. The third term is the primary origin of the wavelength-selective reflection.

This model explains the anisotropic reflection pattern shown in Figure 6.33(a). This is because the reflected light spreads only in the direction perpendicular to the slender platelet, whereas the platelet length l is much longer than the width, so diffraction broadening is limited. For a more quantitative comparison with the experimental

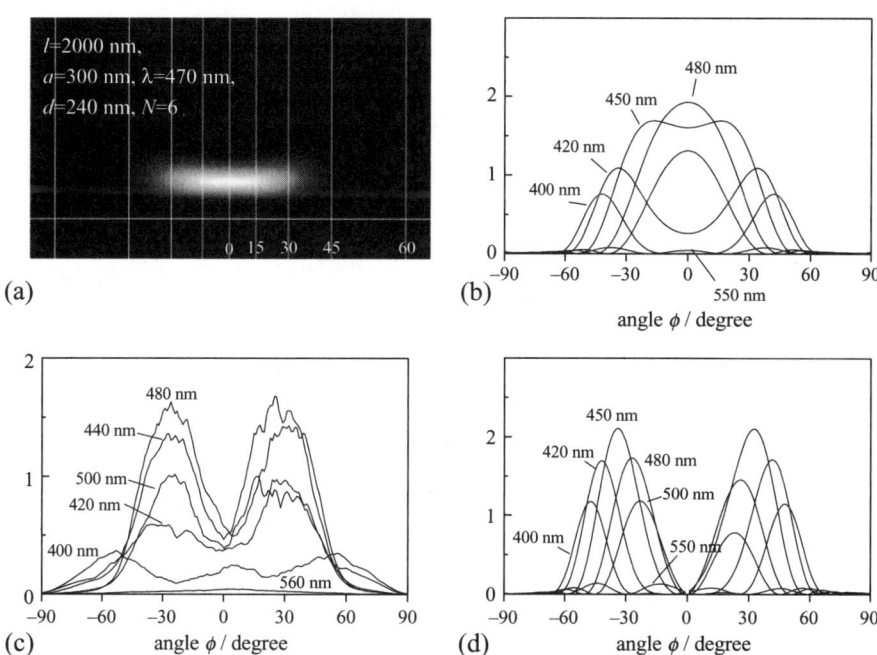

Figure 6.33 Angular dependence of the reflected light. (a) Estimated reflection pattern on the screen calculated on the basis of the 3D discrete multilayer model using Eq. (6-35). Vertical white lines indicate the angles from the screen normal. (b) Angular dependence calculated using Eq. (6-35) in the y–z plane at several wavelengths, where $N=6$, $a=300$ nm, and $d=240$ nm are used. These parameters were chosen for consistency with the electron microscopic observations. Incident light along the z axis, namely $\theta=0$ in Figure 6.32(a), is assumed, and the intensity is plotted against the angle ϕ. (c) Experimental results for a single blue scale of *M. cypris* butterfly. The intensity decreases around the angular origin. (d) Theoretical calculation based on the modified model with alternative heights shown in Figure 6.32(b) with $a=150$ nm and $\Delta z=120$ nm; other parameters are the same as in (b).

results, the angular dependence of the reflected light intensity is calculated for several wavelengths, as shown in Figure 6.33(b), where the incident light direction is along the z axis (normal incidence), and the diffraction intensity is plotted in the y–z plane. The results are in moderate agreement with the experimental results shown in Figure 6.11 (a); a major peak is located at the angular origin for longer wavelengths, and peaks appear at large angles for short wavelengths, in agreement with the experiments.

From these analyses, we can interpret the essential features of the structural color of the *Morpho* butterfly as follows:

1. the narrowness of the branches of the treelike microstructure, which is modeled as thin platelets, causes strong diffraction in a plane perpendicular to the ridges. This is the origin of the anisotropic reflection pattern.
2. Periodically stacked platelets cause optical interference, and, in turn, result in wavelength-selective reflection of blue light.

3. The height irregularity prevents effective interference between the light waves diffracted from different ridges, so diffraction spots do not appear on the reflection side. Conversely, the phase of the transmitted light is less sensitive to the height difference. That is because $k_z - k_{0,z}$ is so small that F_R does not become a simple constant value N_R. As mentioned before, the factor becomes that of the diffraction grating for a case without height irregularity. It causes diffraction spots by selecting a wave vector component along the y direction.

One modification of the discrete multilayer model has been presented by Kinoshita and colleagues (2008); alternating heights of the branches projecting toward the left and right sides of the trunk are introduced into the model, as shown in Figure 6.32(b). The alternating heights tend to cause destructive interference in the normal direction. Thus, the reflected light intensity is predicted to be greatly reduced, as shown in the theoretical calculations in Figure 6.33(d). Such a center-dip angular dependence is qualitatively consistent with the experimental results for a single scale of a few species of *Morpho* butterfly—for example, *M. cypris* (Figure 6.33c) and *M. rhetenor* (Vukusic et al., 1999). However, no such intensity decrease around the normal direction is observed for the ground scales of *M. didius* (Vukusic et al., 1999; Kinoshita et al., 2002a), although alternating heights are observed in electron micrographs of this species. This is one of the remaining mysteries of the *Morpho* butterfly's blue.

The discrete multilayer model helps us to intuitively understand the fundamental optical processes causing the *Morpho* butterfly's structural color. However, it does not consider polarization effects or quantitative reflectance values. Further quantitative analyses have been made recently using computational methods. Zhu et al. (2009) discussed different mechanisms in the angular broadening of the reflection depending on the polarization. Saito et al. (2011) considered the effects of the randomness of the ridge positions in both the lateral and vertical directions. Boulenguez and colleagues (2012) also considered the irregularities found in scale structures with various sizes. The long-lasting mystery of the mechanism of the blue coloration has not been completely solved yet. However, the preceding studies clearly suggest that the treatment of the irregular structure is important for better understanding of the coloration mechanism.

6.5.4 Role of Pigment

The periodicity of the submicron structures is an essential factor in structural color because it can cause wavelength-selective reflection through optical interference. However, studies of the *Morpho* butterfly have shown that the irregularity can also play an important role. In addition, several other factors have been identified that contribute greatly to the coloration. In this section, we describe two such examples found in the *Morpho* species.

As shown in Figure 6.2, the wing of the *M. cypris* butterfly is characterized by distinctive white stripes on the blue color. It is natural to ask how the scale structures of blue and white scales differ. However, both the blue and white scales have been found to have very similar microstructures despite the large color difference. Spectroscopic measurements have confirmed that both types of scale have a reflection peak at a short wavelength. The largest optical difference appears in the amount of optical absorption, as shown in Figures 6.14(c) and (d); the blue part of the wing largely absorbs the light above 550 nm, whereas the absorptance of the white part is much smaller than that of the blue part. In addition, the reflectance is more than 0.3 for wavelengths longer than 550 nm. This higher reflectance at longer wavelengths is considered to be the direct origin of the distinctive whiteness because we see the reflected light. However, it seems difficult for a few-microns-thick scale to have a reflectance as high as 30%. To elucidate the reflection mechanism, the optical properties of all the wing components, single scales, and the wing membrane were thoroughly investigated by careful observations of the wing structure (Yoshioka and Kinoshita, 2006b).

As shown in Figure 6.34(a), the stripe patterns of the ventral and dorsal sides correspond to each other very well; the white stripe is formed as an arrangement of white scales surrounded by blue scales on the dorsal side, whereas the white stripe is surrounded by brown scales on the ventral side. In addition, the wing membrane also appears transparent in the striped part, whereas the other part looks brown. Quantitative measurements revealed that the white scales on both sides have a negligible amount of optical absorption, although the other scales are highly pigmented. Thus, in the white stripes, multiple reflections occur among three layers, the two scale layers and the wing membrane, and the reflectance is greatly enhanced. Conversely, in the blue region, the scales and membrane are highly pigmented, causing a large amount of optical absorption. Thus, the contribution of multiple reflections is almost negligible, resulting in the highly saturated blue color. From these investigations, it was concluded that the white stripe on *M. cypris* is caused by a combination of the microstructure inside the scale and the spatially controlled pigmentation in the three wing layers.

Pigmentation is often employed in natural examples of structural colors. When the pigment is located beneath the color-causing microstructures, it absorbs the transmitted light and prevents it from being reflected back by structures below the color-causing

Figure 6.34 *Morpho* butterflies. (a) Wing color pattern of the ventral (left) and dorsal (center) sides, and naked wing membrane (right) of the *M. cypris* butterfly. In the image on the right, the scales on both sides of the membrane are almost completely removed. ((a) Reproduced by Yoshioka and Kinoshita, 2006b.) (b) Optical micrograph of the wing of *M. didius* showing that the cover scales overlap the ground scales. The layer of cover scales is intentionally disturbed in some parts. (For color version of this figure, the reader is referred to the online version of this chapter.)

microstructure. Thus, the pigmentation enhances the spectral purity of the reflection, greatly enhancing the saturation of the color. Examples have been identified in many species including the peacock feather (Yoshioka and Kinoshita, 2002).

We have seen that anisotropic reflection is one of the major characteristics of the *Morpho* butterfly. However, in one *Morpho* species, *M. didius*, this reflection pattern is modified by overlapping of differently structured transparent scales (Yoshioka and Kinoshita, 2004). In this butterfly, two types of scales exist on the dorsal side forming two scale layers, as shown in Figure 6.34(b); one is a layer of dark blue scales, called ground scales, which exhibit the narrow blue reflection band. The other is a layer of transparent cover scales that overlaps the ground scale layer. The cover scales are often called glass scales because they appear highly transparent. Despite their transparency, the cover scales have a specialized microstructure that modifies the reflection pattern. The ridges of the cover scales are sparse on the bottom scale plate, whereas the bottom plate has an appropriate thickness to cause thin-layer interference at blue wavelengths. Thus, blue light is reflected from the two structural elements inside the cover scales: the treelike microstructure of the ridges and the bottom plate causing single-thin-layer interference. In addition, the reflected light directions in these two reflection mechanisms differ because of the tilt of the branches within the ridges toward the basal plate. These structural characteristics thicken the narrow reflection band when the cover scales overlap the ground scales. It was experimentally demonstrated by the screen projection method that the narrow blue band pattern changes to a diffuse oval pattern (Yoshioka and Kinoshita, 2004).

6.6 Concluding Remarks

This chapter described the structural colors of two insects using basic experiments and analysis. It has been revealed that submicron periodic structure is not the only factor contributing to the coloration; rather, various modifications are introduced to the structure. In the jewel beetle, a few layers near the surface are thicker than the inner layers, and they adjust the phases of the reflected light waves. The scales of the *Morpho* butterfly introduce structural irregularity in the height of the ridges, which results in a narrow reflection band. In addition, a few *Morpho* species have been found to use not only microstructure within the scale but also spatially controlled pigmentation or the macroscopic arrangement of different scales for the colorations. Similar effects of multiscale structure have also been found in the peacock feather (Yoshioka and Kinoshita, 2002).

These examples suggest that the developmental processes are appropriately controlled at different sizes ranging from submicron to even millimeter sizes such that the developed structures overall achieve certain optical effects. It is quite interesting to consider how an established physical mechanism of pattern formation (e.g., a reaction–diffusion system) can be extended or merged with other mechanisms at different sizes to explain the formation of such a multiscale structure. Unfortunately, research in the developmental processes of natural microstructures has not progressed much except for some pioneering work on the butterfly wing scale (Ghiradella, 1978) and tiger beetle (Schultz and Rankin, 1985b). Intense research with respect to this is required.

It is also important to further study the optics of natural structural color in multiscale systems by expanding the range of research subjects. Owing to the recently developed photonic crystal, theoretical methods of treating a periodic system have already been established. However, when optical processes are appropriately combined with those originating from different sized structures, additional effects can be achieved, as observed in the *Morpho* butterflies. Thus, we will likely find unknown, useful, integrated mechanisms in natural structural color that inspire artificial optical applications because so many species of butterflies, birds, and fish with brilliant colors are overlooked without being carefully examined.

Acknowledgments

This study was partially supported by Grants-in-Aid for Scientific Research (nos. 22340121 and 24120004) from the Japanese Ministry of Education, Culture, Sports, Science, and Technology.

References

Adachi, E., 2007. Unexpected variability of millennium green: Structural color of Japanese jewel beetle resulted from thermosensitive porous organic multilayer. J. Morph. 268, 826–829.
Bernard, G.D., Miller, W.H., 1968. Interference filters in the corneas of Diptera. Invest. Ophthalmol. 7, 416–434.
Berthier, S., 2007. Iridescences. Springer, New York.
Berthier, S., Charron, E., Da, A., 2003. Determination of the cuticle index of the scales of the iridescent butterfly *Morpho Menelaus*. Opt. Commun. 228, 349–356.
Born, M., Wolf, E., 1975. Principles of Optics, fifth ed. Pergamon Press, New York.
Boulenguez, J., Berthier, S., Leroy, F., 2012. Multiple scaled disorder in the photonic structure of *Morpho rhetenor* butterfly. Appl. Phys. A: Materials Science and Processing 106, 1005–1011.
Ghiradella, H., 1978. Development of ultraviolet-reflecting butterfly scales: How to make an interference filter. J. Morph. 142, 395–410.
Ghiradella, H., 1991. Light and color on the wing: structural colors in butterflies and moths. Appl. Opt. 30, 3492–3500.
Hariyama, T., Hironaka, M., Horiguchi, H., Stavenga, D.G., 2005. The leaf beetle, the jewel beetle, and the damselfly; insects with a multilayered show case. In: Kinoshita, S., Yoshioka, S. (Eds.), Structural Colors in Biological Systems–Principles and Applications. Osaka University Press, Osaka, pp. 153–176.
Huxley, A.F., 1968. A theoretical treatment of the reflexion of light by multilayer structures. J. Exp. Biol. 48, 227–245.
Kambe, M., Zhu, D., Kinoshita, S., 2011. Origin of retroreflection from a wing of the *Morpho* butterfly. J. Phys. Soc. Jpn. 80, 054801.
Kinoshita, S., 2008. Structural Colors in the Realm of Nature. World Scientific Publishing, Singapore.

Kinoshita, S., Yoshioka, S., Kawagoe, K., 2002a. Mechanisms of structural color in the *Morpho* butterfly: Cooperation of regularity and irregularity in an iridescent scale. Proc. R. Soc. Lond. B 269, 1417–1421.

Kinoshita, S., Yoshioka, S., Fujii, Y., Okamoto, N., 2002b. Photophysics of structural color in the *Morpho* butterflies. Forma 17, 103–121.

Kinoshita, S., Yoshioka, S., 2005. Structural colors in nature: The role of regularity and irregularity in the structure. ChemPhyChem 6, 1442–1459.

Kinoshita, S., Yoshioka, S., Miyazaki, J., 2008. Physics of structural colors. Rep. Prog. Phys. 71, 076401.

Land, M.F., 1972. The physics and biology of animal reflectors. Prog. Biophys. Mol. Biol. 24, 75–106.

Lee, R.T., Smith, G.S., 2009. Detailed electromagnetic simulation for the structural color of butterfly wings. Appl. Opt 48, 4177–4190.

Leertouwer, H.L., Wilts, B.D., Stavenga, D.G., 2011. Refractive index and dispersion of butterfly chitin and bird keratin measured by polarizing interference microscopy. Opt. Exp. 19, 24061–24066.

Lippert, W., Gentil, K., 1959. Über lamellare Feinstrukturen bei den Schillerschuppen der Schmetterlinge vom *Urania*- und *Morpho*-Typ. Z. Morph. Ökol. Tiere 48, 115–122.

Lord Rayleigh, O.M., F.R.S.,1917. On the reflection of light from a regularly stratified medium. Proc. R. Soc. Lond. A 93, 565–577.

Mason, C.W., 1927. Structural color in insects II. J. Phys. Chem. 31, 321–354.

Mossakowski, D., 1979. Reflection measurements used in the analysis of structural colors of beetles. J. Microsc. 116, 351–364.

Nakamura, E., Yoshioka, S., Kinoshita, S., 2008. Structural color of rock dove's neck feather. J. Phys. Soc. Jpn. 77, 124801.

Newton, I., 1704. Opticks: Or, a Treatise of the Reflections, Refractions, Inflections and Colours of Light. Dover Publications, New York.

Noyes, J.A., Vukusic, P., Hooper, I.R., 2007. Experimental method for reliably establishing the refractive index of buprestid beetle exocuticle. Opt. Exp. 15, 4351–4358.

Parker, A.R., 2000. 515 million years of structural color. J. Opt. A: Pure Appl. Opt. 2, R15–R28.

Plattner, L., 2004. Optical properties of the scales of *Morpho rhetenor* butterflies: Theoretical and experimental investigation of the back-scattering of light in the visible spectrum. J. R. Soc. Interface 1, 49–59.

Pouya, C., Stavenga, D.G., Vukusic, P., 2011. Discovery of ordered and quasi-ordered photonic crystal structures in the scales of the beetle *Eupholus magnificus*. Opt. Exp. 19, 11357–11364.

Saito, A., 2011. Material design and structural color inspired by biomimetic approach. Sci. Technol. Adv. Mater. 12, 064709.

Saito, A., Yonezawa, M., Murase, J., Juodkazis, S., Mizeikis, V., Akai-Kasaya, M., et al., 2011. Numerical analysis on the optical role of nano-randomness on the *Morpho* butterfly's scale. J. Nanosci. Nanotechnol. 11, 2785–2792.

Schultz, T.D., Rankin, M.A., 1985a. The ultrastructure of the epicuticular interference reflectors of tiger beetles (Cicindela). J. Exp. Biol 117, 87–110.

Schultz, T.D., Rankin, M.A., 1985b. Developmental changes in the interference reflectors and colorations of tiger beetles (Cicindela). J. Exp. Biol. 117, 111–117.

Srinivasarao, M., 1999. Nano-optics in the biological world: Beetles, butterflies, birds, and moths. Chem. Rev. 99, 1935–1961.

Stavenga, D.G., Leertouwer, H.L., Pirih, P., Wehling, M.F., 2009. Imaging scatterometry of butterfly wing scales. Opt. Exp. 17, 193–202.

Stavenga, D.G., Wilts, B.D., Leertouwer, H.L., Hariyama, T., 2011a. Polarized iridescence of the multilayered elytra of the Japanese jewel beetle, *Chrysochroa fulgidissima*. Phil. Trans. R. Soc. B 366, 709–723.

Stavenga, D.G., Leertouwer, H.L., Marshall, N.J., Osorio, D., 2011b. Dramatic color changes in a bird of paradise caused by uniquely structured breast feather barbules. Proc. R. Soc. B 278, 2098–2104.

Stokes, G.G., 1849. On the perfect blackness of the central spot in Newton's rings, and on the verification of Fresnel's formulae for the intensities of reflected and refracted rays. Cambridge and Dublin Math. J. 4, 1–14.

Trzeciak, T.M., Wilts, B.D., Stavenga, D.G., Vukusic, P., 2012. Variable multilayer reflection together with long-pass filtering pigment determines the wing coloration of papilionid butterflies of the nireus group. Opt. Exp. 20, 8877–8890.

Vukusic, P., Sambles, J.R., 2003. Photonic structures in biology. Nature 424, 852–855.

Vukusic, P., Sambles, J.R., Lawrence, C.R., Wootton, R.J., 1999. Quantified interference and diffraction in single *Morpho* butterfly scale. Proc. R. Soc. Lond. B 266, 1403–1411.

Wilts, B.D., Leertouwer, H.L., Stavenga, D.G., 2009. Imaging scatterometry and microspectrophotometry of lycaenid butterfly wing scales with perforated multilayers. J. R. Soc. Interface 6, S185–S192.

Yeh, P., 1988. Optical Waves in Layered Media. John Wiley and Sons, New York.

Yin, H., Shi, L., Sha, J., Li, Y., Qin, Y., Dong, B., et al., 2006. Iridescence in the neck feathers of domestic pigeons. Phys. Rev. E 74, 051916.

Yoshioka, S., Kinoshita, S., 2002. Effect of macroscopic structure in iridescent color of the peacock feathers. Forma 17, 169–181.

Yoshioka, S., Kinoshita, S., 2004. Wavelength-selective and anisotropic light-diffusing scale on the wing of the *Morpho* butterfly. Proc. R. Soc. Lond. B 271, 581–587.

Yoshioka, S., Kinoshita, S., 2006a. Single-scale spectroscopy of structurally colored butterflies: Measurements of quantified reflectance and transmittance. J. Opt. Soc. Am. A 23, 134–141.

Yoshioka, S., Kinoshita, S., 2006b. Structural or pigmentary? Origin of the distinctive white stripe on the blue wing of a *Morpho* butterfly. Proc. R. Soc. B 273, 129–134.

Yoshioka, S., Nakamura, E., Kinoshita, S., 2007. Origin of two-color iridescence in rock dove's feather. J. Phys. Soc. Jpn. 76, 013801.

Yoshioka, S., Nakano, T., Nozue, Y., Kinoshita, S., 2008. Coloration using higher order optical interference in the wing pattern of the Madagascan sunset moth. J. R. Soc. Interface 5, 457–464.

Yoshioka, S., Kinoshita, S., 2011. Direct determination of the refractive index of natural multilayer systems. Phys. Rev. E 83, 051917.

Yoshioka, S., Kinoshita, S., Iida, H., Hariyama, T., 2012. Phase-adjusting layers in the multilayer reflector of a jewel beetle. J. Phys. Soc. Jpn. 81, 054801.

Zhu, D., Kinoshita, S., Cai, D., Cole, J.B., 2009. Investigation of structural colors in *Morpho* butterflies using the nonstandard-finite-difference time-domain method: Effects of alternately stacked shelves and ridge density. Phys. Rev. E 80, 051924.

Zi, J., Yu, X., Li, Y., Hu, X., Xu, C., Wang, X., et al., 2003. Coloration strategies in peacock feathers. Proc. Natl. Acad. Sci. 100, 12576–12578.

References to Each Chapter

Chapter 1

Haken, H., 1975. Cooperative phenomena in systems far from thermal equilibrium and in nonphysical systems. Rev. Mod. Phys. 47, 67–121.
This paper gives the mathematical background for the phase transition analogy of the cooperative phenomena found in nonequilibrium systems with particular interest in laser oscillation.

Haken, H., 1978. Synergetics. An Introduction. Nonequilibrium Phase Transitions and Self-organization in Physics, Chemistry and Biology. Springer-Verlag, Berlin.
This book is the extension of the preceding paper (Haken, 1975) and develops the phase transition analogy to various cooperative phenomena in physics, chemistry, biology, and sociology.

Chapter 2

Epstein, I.R., Pojman, J.A., 1998. An Introduction to Nonlinear Chemical Dynamics: Oscillations, Waves, Patterns, and Chaos. Oxford University Press, New York.
This book includes numerous examples of nonlinear chemical reactions exhibiting oscillating, chaos, and pattern formation. The authors also provide an overview of oscillation in biological systems and polymers.

Kuramoto, Y., 1984. Chemical Oscillations, Waves, and Turbulence. Springer-Verlag, Berlin.
This book provides mathematical and physical perspectives on the dynamics of chemical waves and turbulence in the self-oscillating fields of a reaction-diffusion system. Synchronization of coupled oscillators is also described based on the phase model.

Pikovsky, A., Rosenblum, M., Kurths, J., 2001. Synchronization: A Universal Concept in Nonlinear Sciences. Cambridge University Press, Cambridge.
This book deals with synchronization of nonlinear oscillators. The background of the synchronization phenomena and the methods of mathematical analyses are described with various examples including density oscillators.

Chapter 3

de Gennes, P.G., Brochard-Wyart, F., Quere, D., 2004. Capillarity and Wetting Phenomena: Bubbles Pearls Waves: Drops, Bubbles, Pearls, Waves. Springer, New York.

This is a fundamental but interesting textbook on surface tension. It shows various kinds of phenomena concerning surface tension. Both theoretical and experimental aspects are described and movies on the enclosing CD are also instructive.

Happel, J., Brenner, H., 1965. Low Reynolds Number Hydrodynamics: With Special Applications to Particulate Media. Prentice-Hall, Englewood Cliffs, NJ.

This is a standard textbook for the low Reynolds number hydrodynamics, which is used in this chapter. It shows a lot of examples of the flow profiles based on the Stokes equations, as well as the fundamental knowledge on theoretical techniques to treat the low Reynolds number hydrodynamics.

Mikhailov, A.S., Calenbuhr, V., 2006. From Cells to Societies: Models of Complex Coherent Action. Springer, Berlin.

The aim of this textbook is to understand the living phenomena in the framework of mathematics and physics. In this book, spontaneous motion is treated theoretically in addition to various pattern formations.

Nepomniashchy, A.A., Velarde, M.G., Colinet, P., 2002. Interfacial Phenomena and Convection. Chapman & Hall/CRC Press, Boca Raton, FL.

In this book, interfacial phenomena related to the surface tension are introduced. They show how to treat the dynamic phenomena such as the Marangoni effect, mainly focusing on hydrodynamics.

Chapter 4

Kano, T., Kinoshita, S., 2007. Viscosity-dependent flow reversal in a density oscillator. Phys. Rev. E 76, 046208.

This paper focuses on the flow-reversal process of a density oscillator. Through detailed experiments, the flow-reversal process was found to depend significantly on the viscosities of the fluids. This finding is well explained by a simple model that takes account of three factors essential for the flow reversal.

Martin, S., 1970. A hydrodynamic curiosity: The salt oscillator. Geophys. Fluid Dyn. 1, 143–160.

This is the first paper that describes the phenomenon of density oscillation. Detailed experimental observations and hydrodynamic analysis of each upflow and downflow are well described.

Pikovsky, A., Rosenblum, M., Kurths, J., 2001. Synchronization: A Universal Concept in Nonlinear Sciences. Cambridge University Press, Cambridge.

This book deals with self-oscillatory systems found in nature, particularly focusing on synchronization phenomena. Backgrounds and analytical methods for the synchronization phenomena are described with various examples including density oscillators.

Chapter 5

Arora, A.K., Tata, B.V.R. (Eds.), 1996. Ordering and Phase Transition in Charged Colloids. Wiley-VCH, New York.

This book deals with topics on ordering phenomena of colloids, including phase transition of charged colloids.

Russel, W.B., Saville, D.A., Schowalter, W.R., 1989. Colloidal Dispersions. Cambridge University Press, Cambridge.
This book is a well-known introductory textbook for colloid science. The book covers fundamentals of colloid physics, including interparticle interaction, Brownian motion, electrokinetic phenomena, sedimentation, diffusion, and non-Newtonian rheology.

Chapter 6

Berthier, S., 2006. Iridescences: The Physical Colors of Insects. Springer-Verlag, Berlin.
This book treats the structural colors of mainly insect species. It contains many colorful photographs of butterflies and beetles, and also intuitive illustrations of optical processes. In addition, many electron micrographs can be found that clearly show how elaborate the microstructures of the animals can be. It is characteristics to this book that a brief description is given about the developmental processes of butterflies and beetles from an egg to an adult.
Kinoshita, S., 2008. Structural Colors in the Realm of Nature. World Scientific Publishing, Singapore.
This book deals with various structural colors found in natural systems. It covers the basic optical processes to detailed coloration mechanisms in some species of insects. In appendixes, numerous butterfly and bird species with structural colors are carefully listed with many references, which would be quite useful for those who are going to start researches. In addition, detailed derivations of some mathematical formula are included.
Kinoshita, S., Yoshioka, S. (Eds.), 2005. Structural Colors in Biological Systems—Principles and Applications. Osaka University Press, Osaka.
This book consists of 19 monographs about structural color that were written by pioneering researchers in this field. The research subjects include various animal species including insects, birds, and fish. In addition, several monographs are about the applications of structural color to textile and cosmetic industries, and to a lasing phenomenon.
Journal of Royal Society of Interface 6 (2), 2009.
This special issue of the journal contains the papers written by the famous researchers in this field who attended a 2008 international meeting, Iridescence: More Than Meets the Eye. As well as an introduction of this meeting and the current understanding of the iridescence, 12 research articles are included about the coloration mechanisms of bird feathers, insects, and cephalopods. One paper summarizes the physical methods to study the structural color. The papers are available from the journal's website, http://rsif.royalsocietypublishing.org/.
Forma 17 (2), 2002.
Japanese researchers contribute to this issue of the journal that focuses on structural color. The subjects are, for example, the antireflection properties of the wing of a moth, variable colors found in fish, and mimicry in the microstructure. The papers are available online from the website http://www.scipress.org/journals/forma/index.html.

Index

Note: Page numbers followed by *f* indicate figures and *t* indicate tables.

A
Alder transition, 165–166
Ambient illumination, 201–202, 204–205, 204*f*

B
Belousov-Zhabotinsky (BZ) reaction, 12–17, 61–83, 106–111, 117–118
  Brusselator model, 12–14
  Marangoni flow, 95
  mathematical model, 117–118
  Oregonator model
    chemical oscillation, origin of, 15
    nullcline, 16
    stationary point, 16
  oscillation
    absorption spectrum, ferroin-catalyzed, 64–65
    CSTR, 65–66, 66*f*
    ruthernium–bipyridine complex, 65
    stirring condition, 66
  oscillations and spatiotemporal patterns, 61, 62*f*
  oscillations and traveling waves, 95
  reactants color, 119–120
  reaction mechanism and numerical simulation
    FKN mechanism, 68–69
    limit cycle oscillation, 71–72
    Oregonator model (*see* Oregonator model)
    trajectories of, 72, 73*f*
  rotating spiral wave, 61, 74*f*
  self-oscillatory systems, 119–120
  spiral wave and scroll ring, 95, 96*f*
  suspended droplet swims, 95
  synchronization
    analytical method, 75–76
    chemical oscillation, 75
    coupled BZ oscillators, 77–81
    coupling function, 76–77
    target pattern, 95, 96*f*
    time course of, circular wave, 61, 74*f*
  wave propagation and pattern formation, 66–68
    reaction–diffusion equations, 72–75, 74*f*
    rotating spiral waves, 66–67, 67*f*
    target pattern, 66–67, 67*f*
Brusselator model, 12–14, 13*f*

C
Camphor boat, 93
Charged colloidal crystals, 140–141, 168, 175*f*, 176
Colloidal crystals, 21–28, 165–198
  anisotropic particles
    kagome lattice structure, 194–195
    types of, 194, 194*f*
  binary colloids, 195
  charged colloidal crystals, 175–176, 175*f*
  charge-induced crystallization, silica
    effective charge number, 182
    experimental parameters, 181–182
    NaOH and NaCl concentrations, 181–182
    surface charge number tuning, 182*f*
    theoretical phase diagram, 182–183, 183*f*
  colloidal dispersion system
    Bragg spots, 170
    colloidal particles, 166
    material properties, 166
    opal-type colloidal crystal, 167, 167*f*
    particle positions structure identification, 169–170, 169*f*
    radial distribution function, 168, 169*f*
    volume fraction, 167
  colloidal samples
    metal oxides, 176
    purification, 176–177
    silica particles, 176–177
    size distribution, 176
    strong acid groups, 176

Colloidal crystals (*Continued*)
  electrostatic (Coulomb) interaction
    cations and anions density, 172–173
    charged colloidal phenomena, 171–172, 172f
    charge density, 173–174
    charge-neutrality condition, 172
    charge number, 174–175
    Fourier transformation, 173–174
    interaction energy, 174–175
    linear Poisson-Boltzmann equation, 173, 175
    nonlinear Poisson-Boltzmann equation, 172–173
  entropy, 165–166
  gel immobilization
    Bragg wavelength tuning, 189, 190f
    Py diffusion, 189, 189f
  impurity exclusion
    particle size, 189–190
    PS particles, 190–191, 192f
    unidirectional crystallization, 191–193, 192f, 193f
  microscopy, 181
  modification, fluorescent molecules, 195
  opal-type colloidal crystals, 193–194
  order-formation process
    Bragg condition, 21f, 23
    2D arrays formation, 25–26
    Debye screening, 22
    experimental preparation, 23
    experimental setup, 24, 24f
    functional groups dissociation, polystyrene sphere, 21–22, 22f
    initial NaCl concentration, 25, 25f
    ion concentration variation, 22, 22f
    light illumination, 21–22
    monolayer colloidal array collapse, 26, 27f
    particle volume, 22–23
    two-dimensional colloidal crystal growth, 24f, 25
  preparation of
    aqueous dispersion, 167–168
    charged colloidal crystals, 168
    opal crystals, 168
  self-assembly, structural patterns, 195
  spectroscopy
    Bragg wavelengths, 180–181

    transmission and reflection spectra, 179–180, 180f
    visible and near-infrared reflection spectra, 179–180, 180f
  unidirectional crystallization
    atomic and molecular crystals, 183–184
    base diffusion, 184–185
    optical quality, 183
    temperature gradient, 185–189
  van der Waals interaction
    coagulation, 170–171, 171f
    Derjaguin-Landau-Verwey-Overbeek theory, 171
    interparticle magnitude, 170
  X-ray scattering
    body- and face-centered cubic lattices, 177–179
    diffraction conditions, 177–179
    form factor, 179
    Miller indices, 178
    ultra-small-angle X-ray scattering patterns, 177, 178f
Confocal laser scanning microscopy (CLSM), 181, 190–191, 192f, 193f
Continuous-flow stirred tank reactor (CSTR), 12, 65–66, 66f
Coulomb interaction, colloidal crystals
  cations and anions density, 172–173
  charged colloidal phenomena, 171–172, 172f
  charge density, 173–174
  charge-neutrality condition, 172
  charge number, 174–175
  Fourier transformation, 173–174
  interaction energy, 174–175
  linear Poisson-Boltzmann equation, 173, 175
  nonlinear Poisson-Boltzmann equation, 172–173
Coupled Belousov-Zhabotinsky (BZ) oscillators
  in-phase and out-of-phase synchronization, 77, 78f, 78t
  phase difference and coupling functions, 79, 79f

D

Debye screening, 21–23, 22f, 173
Density (saltwater) oscillators, 18–20, 119–162
  biological membranes, 126

Index

coupled density oscillators, 124–125, 125f
electrical potential, 126
experimental procedure and oscillation trend
  charge-coupled device (CCD) camera, 128–129
  electrical potential, 128–129
  fluid temperature, 127–128
  laser displacement meter, 128–129
  long-time behavior, 130f, 131
  magnified view, 129–131, 130f
  oscillation amplitude, 129–131, 130t, 131f
  temporal evolution, height, 129, 129f
  up- and downflow branches, 131, 131f
experimental setup, 126, 127, 127f
flow-reversal process, 126
  downflow velocity, 149–150
  flow-reversal model, 149–150, 150f
  heavy-fluid surface height, 155, 156f
  hydrodynamic analysis, 141–143
  hydrostatic pressures, 152
  intuitive interpretation, 158, 158f
  nondimensionalized variables, 154
  PL_1 and PL_2 planes, 152–153, 153f
  Rayleigh-Taylor instability, 140–141
  viscosity-dependent (*see* Viscosity-dependent flow reversal)
hydrodynamic analysis, upflow and downflow branch
  3D Hagen-Poiseuille flow, 133–134
  head loss, 133–134
  heavy fluid, 133–134, 134f
  hydrostatic equilibria, 134f, 135–136
  hydrostatic pressure, 133–134
  light fluids, 134–135, 134f, 135f
  Navier-Stokes equation, 131–133
  pipe axis, 131–133, 132f
oscillation process, 124, 124f
phenomenological model
  coupled density oscillator, 140
  heavy-fluid surfaces, 139–140
  hydrodynamic simulation, 137–138
  oscillator number, 139–140
  oscillatory behavior, 137, 138
  Rayleigh equation, 137
  vector fields, 138, 139f
relaxation oscillations, 121–123
salt oscillator, 123, 123f, 124

self-oscillatory phenomena, 119–121
synchronization, 125–126
taste-sensing mechanisms, 126
three-dimensional Hagen-Poiseuille flow, 162–163
two-dimensional Hagen-Poiseuille flow, 163
two oscillators coupled, 125–126, 125f
Derjaguin-Landau-Verwey-Overbeek (DLVO) theory, 24, 171
3D Hagen-Poiseuille flow, 133–134
Directional illumination, 201–203, 204–205, 207
Driven spontaneous motion
  drifting droplet, aqueous surface
    camphor boat, 93
    ethanol, 92
    floating alcohol droplet, 92
    Marangoni effect, 92
    motile cellular fragments, 93
    pentanol droplet motion, 93–94, 94f
    pentanol molecules, 92, 93f
  gliding droplet, glass surface
    glass surface, 90
    oil droplet, 88–90, 91, 91f
    reactive wetting, 92
    tears of wine, 89, 89f
  swimming droplet, aqueous phase
    advection-reaction-diffusion equation, 97
    Belousov-Zhabotinsky (BZ) reaction, 95, 96f
    Marangoni flow, 95
    Millipore-Q system, 96–97
    petri dish, 95
    Stokes equation, 97
Droplets dynamics
  active matter, 85–86
  BZ reaction, 117–118
  hydrodynamics for spontaneous motion
    force-free condition, 103–104
    hydrodynamic equation, 103–104
    inhomogeneous surface tension, 102–103
    Navier-Stokes equation, 97–98
    reaction-diffusion systems, 105
    shear stress, 103–104
    spherical and polar coordinates, 98–99, 99f

Droplets dynamics (*Continued*)
　steady-state distribution, 103
　Stokes equation, 98–99
　Stokes flow, 99–101
　streaming flow, 104*f*
　stress tensor, 97–98
　surface tension, frame, 101–102
　viscosity, 97–98
　non-Newtonian behavior, 112
　pattern formation
　　BZ droplet motion, 106–107
　　co-moving frame and lab frames, 108, 110*f*
　　flow profiles, 108–111, 109*f*, 112*f*
　　Legendre polynomials, 107
　　scroll ring, 108, 110*f*
　　spatiotemporal plot, chemical wave, 107, 108*f*
　　spiral wave, 108–111, 111*f*
　Stokes equation, 116–117
　surface tension (*see* Surface tension)

F
Fabry-Perot interference, 230–231
Field, Körös and Noyes (FKN) mechanism, 14, 68–70
Fraunhofer diffraction, structural color
　designed pattern, 240*f*
　diffracted wave amplitude, 238–239
　geometry of, 237–238, 237*f*
　intensity pattern, 238–239
　inverse Fourier transformation, 239–240
　Kirchhoff's formula, 237–238
　light source, 237–238
　narrow slit, 238
　single slit, 238–239, 239*f*

H
Hard-sphere colloid system, 165–166, 166*f*
Harmonic oscillator
　mechanical, 3, 3*f*
　orbits, phase space, 3–4, 3*f*
　time variation, 3, 3*f*
Hopf bifurcation, 11, 14, 54
Huxley's method, 230

I
Inhomogeneous surface tension, 102, 102*f*, 104–105, 104*f*
Iridescence, 201, 204–205, 204*f*

J
Jewel beetle, structural color
　ambient illumination, 204–205, 204*f*
　Chrysochroa fulgidissima, 204
　directional illumination, 204–205
　electron micrograph, 205–206, 205*f*
　integrated reflectance spectra, 216–217, 216*f*
　iridescence, 204–205, 204*f*
　multilayer interference theory
　　angle-and polarization-dependent reflectance spectra, 234–235
　　cuticle layers, 237
　　destructive interference, 236
　　layer thickness, 234–235, 236
　　p-polarization, 236
　　quarter-wave stack model, 235–236
　　refractive indices, 234–235, 234*f*
　　spectral features, 235
　　s-polarization, 236
　optical micrographs, 205–206, 205*f*
　θ–2θ scan measurement
　　angle-and polarization-dependent reflectance spectra, 213–214, 214*f*
　　angle dependence, 213
　　incident angles, 212–213, 213*f*
　　p-polarization, 214
　refractive index
　　high-index layer, 218
　　natural multilayer systems, 217–218

L
Laplace pressure, 101
Leidenfrost droplets, 85–86
Limit cycle, 7–11, 8*f*, 10*f*, 13, 15–17, 71–73, 71*f*, 73*f*, 76, 120–121, 120*f*, 138–139, 139*f*
Linear Poisson-Boltzmann equation, 173
Liquid immersion experiment
　jewel beetle, 207
　Morpho wing color change, 206, 206*f*, 207
　peacock feather, 206*f*, 207
Lotka-Volterra model
　activator-inhibitor system, 5–6
　motion, phase space, 5*f*, 6
　prey-predator dynamics, 4–5, 5*f*
　time variations in, 5, 5*f*

M

Marangoni effect
  adsorption/desorption process, 95, 97
  stress field induces flow, 88
  surface tension, 88, 92
  tears of wine, 89
  thermal gradient, 88
Millipore-Q system, 96–97
Morpho butterfly, structural color
  angle-dependent reflection
    anisotropic reflection, 212
    diffraction ring, 210f
    fiber rotation method, 212
    imaging scatterometry, 212
    optical interference, 212
    quantitative characterization, 210–212, 211f
    reflected light intensity, wavelengths, 211f
    screen projection method, 207, 208f
    single-scale examination, 208–209, 209f
    transmission pattern, 209–210
  diffraction grating
    anisotropic reflection pattern, 244–245, 245f
    discrete multilayer model, 242–243, 243f, 246
    height irregularity, 246
    intensity of reflected light, 243–244
    multislit diffraction grating factor, 243–244
    optical interference, 245
    reflected light wave amplitude, 243–244
    reflection intensities, 244
    wavelength-selective reflection, 242
  directional illumination, 202–203, 202f
  optical microscope, 203
  pigment role
    anisotropic reflection, 248
    blue and white scales, 246–247
    color-causing microstructures, 247–248
    ventral and dorsal sides, 247, 247f
  reflectance and transmittance measurement, 215–216, 215f
  refractive index
    cuticle, 217
    polarization-dependent effective, 217
    wavelength dependence, 217
  ridges, 203–204

  scanning and transmission electron micrographs, 203–204, 203f
Mullins–Sekerka instability, 29–34, 34f
Multilayer interference, structural color
  ideal and nonideal multilayer systems, 231–234
  interference conditions
    interference condition, 227–228
    multilayer system, 228f
    path length difference, 227
    reflectance and transmittance spectra, 227
  jewel beetle
    angle-and polarization-dependent reflectance spectra, 234–235
    cuticle layers, 237
    destructive interference, 236
    layer thickness, 234–235, 236
    p-polarization, 236
    quarter-wave stack model, 235–236
    refractive indices, 234–235, 234f
    spectral features, 235
    s-polarization, 236
  reflectance spectrum
    Fabry-Perot interference, 230–231
    higher-order interference peaks, 231
    Huxley's method, 230
    quarter-wave stacks, 229–230, 229f
    recursive method, 228–229, 229f
    reflection paths, 228–229
    two-layer system, 228–229

N

Navier-Stokes equation, 19, 97, 131–133
Neon lamp, 17–18, 122
Neon-lamp oscillator, 17–18
Nonequilibrium phenomena
  order and fluctuation, 51–55
  order-formation process (see Order-formation process)
  oscillatory phenomena (see Oscillatory phenomena)
  pattern formation
    color-producing nanostructures, 41–51
    crystal growth instability, 29–34
    turing pattern, 35–41
Nonlinear Poisson-Boltzmann equation, 172–173

O

Opal, 46, 168
Order-formation process
  colloidal crystals
    Bragg condition, 21f, 23
    2D arrays formation, 25–26
    Debye screening, 22
    experimental preparation, 23
    experimental setup, 24, 24f
    functional groups dissociation, polystyrene sphere, 21–22, 22f
    initial NaCl concentration, 25, 25f
    ion concentration variation, 22, 22f
    light illumination, 21–22
    monolayer colloidal array collapse, 26, 27f
    particle volume, 22–23
    two-dimensional colloidal crystal growth, 24f, 25
Oregonator model, 14–17, 69–72, 106
  chemical oscillation, origin of, 15
  elementary reactions, 70
  nullcline, 16
  stationary point, 16
Oscillatory phenomena
  Belousov-Zhabotinsky reaction
    Brusselator model, 12–14
    Oregonator model, 14–17
  harmonic oscillator
    mechanical, 3, 3f
    orbits, phase space, 3–4, 3f
    time variation, 3, 3f
  Lotka-Volterra model
    activator-inhibitor system, 5–6
    motion, phase space, 5f, 6
    prey-predator dynamics, 4–5, 5f
    time variations in, 5, 5f
  relaxation oscillation
    neon-lamp oscillator, 17–18
    saltwater oscillator, 18–20
  Van der Pol oscillator
    basic equation for, 7
    Hopf bifurcation, 11
    Liènard's theorem, 7–8
    limit-cycle orbits, 7, 8f
    linear stability analysis, 10
    nullcline, 8
    phase space, orbits, 8, 9f
    stationary point, 10–11
    time variation, 8, 10f
    vector fields, 8–9, 10f

P

Pair correlation function, 168, 169f
Pattern formation
  color-producing nanostructures, 41–51
    developmental studies on, 47–51
    diffraction grating, 44–46
    light scattering, 47
    photonic crystals, 46–47
    thin-layer interference, 42–44
  crystal growth instability
    directional solidification, 29, 29f
    energy conservation, 30
    Gibbs–Thomson equation, 30
    liquid–crystal interface modulation, 32, 32f
    Mullins–Sekerka instability, 29–34, 34f
    temperature gradient, 31, 32f
    types of, 29–30
  turing pattern
    in biological systems, 38–41
    in chemical systems, 37–38
    turing instability, 35–37
Phase-transition analogy
  adiabatic approximation, 52–53
  atomic polarization, 52–53
  functional shape, 53–54, 54f
  Hamiltonian system, 52
  Markovian type, 53
  phenomenological equations, 51

Q

Quarter-wave stack model, 235–236

R

Radial distribution function, 169f
Rayleigh equation, 137
Rayleigh–Taylor instability, 140
Reactive wetting, 92
Relaxation oscillation, 17–20, 121–123
Ridges, 203–204
Ruthernium–bipyridine complex, 65

S

Saffman-Taylor instability, 143–144
Salt oscillator, 123, 123f, 124
Saltwater oscillator, 18–20

Self-oscillatory phenomena, 119–121
Static structure factor, 169–170
Stokes equation, 116–117
Structural color, 41–51, 199–251
  diffraction grating
    diffracted light amplitude, 240–242
    interference condition, 240–242
    multiple slits, 241f
    phase factor, 240–242
  Fraunhofer diffraction
    designed pattern, 240f
    diffracted wave amplitude, 238–239
    geometry of, 237–238, 237f
    intensity pattern, 238–239
    inverse Fourier transformation, 239–240
    Kirchhoff's formula, 237–238
    light source, 237–238
    narrow slit, 238
    single slit, 238–239, 239f
  Fresnel's equations
    amplitude reflectance and transmittance, 218–221, 220f
    energy reflectance, 221
    Maxwell's equations, 218–220
    polarization states, 218–220
    refractive index, 218–220, 219f
    Snell's law, 218–220
  illumination, 201–202
  jewel beetles (*see also* Jewel beetle, structural color)
    ambient illumination, 204–205, 204f
    Chrysochroa fulgidissima, 204
    directional illumination, 204–205
    electron micrograph, 205–206, 205f
    iridescence, 204–205, 204f
    optical micrographs, 205–206, 205f
  liquid immersion experiment, 206–207
  Morpho butterfly (*see* Morpho butterfly, structural color)
    directional illumination, 202–203, 202f
    optical microscope, 203
    ridges, 203–204
    scanning and transmission electron micrographs, 203–204, 203f
  multilayer interference
    ideal and nonideal multilayer systems, 231–234
    interference conditions, 227–228
    reflectance spectrum, 228–231
  single thin-layer interference
    interference conditions, 221–223
    reflectance spectrum, 223–224
    thickness variation, 225–227
Surface tension
  air-liquid interfaces, 86
  driven spontaneous motion (*see* Driven spontaneous motion)
  fluid-fluid surface, 88
  free-energy cost, 86
  Marangoni effect, 88
  measurement, 87, 87f
  solid-liquid interface/solid-air interface, 86
  thermodynamic variables, 88
  Young's equation, 86

T

Tears of Wine, 89
Three-dimensional Hagen-Poiseuille flow, 162–163
Two-dimensional Hagen-Poiseuille flow, 163

V

Van der Pol oscillator, 7–11, 120–123
  basic equation for, 7
  Hopf bifurcation, 11
  Liènard's theorem, 7–8
  limit-cycle orbits, 7, 8f
  linear stability analysis, 10
  nullcline, 8
  phase space, orbits, 8, 9f
  stationary point, 10–11
  time variation, 8, 10f
  vector fields, 8–9, 10f
Viscosity-dependent flow reversal
  asymptotic values, 144–146, 146f
  densities, 144, 145f
  down- to upflow, stereomicroscope, 148, 148f
  experimental setup, 144, 144f
  fluid dynamics, 143–144
  fluid viscosity, 148
  heavy fluid viscosity, 144–146, 145f
  Saffman-Taylor instability, 143–144
  temporal evolution of fluid surface height, 144–146, 145f
  temporal evolution of intrusion, light fluid, 148, 149f
  time constants, 144–146, 146f
  viscosity-dependent behavior, 149